Theories and Technologies of Hypervelocity Shock Tunnels

Learn to design and improve state-of-the-art aerodynamic ground testing facilities in this comprehensive reference book, with particular focus on high-enthalpy shock tunnels. In the book, readers will discover how to extend test time with brand new concepts and duplicate real hypersonic flight test conditions, including the latest advances in detonation-driven high-enthalpy shock tunnels.

Through a systematic approach, the book describes technologies for a variety of different drivers in hypersonic and high-enthalpy shock tunnels. The fundamental theories for hypersonic and high-enthalpy shock tunnels are described step-by-step, with examples throughout, providing an accessible introduction. Built on years of real-world experience, this book examines in detail the advantages and challenges of improving test flow qualities, including increasing total pressure and enthalpy, model scale amplification, and test time extending for different types of shock-tunnel drivers. This is an ideal companion handbook for aerospace engineers as well as graduate students.

Zonglin Jiang is a professor at the Institute of Mechanics, Chinese Academy of Sciences (CAS), a distinguished fellow at the International Shock Wave Institute (ISWI), a member of Russian Academy of Natural Sciences, and an associate fellow at the American Institute of Aeronautics and Astronautics (AIAA). He has been internationally recognized for his universal framework of detonation initiation and propagation, the theory and technology for detonation-driven hypervelocity shock tunnels, the standing oblique detonation ramjet engine, and the dispersion-controlled dissipative shock-capturing scheme, and rewarded the 2016 AIAA Ground Testing Award, the National Technology Invention Prize of China, the Outstanding Scientific, and Technology Achievement Prize of CAS.

Randy S. M. Chue, formerly Associate Professor at Nanyang Technological University, Singapore, has worked extensively in the analysis and optimization of various wind tunnels for testing scramjets and reentry vehicles, such as the NASA HyPulse and blowdown facilities located at the General Applied Science Laboratory where he was the chief engineer for computational fluids. His teaching and research were focused on propulsion.

Theories and Technologies of Hypervelocity Shock Tunnels

ZONGLIN JIANG
Chinese Academy of Sciences, Beijing

RANDY S. M. CHUE
Nanyang Technological University, Singapore (Retired)

CAMBRIDGE
UNIVERSITY PRESS

CAMBRIDGE
UNIVERSITY PRESS

Shaftesbury Road, Cambridge CB2 8EA, United Kingdom

One Liberty Plaza, 20th Floor, New York, NY 10006, USA

477 Williamstown Road, Port Melbourne, VIC 3207, Australia

314–321, 3rd Floor, Plot 3, Splendor Forum, Jasola District Centre, New Delhi – 110025, India

103 Penang Road, #05–06/07, Visioncrest Commercial, Singapore 238467

Cambridge University Press is part of Cambridge University Press & Assessment, a department of the University of Cambridge.

We share the University's mission to contribute to society through the pursuit of education, learning and research at the highest international levels of excellence.

www.cambridge.org
Information on this title: www.cambridge.org/9781316516751

DOI: 10.1017/9781009030991

First published 2023

A catalogue record for this publication is available from the British Library

A Cataloging-in-Publication data record for this book is available from the Library of Congress

ISBN 978-1-316-51675-1 Hardback

Contents

Preface

Hypersonic test facilities have been in development for decades with the aim of meeting the requirements for hypersonic flight, which is a beauteous dream of many scientists and engineers who are dedicated to aerospace science and engineering. The hypervelocity shock tunnel is one class of hypersonic test facility and belongs to the family of shock tunnels. The word "hypervelocity" is adopted here to distinguish it from the "hypersonic" in the experimental methods, in which a low sound speed is approached with a low-temperature test gas medium to obtain high Mach number test flows. In the "cold" hypersonic flow generated in this way, the total test flow temperature is significantly lower than that in true hypersonic flight, during which a thermochemically reacting gas flow dominated by the local flow temperature is formed around a flying vehicle. The reactions include air dissociation, ionization, and recombination near the vehicle surface. It is the thermochemically reacting nature that makes hypersonic flows stand out from supersonic and subsonic flows because the mechanisms for energy conversion associated with adiabatic compression and viscous energy dissipation need further investigation.

Since the 1950's, hypersonic flight tests have revealed various problems that had not been predicted either through analysis and computation or through wind tunnel ground testing. Such experiences led to a great deal of effort on improving both analytical techniques and ground-testing methods. This book is devoted to hypervelocity shock tunnels not only because the hypersonic test facilities developed so far are too broad to cover, but also because the investigation into the hypersonic flow chemistry is a more challenging research topic than others. In addition, the shock tunnel is simple to construct, intuitive to operate, easy to extend to large scales, and also has a low thermal load, which is a big issue in the development of wind tunnel test facilities.

This book is concerned only with high-enthalpy shock tunnels that are capable of generating hypervelocity air flows where the chemistry can be reasonably simulated. And also, the theories and techniques for large-scale and long-test-duration shock tunnels are emphasized and discussed in detail. The large-scale test flow is of fundamental importance for high-enthalpy flow tests because the test models are not scalable in many cases where chemical reactions dominate. Extending the test duration is also significant, not only because the long test duration can extend the applicability of shock tunnels, but also because it allows for the challenging work on supersonic combustion exploration, aerodynamic forces and moments measurement, and experimental data accuracy improvement to be carried out. It is believed that the hypervelocity shock tunnel is the primary enabler both for the development of hypersonic vehicles and the better understanding of high-temperature flow physics.

We intend this book to be accessible to those researchers with a background in thermodynamics, aerodynamics, and aerospace science and engineering, especially for graduate students, scientific researchers, and professors who work on hypersonic and high-temperature gas dynamics. A chapter on shock wave relations and aerothermodynamic states has been added to the book to render it readable for beginners. The topics for each chapter, the order of their organization, the selection of theoretical expressions, and the experimental data interpretations are bound to the authors' views, which reflect our more than 30 years' research experience on the physical understanding of hypersonic research practices. Therefore, the book is a specialized one because there are some attempts in it not only to explore new phenomena, new mechanisms, new theories, and new techniques, but also to approach some puzzles and unknowns, concerning both theoretical and technical aspects, that remain under exploration.

It is difficult to recall the exact time when the author first became interested in high-enthalpy shock tunnels, but several events still remain clear in my memory. The first one was the discussion with Professor Kazuyoshi Takayama in the year 1995 when he was planning to reconfigure Professor I. I. Glass' shock tube into a detonation-driven shock tunnel for high-enthalpy flow generation. Professor Takayama wanted to install a one-meter mirror in order to visualize the one-meter-long hypersonic boundary layer to observe its nonlinear development on a large scale. Later, Professor Takayama and I visited Dr. Katsuhiro Itoh at the National Aerospace Laboratory in Kakuda, Japan, and I was greatly surprised by the scale of the High Enthalpy Shock Tunnel (HIEST), the biggest free-piston-driven high-enthalpy shock tunnel in the world at the time. It was a big step forward for generating hypersonic flow for solving engineering problems. In 1996, Professor Hongru Yu visited Tohoku University in Sendai, Japan, and reported his new progress on the detonation-driven shock tunnel. The promising experiment on the detonation driver was performed in the TH2-D shock tunnel in Aachen, Germany, by Professors Yu and Hans Grönig as a joint research project between the Shock Wave Laboratory, RWTH Aachen University, and the Institute of Mechanics, Chinese Academy of Sciences. In one faculty member party held by the College of Engineering, Tohoku University, I met with Professor Chul Park who was a professor at the College of Engineering at that time. When we talked about high-enthalpy flow, Professor Park told me that high-enthalpy flow research is like a black hole: as you put more in it, you get less out. However, your gains may be huge once you do eventually get something out of it. I believe he was right, and this was the beginning of my research on hypersonic flow.

I would like to express my sincere thanks to Dr. Randy S. M. Chue (formerly of Nanyang Technological University, Singapore), my co-author of this book. Randy had not only completed his chapters, but also revised other chapters to make them readable. Without his help, I cannot imagine how I could have completed the monograph in well-written English. I met Randy and his wife Patricia in Sendai in 1995 and we have been friends since then. Randy is an earnest scholar and also an excellent expert in detonation and hypersonic flow research. We collaborated very well, and I sincerely hope we have more opportunities for collaboration in the future. I would also like to express my thanks to my colleagues, Professors Hu Z. M., Wang C., Wang Y. P., Liu Y. F., Han G. L., Yuan C. K., Li J. P., and Luo C. T. who contributed a lot to this book, especially in manuscript proofreading and revising, figure-replotting, and formula reformatting. Without their dedication, I am sure that the book would never have been ready for publication. Finally, I would like to acknowledge the contribution of my family, Li, Peng, Xiaolei, and Chenyue, in creating an enjoyable and loving environment for working and writing. It was essential to this book.

Zonglin Jiang

Zonglin JIANG

1 Introduction

The aspiration to fly faster and higher has always been the driving force behind the progress of aeronautics. This desire stems from the practical need for faster modes of transportation as well as humanity's innate curiosity to explore the unknown. To travel to new worlds, humans or their devices must transcend the confines of this planet.

Hypersonic flight not only draws the imagination of space travel to exotic realms and peaceful exploration of space but also evokes dire portents of military exploitation. Machines that are developed to fly at hypersonic speeds must first be tested on the ground in order to evaluate their designs. This book is about ground testing of hypersonic flight vehicles. We will define more precisely what hypersonic means later in the chapter but suffice here to say that it refers to flight at velocities much greater than the speed of sound. Because of the high speeds encountered, hypersonic flight involves extreme flow conditions that are very difficult to replicate in ground testing facilities.

Hypersonic flight occurs when a missile, rocket, aircraft, or spacecraft travels through the atmosphere at the extreme high-speed end of the flight spectrum. This can occur when an intercontinental ballistic missile reenters Earth's atmosphere. A similar situation occurs when a spacecraft reenters Earth's atmosphere from a low Earth orbit (LEO) or when it returns from the Moon or other celestial bodies in the solar system.

For Earth-orbiting spacecraft, the reentry into Earth's atmosphere poses a major challenge. The orbital velocity of LEO spacecraft is approximately 7.5 km/s (or Mach 25). When the spacecraft reenters the atmosphere at this speed, the shock-heated gas around the vehicle results in high heating on the vehicle's surface. Therefore, it is vital to design the thermal protection system (TPS) for the vehicle's survival during reentry. However, there is little room for safety factors and conservative designs, as these spacecraft have very stringent weight requirements. A conservative TPS design would mean that the heat shield is too massive for any payload to be carried.

Spacecraft returning from Moon missions will reenter Earth's atmosphere at higher velocities than those from LEOs. The Apollo command modules that carried the returning astronauts from the Moon reentered Earth's atmosphere at approximately 11 km/s (or Mach 36). For spacecraft returning from Mars, the reentry velocity would be as high as 13 km/s (over Mach 40). These far exceed the reentry velocities of LEO spacecraft, and thus the heating problem is also more severe.

Hypersonic flight also occurs when a spacecraft enters into the atmosphere of another planet or moon that has an atmosphere of its own. The entry velocity into Mars' atmosphere is typically around 6 km/s. The entry velocity into Jupiter's atmosphere is much higher. While the entry velocity at Mars is no higher than that for Earth reentry from LEO, the Martian atmosphere presents other challenges. Because of the very low atmospheric pressure and density in the Martian atmosphere, it is difficult to use aerodynamic drag to rapidly decelerate and land the entry spacecraft safely. This is especially an issue for landing much heavier manned spacecraft in future missions.

Another occurrence of hypersonic flight is when an aircraft cruises in the atmosphere at a hypersonic speed, or when a trans-atmospheric vehicle is being accelerated from subsonic speeds through the atmosphere in order to reach the orbit. For the latter, currently, access to space is accomplished using rockets. Ascending rockets encounter hypersonic conditions at relatively high altitudes near the fringe of the atmosphere. In the future, if the accelerator vehicle is propelled by air-breathing propulsion rather than by rocket, it would need to stay in a sufficiently dense part of the atmosphere to ingest adequate air for the generation of thrust, thereby lengthening the duration and intensifying the exposure to hypersonic conditions.

Before we delve into the subject of hypersonic ground testing facilities, we will first look at what kind of hypersonic flights we expect to see in the future. These provide good indications of the needs that are called for in testing them. It is only after a vehicle design has been carefully evaluated in ground facilities before flight testing can proceed. Indeed, the development of hypersonic flight vehicles is very much paced by the ability to test them on the ground.

1.1 Hypersonic Flight for the Future

1.1.1 Earth-Orbiting Spacecraft

Although still infrequent, the number of space flights is steadily increasing. Most of the flights are for access to the Earth orbits, with the majority being the launching of commercial satellites, such as those for communication, Earth observation, navigation, and air traffic tracking. Space flights are also regularly needed to ferry supplies to the International Space Station (ISS) and to transport crew members to and from it. Increasingly, scientists are exploiting space-based platforms to avoid the distortion of Earth's atmosphere by putting telescopes in the orbit. The Hubble Space Telescope is the world's first space-based optical telescope that has obtained amazing images of distant universe. The Transiting Exoplanet Survey Satellite (TESS) is another space telescope that was launched in April 2018 to search for exoplanets.

Today, besides the major space powers (United States, Russia, and Europe), more and more countries are developing the capability to construct and launch spacecraft. For example, China already launched more rockets into the orbit than any other country in 2018. Furthermore, spaceflight is no longer the sole purview of government space agencies. Private space transportation companies now develop and launch their own spacecraft, such as the SpaceX Dragon and Orbital-ATK (now Northrop Grumman) Cygnus, which are being used to deliver supplies to the ISS (Arney 2016).

The number of space launches will further increase as the need to place satellites into orbit continues to grow. The rise in commercial demand is particularly remarkable. Currently, most communication satellites are placed in geostationary Earth orbit (GEO). However, because of the long distance between the satellite and the ground, it takes a relatively long time (0.24 s) for the radio signals to travel to the satellite and back. The long delay results in high latency. With the rapid growth in Internet usage, there is a great desire to place communication satellites in LEOs. But because a low-orbit satellite can only serve a small local region, many such satellites are required to cover the entire globe. Since the 2010s, the private sector has invested heavily to build constellations of small satellites for communication and other applications. As examples, Amazon plans to launch a constellation of thousands of LEO internet satellites to provide low-latency, high-speed broadband network connectivity for users around the world. SpaceX's Starlink mission is already in the process of deploying its own mega-constellation of up to 12,000 LEO satellites to provide worldwide internet connections (Hofactor 2020a).

Currently, rockets have been used exclusively for launching vehicles to space. However, rockets are inefficient and expensive as they are nonreusable and the fact that they must carry the oxidizer means that there is little mass left for carrying payloads. The tremendous growth in the demand for commercial satellites is an impetus to develop alternative methods for conducting space launches. Already, private companies such as SpaceX and Blue Origin are making progress in reusable rockets (Klotz 2017). The SpaceX Falcon 9 is a partially reusable two-stage rocket launcher. After separating from the second stage, the first stage is capable of landing itself vertically on land or at sea, which will then be refurbished for reuse. Falcon 9 has already been used to launch the Dragon spacecraft to the ISS (Hadhazy 2017). Blue Origin's New Shephard rocket is also a two-stage rocket launching system capable of vertical takeoff and vertical landing so that the first stage can be reused (Klotz 2017).

Reusability together with small-size satellites allows Falcon 9 to launch 60 satellites in a single launch. However, the inherent disadvantage of needing to carry its own oxidizer still remains. To further improve launch efficiency and reduce cost, future technology will need to exploit hypersonic air-breathing propulsion.

1.1.2 Moon Exploration

In the field of science, there is also a growing interest for space exploration, and the Moon is a prime destination. Since Apollo 17 astronauts last stepped on the Moon in 1972, only robotic missions have taken place, and there have not been manned return flights. But there is a great interest in further exploring the Moon and using it as a base to explore Mars.

Besides the United States, there are already a number of countries that have executed or are planning missions to go to the Moon. India launched its first lunar probe, Chandrayaan-1, in October 2008, which started orbiting the Moon in November

2008. The probe's Terrain Mapping Camera was used to produce a high-definition map of the Moon. The Moon Mineralogy Mapper (M^3), a NASA instrument on board the orbiter, has confirmed the magma ocean hypothesis that the Moon was once completely molten (*Economic Times* 2009). The Miniature Synthetic Aperture Radar (Mini-SAR), another NASA instrument, detected the existence of water across wide expanses of the Moon's polar regions, which was later confirmed by the M^3 infrared data (NASA 2009). A second mission, the Chandrayaan-2 orbiter and lander, was intended as a follow-up to Chandrayaan-1's discovery of water and to identify potential water sources that may support humans to set up a base on the Moon. It was launched in July 2019 and reached the Moon's orbit in August 2019. However, the landing was unsuccessful, and the lander crashed on the Moon's surface (Singh et al. 2019).

Between 2007 and 2020, China successfully landed three spacecraft on the surface of the Moon. Chang'e-3 landed on the Moon on December 14, 2013 (Barbosa 2013) and became the first spacecraft to soft-land on the Moon since NASA's last manned mission, Apollo 17, on December 19, 1972, and the former Soviet Union's unmanned spacecraft, Luna 24, on August 22, 1976. Chang'e-3 carried a lunar rover, Yutu, which was designed to conduct surface topology and geology surveys, as well as astronomical observations. Chang'e-4 was originally built as a backup for Chang'e-3 and became available after Chang'e-3 landed successfully in 2013. On January 3, 2019, Chang'e-4 achieved the milestone of the first landing on the far side of the Moon (Hadhazy 2019). The lunar far side is distinctly different from the near side. It has relatively few lunar maria, which were formed by volcanic eruptions, but has more craters. The selected landing site is within the von Karman crater in the South Pole–Aitken basin, which is itself the oldest, largest, and deepest impact crater on the Moon. The basin is of particular interest because the powerful impact that created it 3.9 billion years ago may have ejected a vast amount of deep mantle material that remain undisturbed since the impact (Hadhazy 2019). Examining the exposed material may provide insights into the formation of the Moon and the earth as well.

Another significance of the Chang'e-4 mission is the establishment of the communication link between the earth and the Moon's far side. Because the Moon's body blocks direct communication between the earth and the Moon's far side, it had impeded lunar far-side missions. To overcome this, a relay communication satellite, Queqiao, was launched ahead of the Chang'e-4 spacecraft to a halo orbit near the Earth–Moon Lagrange point L_2, about 60,000 km beyond the Moon (Hadhazy 2019). The relay satellite provides line-of-sight links with both the Moon's far side and the earth, which may serve as a key space infrastructure for the international community, thereby opening up further opportunities for the exploration of the lunar far side.

Chang'e-5 is a lunar sample return mission that was launched on November 23, 2020. Landing took place on December 1, 2020, in the Mons Rümker region of Oceanus Procellarum. Samples were collected on the lunar surface as well as by drilling up to 2 meters into the lunar regolith. Past orbital observations indicate that the area sampled has a relatively young geological age. Determination of the samples' age and composition would yield insights into late-stage geological activities on the

Moon. Following sample collection, the ascender module lifted off from the lunar surface on December 3 and rendezvoused with the orbiter on December 5. On December 16, the return capsule reentered Earth's atmosphere at a velocity of 11 km/s and landed successfully (eoPortal Chang'e-5 n.d.).

A notable recent example of lunar mission is the Israeli Beresheet that was planned to land on the Moon in 2019. The spacecraft was designed and built by SpaceIL, a private nonprofit organization. The ambitious mission sought to make history as the first ever privately funded lunar mission. The tiny Beresheet was launched atop a SpaceX Falcon 9 rocket in February 2019. The spacecraft reached the intended landing site in April 2019, but because of engine malfunction, it was crashed upon landing (Aharonson et al. 2020). Nevertheless, the mission was a partial success for completing the long journey and coming so close to landing on the Moon successfully.

Another private company, Blue Origin, has proposed a plan for commercial Moon landing using a new Moon lander Blue Moon. Blue Origin envisions the lander to serve as an early step to develop an infrastructure for lunar exploration and pave the way for the creation of space colonies on the Moon (Hofactor 2020b). Clearly, the stage is set to hasten the exploration of the Moon, with crew return flights to occur eventually. This will fuel the need for designing the reentry capsules for such missions.

1.1.3 Near-Earth Objects

Near-Earth objects (NEOs) are comets and asteroids that have orbits that bring them into Earth's neighborhood. NEOs are the primitive leftover building blocks of the solar system. They are of particular interest because their state has remained relatively unchanged since the formation of the solar system some 4.6 billion years ago. They can provide valuable information on the chemical composition of the primordial mixture from which the planets were originally formed.

There have already been a number of flyby missions to explore NEOs since 1978. NASA's Near Earth Asteroid Rendezvous Shoemaker (NEAR Shoemaker) became the first spacecraft to land on an asteroid called Eros on February 12, 2001 (McCurdy 2005). Stardust, another NASA mission, is of significance here as it was the first to return samples from a comet to Earth, and thus hypersonic reentry flight. The Stardust spacecraft encountered Comet Wild 2 on January 2, 2004. During the flyby, Stardust deployed the Sample Collection Plate to collect dust grains from the coma of the comet (Siddiqi 2018). On January 15, 2006, the Sample Return Capsule successfully separated from Stardust and reentered Earth's atmosphere at a velocity of 12.9 km/s, the fastest reentry speed into Earth's atmosphere ever achieved by a man-made object (Desai et al. 2006).

Hayabusa was a robotic spacecraft developed by the Japan Aerospace Exploration Agency (JAXA) to collect samples from the small near-Earth asteroid 25143 Itokawa and return them to Earth. The spacecraft was not intended to land on the asteroid but simply to touch the surface with its sample-capturing device and then moves away. In November 2005, Hayabusa retrieved samples from Itokawa (Grinstead et al. 2011).

The reentry capsule carrying the collected samples reentered Earth's atmosphere in June 2010. At a reentry velocity of 12.2 km/s, it became the second fastest reentry into Earth's atmosphere after NASA's Stardust reentry (Grinstead et al. 2011).

Hayabusa 2 is JAXA's second asteroid sample return mission. It was launched on December 3, 2014, and it arrived at asteroid 162173 Ryugu on June 27, 2018. On February 21, 2019, the spacecraft approached the surface of Ryugu to conduct surface sampling (eoPortal Hayabusa-2 n.d.) A 5-g tantalum projectile was fired at 300 m/s into the asteroid's surface, and the ejected particles were collected by a catcher in the spacecraft (JAXA 2019). On April 5, 2019, subsurface sampling was initiated by firing a 2-kg copper projectile at 2 km/s to excavate deeper materials that have not been subjected to space weathering. The spacecraft returned to the impact site on April 25, 2019, to confirm that a crater was created. On July 11, 2019, the spacecraft touched down on the asteroid and collected samples from inside the crater. The spacecraft started its return journey to Earth on November 13, 2019. On December 6, 2020, the sample return capsule executed reentry to Earth at a velocity of 12 km/s (eoPortal Hayabusa-2 n.d.)

The NASA OSIRIS-REx (Origins, Spectral Interpretation, Resource Identification, Security, Regolith Explorer) is an asteroid study and sample-return mission. If successful, it will be the first spacecraft from the United States to return samples from an asteroid. It was launched in September 2016 and reached the proximity of the asteroid Bennu in December 2018. Sample collection was completed in October 2020 and the reentry to Earth is expected to occur in September 2023 (NASA 2020). Besides being a primitive asteroid that has not significantly changed since its formation several billion years ago, it is also a potentially hazardous asteroid, as it has a small probability of impacting Earth in the late twenty-second century. The samples will allow the determination of the asteroid's physical and chemical properties. Another mission objective is to study the effect of nonuniform sunlight heating on the asteroid's rotating body and its impact on Bennu's orbit. The findings will be used to develop better estimates of the asteroid's impact probability with Earth (Lauretta et al. 2017).

Besides NASA and JAXA, the European Space Agency (ESA) has sent the Rosetta spacecraft to study Comet 67P/Churyumov–Gerasimenko. Its lander module, Philae, performed the first successful landing on a comet on November 12, 2014 (Costa et al. 2016). China is proposing to send a probe to collect samples from asteroid 2016 HO3 (Kamo'oalewa). The probe will fly back to the proximity of Earth, and a reentry capsule will be released to return the samples to Earth. After that, the probe will resume its journey to the main asteroid belt to explore Comet 133P (Gibney 2019).

1.1.4 Mars Exploration

There is a lot of interest to explore and study Mars, the most accessible planet in the solar system. There have already been a number of missions that flew robotic devices to orbit and land on Mars. As of February 2021, the United States had successfully deployed five rovers (Sojourner, Spirit and Opportunity, Curiosity, and Perseverance) to Mars. The latest rover was Perseverance of NASA's Mars 2020 mission. It was launched on July 30, 2020, and landed on Mars on February 18, 2021, in Jezero Crater, a site that was once an ancient lake. The collected rock and soil samples could potentially be returned to Earth in future missions (NASA Mars 2021, NASA Perseverance 2020). NASA's Mars 2020 mission was joined by two other Mars missions. On July 20, 2020, the United Arab Emirates Space Agency's Hope spacecraft was launched from Tanegashima Space Center, Japan, with a Japanese rocket H-IIA. It entered Mars orbit on February 9, 2021, to study the atmosphere of Mars for two years (Abbany 2021). China's Tianwien-1 was launched from the Wenchang Spacecraft Launch Site on July 23, 2020, and entered Mars orbit on February 10, 2021. The spacecraft consists of an orbiter and a lander, with the latter carrying a rover, Zhurong, that was scheduled to land in May 2021 to explore the Mars surface (Abbany 2021). On May 14, 2021, Zhurong successfully landed on Utopia Planitia – the vast Martian plain where the NASA Viking 2 spacecraft previously landed in the 1970s (Planetary Society n.d.).

The scientific goals for these missions are to determine whether Mars has ever had and has still water, and whether it has chemicals necessary for living organisms to grow. These questions cannot be answered easily using the instruments transported to Mars. Thus, there is a strong desire to return Martian samples to Earth for detailed analysis. The ultimate goal is to send humans to explore Mars and return them back safely.

Although Mars exploration has been a goal of national space programs since the 1960s, necessary technologies to accomplish manned missions are still being developed. While NASA does not yet have definite plans for a human Mars mission, SpaceX is developing the Falcon Heavy rocket and is planning to use it for crewed Mars missions (SpaceX Falcon 2021). We can expect the increase in the need for the design of vehicles for atmospheric entry into Mars and reentry to Earth from Mars exploration missions.

1.1.5 Other Planets and Moons

There are other interesting planets and moons to explore. For example, Titan, which is Saturn's largest moon and is the second-largest moon in the solar system after Jupiter's Ganymede. It is the only planetary body other than Earth in our solar system with a thick nitrogen atmosphere. It has a surface pressure about 1.5 times that on Earth. Methane is the second-most abundant atmospheric constituent on Titan, but as it is constantly destroyed by solar UV, cosmic rays, and electrons from Saturn's magnetosphere (Owen et al. 1997), it leads to the question of how it gets replenished in Titan (Lebreton et al. 2005). Are there hydrocarbon oceans that serve as reservoirs for Titan's atmospheric methane? Another significance of Titan is that its atmosphere is similar to that of early Earth when free molecular oxygen had not existed yet. The absence of molecular oxygen is known to be beneficial for the origin of life. Therefore, Titan can serve as a planet-sized laboratory of a primitive Earth atmosphere to study how life began (Owen et al. 1997).

The Huygens probe was part of the Cassini–Huygens mission to investigate Titan. It was released from the Cassini orbiter and headed toward Titan on December 25,

2004. On January 14, 2005, the probe entered the top layers of Titan's atmosphere at an altitude of approximately 1,400 km with a speed of 6 km/s. It then decelerated rapidly to 400 m/s (approximately Mach 1.5) at an altitude of about 155 km in less than 5 min, at a point where the main parachute was deployed. The descent took about 2.5 hours before it landed successfully on Titan (Lebreton et al. 2005).

Although the Cassini–Huygens spacecraft had spent 13 years orbiting and observing Saturn, and its journey to Saturn included flybys of Venus, the asteroid 2685 Masursky, Jupiter, and Saturn's other moons (Phoebe, Enceladus, Iapetus, Rhea, Hyperion, and Dione), its entry into Titan's atmosphere was the highlight as it was the first ever landing in the outer solar system and on a moon other than that of Earth (Lebreton & Matson 2002). Huygens' data provided valuable information on the physical properties of Titan's atmosphere as well as the features of its surface. However, much more is needed to understand this moon, and the answers to the many new and remaining questions will have to wait for future exploration efforts.

Another interesting planet to explore is the gas giant Jupiter that is the largest planet in the solar system. On December 7, 1995, the Galileo probe descended into the Jovian atmosphere. At 48 km/s, the atmospheric entry into Jupiter is the most severe for any human-made spacecraft. During the descent, the probe collected data on Jupiter's atmospheric conditions and composition for 61 min until the transmitter failed due to the enormous atmospheric pressure (Siddiqi 2018). Much longer missions to study Jupiter have also been proposed. One of them uses an autonomous supersonic aircraft powered by a nuclear ramjet engine that can be transported in a spacecraft similar to the size of that which performed the Galileo mission (Maise et al. 2003, Sforza 2012). After arriving at Jupiter, an entry capsule carrying the nuclear ramjet will enter into the Jovian atmosphere. When the capsule decelerates to about Mach 3, the ramjet separates from it, and then the nuclear engine starts and uses the unlimited hydrogen in the Jovian atmosphere as a propellant to propel the vehicle that would then fly continuously and collect data for months. Hydrogen is an excellent propellant because of its low molecular weight, and so the nuclear ramjet is ideal for exploration in the two largest gas giants, Jupiter and Saturn, with both having hydrogen as their main atmospheric constituent. However, because essentially any gas could be used as a propellant, the nuclear ramjet could operate in any planetary atmosphere.

1.1.6 Suborbital Hypersonic Flight and Air-breathing Propulsion

We now turn our attention to suborbital hypersonic flight. Entries into Earth's and other celestial bodies' atmospheres are not the only conditions that involve hypersonic flight. Hypersonic flight also occurs when an aircraft travels within the atmosphere at hypersonic speeds. Although rockets are currently used to launch spacecraft to space, they only reach hypersonic speeds as they approach the upper edge of Earth's atmosphere.

A major disadvantage of rockets is that they must carry the oxidizer on the vehicle, which means that the amount of payload that can be carried is severely limited.

To overcome this, an air-breathing propulsion system that utilizes the surrounding air, which is plentiful in the atmosphere, would free up valuable vehicle mass and space for carrying payloads. Such air-breathing propulsion systems could use a combination of turbine engine, ramjet, scramjet, and rocket to accelerate the aircraft-like vehicle gradually from the ground to hypersonic speeds. As the vehicle reaches the upper atmosphere where the air is too thin to support air-breathing propulsion, it transitions to rocket power for the final orbit insertion.

Another application of hypersonic flight is suborbital cruise. In this regard, an air-breathing engine would propel the aircraft to cruise at hypersonic speeds over a long distance. Although rockets can also cruise at high Mach numbers in an atmosphere, their range is limited because of the burden of carrying their own oxidizer.

Whether air-breathing engines are used for accelerating vehicles to space or for suborbital cruise, the hypersonic vehicles involved have some key differences from those used for atmospheric entry. For atmospheric entry, the main objective is to decelerate the vehicle as rapidly as possible from orbital or superorbital speeds down to subsonic values. The vehicle shape is, therefore, designed for high drag and for minimizing extreme aeroheating. This results in blunt-shaped bodies, especially when convective heating dominates over radiative heating.

On the other hand, hypersonic vehicles propelled by air-breathing engines would have to stay in the much denser part of the atmosphere for a much longer time, as a large amount of air is needed to be ingested to generate thrust. Thus, hypersonic atmospheric flights can also experience severe aeroheating just as for flights during atmospheric entry. Furthermore, because hypersonic cruise vehicles must also have high aerodynamic performance, that is, good lift–drag characteristics, the body shape would need to be slender rather than blunt, which allows the extreme aeroheating to have a larger impact on the vehicle.

The supersonic combustion ramjet (scramjet) has been studied since the 1960s. The first aircraft that achieved hypersonic speed was the X-15. This rocket-powered plane was developed to study hypersonic flight and was to be used as a test bed to test hypersonic air-breathing engines. The X-15 reached a speed of Mach 6.7 in 1967, but the program was canceled in 1968 (Urzay 2018). The most ambitious hypersonic program that involves air-breathing propulsion was the National Aero-Space Plane (NASP) in the United States to create a scramjet-powered single-stage-to-orbit space plane. The program started in the late 1980s but was canceled in 1993 with no test flight flown (Urzay 2018). A milestone was achieved in 2004 when NASA successfully flew a self-propelled scramjet-powered unmanned vehicle, X-43A, at Mach 10 (Urzay 2018). As the scramjet operated only for about 10 s during the test flight, the X-43A does not represent a true aerospace plane that can cruise long distances through the atmosphere. The next major program in the United States to progress scramjet-powered flight was the X-51, which was started in 2004. One of the program's goals was to extend the operation of scramjet–powered flight from 10 s to several minutes. In May 2013, an unmanned X-51A successfully flew at Mach 5.1. The scramjet engine burned for 210 s to accelerate the vehicle to Mach 5.1, in the longest ever mission for a vehicle of its kind (Urzay 2018).

Another type of hypersonic cruise is that of the hypersonic glide vehicle (HGV). It is a class of suborbital flight vehicle that is boosted by a rocket to the high upper atmosphere, which then employs aerodynamic lift to glide over a long range at high altitudes before reentering the atmosphere. The DARPA Falcon (Defense Advanced Research Projects Agency's Force Application and Launch from CONtinental United States) program test flew two Hypersonic Technology Vehicle 2 (HTV-2): one in 2010 and another in 2011. The HTV-2 is an experimental HGV capable of flying at Mach 20 near the edge of the atmosphere (Malik 2012). As gliders, the HTV-2s did not have propulsion. However, technologies being tested have great relevance to scramjet-propelled hypersonic flight. In contrast to other reentry vehicles, the HTV-2 has a slender shape to improve its aerodynamic performance in order to achieve the flight range. Both this and the long flight duration pose a major challenge regarding aeroheating. Since 2014, China has tested various hypersonic gliders, such as the DF-ZF, reaching speeds between Mach 5 and Mach 10 (Wood & Cliff 2020). Russia also tested the Avangard HGV from 2015 to 2018, which is designed to operate at Mach 20 at the edge of the atmosphere (Button 2018).

Up until the 2000s, the development of hypersonic atmospheric flight had not been sustainable, and developmental efforts were followed by periods of stagnation. However, research and development seems to have intensified again in the 2010s. The scramjet is a high risk, but potentially a high pay-off, technology. The countries that can first master this technology and exploiting it in weapons systems will have a significant advantage over the others. Thus, it is not surprising to find that countries are racing to progress the readiness of hypersonic air-breathing propulsion.

In the United States, Lockheed-Martin is reportedly developing a hypersonic SR-72 vehicle capable of reaching Mach 6 as a successor of the retired reconnaissance plane SR-71 (Norris 2013). Reaction Engines in the United Kingdom is testing the Synergetic Air Breathing Rocket Engine (SABRE) hybrid air-breathing rocket engine that would rapidly precool the incoming air with liquid hydrogen to allow a gas turbine engine to continue working up to Mach 5.5, at which point, the engine would transition to rocket mode and accelerate the aircraft to orbit (Odom & Johnston 2018). The Indian Space Research Organization (ISRO) conducted an experimental hydrogen-fueled Mach 6 scramjet engine that operated for about 5 s in 2018 (Odom & Johnston 2018). In Japan, JAXA plans to test an experimental airplane, HIMICO (High-Mach Integrated Control Experiment), which uses liquid hydrogen to precool the incoming air in order to extend the operability of conventional turbine engines to Mach 5 (Odom & Johnston 2018). Brazil plans to flight test a prototype of the 41-X unmanned hypersonic airplane with a scramjet engine at an altitude of 30 km and a Mach number of 7 (Martos et al. 2017).

Intense research is currently being undertaken in various countries to develop the scramjet into a mature technology. The initial application of hypersonic air-breathing propulsion will always be in weapons of destruction, but eventually, the peaceful application of it in the next generation transport would hopefully follow to benefit all mankind.

1.2 Hypersonic Aerothermodynamics: Key Issues and Research Progress

A selected list of key issues related to hypersonic flow and the development of hypersonic flow theory will be briefly discussed in this section. The list is not intended to be exhaustive, and the introductory remarks in Sections 1.2.1 to 1.2.5 below serve mainly to bring out the significance and challenges of the hypersonic flow regime rather than to explore the theory in-depth. Examples of experimental studies that have been carried out in test facilities on some of the topics will be discussed in subsequent chapters in the book.

1.2.1 Hypersonic Flow Defined

A flow is termed hypersonic when the freestream velocity U_∞ is much larger than the speed of sound a_∞. Using the definition of Mach number, which is the ratio of the velocity to the sound speed in the freestream, we have

$$M_\infty = \frac{U_\infty}{a_\infty} \gg 1. \tag{1.1}$$

The Mach number is a measure of the square root of the kinetic energy of the gas to its thermal energy. For hypersonic flows, the freestream kinetic energy is much larger than the thermal energy of the freestream fluid particles. In particular, the term "hypersonic" is used to distinguish those flows in which the velocity is so large that the conditions behind the shock waves on a body are such that the gas is no longer inert. As the freestream Mach number increases beyond 1.0, shock waves form over the body. The rise in temperature behind the shock waves, which increases with the freestream Mach number, eventually causes the gas molecules to be vibrationally excited. As the Mach number increases further, the chemical bonds that bind the molecules together start to break, leading to dissociative reactions. At even higher Mach numbers, ionization occurs. These processes are often collectively referred to as "real-gas effects."

Although it is commonly accepted that "hypersonic" refers to speeds greater than about Mach 5, there are better quantitative measures to distinguish it from the supersonic regime. Indeed, it is rather imprecise to define the threshold of "hypersonic" based on a Mach number, because the phenomena that characterize hypersonic flight depend on the flow conditions as well as on the body shape and its orientation.

In order to better define the hypersonic regime, we will examine two of its characteristics. The first is the hypersonic Mach number independence principle. The principle of hypersonic Mach number independence was first derived by Oswatitsch (1951, 1980) for a calorically perfect gas with no viscosity. The derivation of the principle can be readily found in the literature, such as the book by Anderson (2006). It can be shown that at very large freestream Mach numbers, certain nondimensional aerodynamic quantities, such as pressure coefficient, lift and wave-drag coefficients, and flow field structure become independent of the Mach number. A key parameter that characterizes Mach number independence is

$$M_\infty^2 \sin\beta \gg 1, \tag{1.2}$$

where β is the local shock-wave angle. Equation (1.2) shows that hypersonic Mach number independence begins earlier for blunt bodies than for slender ones. For blunt bodies, sin β is large, so that Eq. (1.2) is satisfied at lower M_∞ than for slender bodies. The parameter also depends on the orientation of the flight body. For a slender body at a high angle of attack, sin β is large enough so that it fulfills the Mach number independence principle at a lower Mach number than at zero angle of attack.

While Oswatitsch's principle does not include viscous and high-temperature reacting gas effects, ballistic range measurements have demonstrated that the drag coefficients for the sphere and cone–cylinder configurations do approach their respective constant values at high Mach numbers (Charters & Thomas 1945, Stevens 1950, Hodges 1957, Cox & Crabtree 1965).

Another important feature of hypersonic flow is the occurrence of real-gas effects. At high Mach numbers, the temperature downstream of the bow shock is sufficient to cause vibrational excitation, dissociation, and even ionization. Since the post-shock temperature for a normal shock is higher than that for oblique ones, and as blunt bodies have a larger portion of the surfaces exposed to the normal shock, they experience reacting gas effects at lower Mach numbers than slender bodies. For air at a pressure of one standard atmosphere, molecular oxygen begins to dissociate at about 2,500 K. At 4,000 K, molecular oxygen is mostly dissociated, and molecular nitrogen begins to dissociate. At about 9,000 K, molecular nitrogen is mostly dissociated and ionization begins (Anderson 2006). Thus, at standard sea level freestream conditions, reacting gas effects would start at a temperature of approximately 2,500 K, which would correspond to a post-shock temperature for the freestream Mach number of about 8. However, dissociative reactions also depend on the pressure and therefore on the flight altitude. At lower pressures or high altitudes, dissociation occurs at lower temperatures and therefore lower Mach numbers.

A more illustrative picture of the level of real-gas effects experienced during flight can be obtained by relating it to the flight speed. Consider a flight body flying in Earth's atmosphere at a certain Mach number that is larger than 1, giving rise to a bow shock in front of the body. As the flight Mach number increases, the temperature of the shock-heated air will eventually be sufficiently high to cause chemical reactions to occur. The chemical composition of the air behind the normal portion of the shock can be calculated readily using a chemical equilibrium analysis (see, e.g., McBride & Gordon 1996). Shown in Figure 1.1 are the species mole fractions of the air behind a normal shock at freestream Mach numbers ranging from 1.5 to 40. For simplicity, a flight altitude of 30 km is assumed, corresponding to an ambient pressure of 1,200 Pa, temperature of 226 K, and sound speed of 302 m/s. At Mach numbers below 5, the air is quite inert, comprising N_2 and O_2 with mole fractions of 79% and 21%, respectively. At about Mach 6, dissociation of the N_2 and O_2 molecules starts and a small amount of NO is formed. At about Mach 8, a significant dissociation of O_2 to O begins and is completed near Mach 15, at which point a significant dissociation of N_2 to N

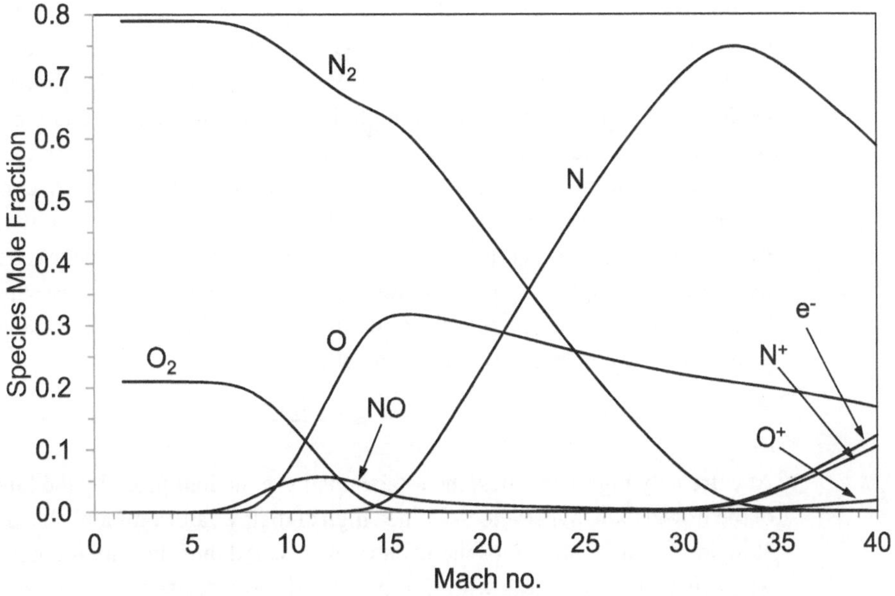

Figure 1.1 Species mole fraction of air behind a normal shock at different freestream Mach numbers at an altitude of 30 km.

starts. By about Mach 30, N_2 is completely dissociated and ionization begins, giving rise to free electrons e^- and ions N^+ and O^+. It is of interest to note three flight Mach numbers – 25, 36, and 40. Mach 25, or a flight velocity of about 7.5 km/s, is the typical reentry velocity from LEOs. As can be seen from Figure 1.1, there is a complete dissociation of O_2 and a significant dissociation of N_2 at this speed. Mach 36, or approximately 11 km/s, is encountered during a flight's reentry into the Earth's atmosphere following lunar missions. At this Mach number, N_2 is completely dissociated, and ionization has started. For spacecraft returning from missions to comets and asteroids, the Earth reentry Mach number is around 40 or a velocity of 12 km/s, at which point ionization becomes appreciable.

Since Figure 1.1 is obtained for a fixed flight altitude, and thus the freestream pressure, one should bear in mind that real-gas effects generally increase at lower pressures. A flight velocity–altitude map with regions of vibrational excitation, dissociation, and ionization superimposed over the map has been computed by Tauber et al. (1987). It provides the same information as Figure 1.1 but adds details regarding variations with the flight altitude. Note also that the intensity of real-gas effects depends on the vehicle's shape and orientation as well. For shock-wave angles smaller than that for a normal shock, as encountered on slender bodies, the post-shock temperature will be lower, leading to weaker real-gas effects. Nevertheless, the general conclusion is that a true quantitative description of the flow encountered in hypervelocity flight must include real-gas effects.

Note that "cold hypersonic" flows can be generated by cooling the fluid at a medium to very low temperature to decrease its sound speed so that high Mach numbers can be reached with relatively low velocities. However, such high Mach number flows must be differentiated from those in real flight because the low temperature in cold hypersonic flows precludes chemical reactions and the associated real-gas effects.

Other commonly used terms to describe extremely high-velocity conditions are "hypervelocity" and "high enthalpy," both avoid the ambiguity of low-temperature hypersonic flows. "Hypervelocity" represents both high Mach number and high velocity. "High enthalpy," or more appropriately high total enthalpy, essentially means high velocity. This can be seen in the definition of total or stagnation enthalpy, which is the sum of static enthalpy and kinetic energy:

$$h_t = h + \frac{U_\infty^2}{2}. \tag{1.3}$$

At extremely high velocities, the total enthalpy h_t is dominated by the kinetic energy of the flow. Both "hypervelocity" and "high enthalpy" are commonly used in the literature in the context of atmospheric entry events, and thus they are more precise terms when characterizing flows with significant real-gas effects.

The preceding discussions on Mach number independence and reacting gas effects indicate that the Mach number is no longer the best parameter to characterize hypersonic flows. At large Mach numbers, the kinetic energy far exceeds the thermal energy. A more suitable reference energy quantity, which reflects that the substantial reacting processes are taking place, can be utilized to construct a different nondimensional parameter. For example, for reentry from LEOs, the dissociation energy of nitrogen could be used as a reference quantity. However, this is not a widely accepted practice.

1.2.2 Stagnation Region

Consider a flight vehicle flying at a very high velocity. When viewed in the reference frame of the vehicle, as the incoming air is decelerated and brought to rest near the vehicle's surface, the flow kinetic energy is converted into thermal energy, thus heating both the airflow and the vehicle body. The shock wave formed around the body heats the airflow, while the intense friction in the boundary layer heats the body.

Temperatures and surface pressures are usually the greatest at the stagnation point, which is downstream of the normal portion of the bow shock wave over the vehicle. This leads to extremely high heat transfer to the vehicle.

The heat transfer in the stagnation region became an important problem early on in the development of hypersonic flow theory. In the 1950s, aerodynamic heating was a major challenge for the development of intercontinental ballistic missiles (ICBMs) as ICBMs have to endure extreme heating during reentry into the atmosphere. It was initially thought that a reentry body should have a conical sharp nose with a pointed tip, much like an artillery shell. However, with the sharp-nose cone, a conical shock wave is attached that is relatively weak and that does little to heat the airflow, whereas the severe heat transfer from the boundary layer into the vehicle's body could melt the sharp nose and even destroy the vehicle structure.

Harvey Allen came up with a revolutionary idea that by diverting more of the reentry energy into the airflow, there is less left to heat the vehicle. This is achieved by using a blunt nose that generates a strong detached shock wave standing ahead of the vehicle, so that most of the heat goes off the surface and into the flowfield, not into the vehicle.

Allen's blunt-body analysis (Allen & Eggers 1953) used simplified assumptions that neglected chemical reactions and radiation to show that the maximum convective heat transfer rate q_{conv} is inversely proportional to the square root of the leading-edge nose radius at the stagnation point:

$$q_{\text{conv}} \propto \frac{1}{\sqrt{R_N}}. \tag{1.4}$$

Thus, to reduce the stagnation-point heat transfer rate, the nose radius R_N should be as large as possible. The blunt-body concept led to the choice of the manned reentry capsule shapes for the Mercury, Gemini, and Apollo vehicles. Indeed, all successful reentry bodies have been blunt because of this concept. Allen's original report (Allen & Eggers 1953) was classified as it was intended for ballistic missile applications, and it became declassified and publicly released in 1958 (Allen & Eggers 1958).

The chemically reactive nature of hypersonic flows can have a profound influence on the heating of the vehicle's surface. Fay and Riddell (1958) formulated the equations to calculate the convective heat transfer rate at the stagnation point of a blunt body in chemically reacting air. Fay and Riddell's theory was derived using laminar boundary-layer equations applied to the stagnation region of spherical and cylindrical geometries. The equations include different possible scenarios: chemical equilibrium and frozen boundary layers, and catalytic and non-catalytic walls.

The stagnation-point heat transfer obtained from Fay and Riddell's equations could be used both to verify the validity of wind-tunnel measurements and also computational fluid dynamics (CFD) solutions. It is also often used as a denominator to construct a nondimensional heat transfer ratio at different locations along the vehicle's surface. This dimensionless parameter could then be used to extrapolate wind-tunnel model measurements to actual flight conditions.

Fay and Riddell (1958) noted that while the process of dissociation and recombination is important for determining the thermodynamic state of the air throughout the flowfield, the effect on heat transfer is secondary. Also, the result is dominated by the freestream flow properties, with only weak dependence on the condition of the wall's surface. Thus, an approximate value for the stagnation-point convective heat transfer rate for a sphere can be computed using the following simple equation (Bertin 1994):

$$q_{\text{conv}} = \frac{C}{\sqrt{R_N}} \rho_\infty^{0.5} U_\infty^3, \tag{1.5}$$

where ρ_∞ and U_∞ are the freestream density and velocity, respectively, R_N is the nose radius, and C is a constant. For reentry conditions into Earth, Tauber (1989) obtained an identical approximate equation where C equals 1.83×10^{-4} when SI units are used and the resulting heat transfer rate is in W/m^2. Similar correlations have been obtained by other authors with slightly different values of the exponents for the density and velocity. Equivalent equations for different atmospheric gases in other planets have also been obtained by various authors, such as Marvin and Deiwert (1965), Sutton and Graves (1971), and Justus and Braun (2007).

The significance of Eq. (1.5) is that the convective heat transfer rate at the stagnation point scales inversely with the square root of the nose radius, as originally pointed out by Allen and Eggers (1953). The larger the nose radius, the smaller the stagnation-point heat transfer rate. To minimize convective heating, blunt forebody shapes with large nose radii have been used on reentry vehicles.

For many problems of interest, such as those for suborbital flight and the reentry from LEOs, the heat transfer is dominated by convective heating. However, when the flight speed reaches about 10 km/s, radiative heating becomes significant and will eventually exceed convective heating as flight speed increases further. For example, radiation was the dominant mode of heat transfer for the Galileo probe that entered Jupiter's atmosphere at a velocity of about 50 km/s.

For an emitting and nonabsorbing gas (transparent gas), the stagnation-point heat transfer rate due to radiation can be approximated by (Anderson 2006)

$$q_{\text{rad}} = \frac{E}{2} R_N \frac{\rho_\infty}{\rho_s}, \qquad (1.6)$$

where E is the total energy emitted by the gas in all directions per second per unit volume, ρ_∞ and ρ_s are, respectively, the density in the freestream and behind the normal shock, and R_N is the nose radius. Equation (1.6) shows that the radiative heat transfer is directly proportional to the nose radius, whereas in Equation (1.5), the convective heat transfer is inversely proportional to the square root of the nose radius. Thus, the design of superorbital reentry vehicles would need to find a compromise between the two phenomena. To minimize convective heating, the nose radius should be made large, while to minimize radiative heating, the nose radius should be small.

With the advent of modern computers and the advances of CFD, a detailed solution of the stagnation region flowfield can be readily obtained. To achieve high-fidelity solutions, it is necessary to have accurate modeling of the physical and chemical processes at the temperatures encountered during hypersonic flight. The validation of these models requires experimental data. However, it remains extremely difficult to replicate the flow conditions in ground testing facilities.

1.2.3 Viscous Interactions

Localized high heating can also occur because of viscous effects. There are two important viscous interactions in hypersonic flow. The first is associated with the interaction between the boundary layer and the inviscid flowfield. In classical low-speed boundary-layer theory, the boundary-layer thickness is assumed to be small so that viscous effects are confined to the thin layer near the surface. The outer inviscid flow remains unaffected by the presence of the boundary layer. In the hypersonic regime, the flow deceleration from the freestream to the boundary layer converts a large amount of the freestream kinetic energy into thermal energy. If the static pressure remains uniform across the boundary layer, the density becomes very low in the boundary layer, thus making it exceptionally thick. The presence of the boundary layer can weakly or strongly affect the inviscid outer flow, which in turn feeds back to alter the characteristics of the boundary layer. Such interaction between the outer and viscous flows is referred to as pressure interaction.

The second interaction is the shock-wave/boundary-layer interaction that is due to the impingement of a strong shock wave on a boundary layer. Because of such interaction, the boundary layer is disturbed and may even separate depending on its thickness and the strength of the shock.

Shock-on-shock interaction is part of the shock-wave/boundary-layer interaction family, and it can cause high localized heating. Early studies have reported a large increase in the local heat transfer rate as an extraneous shock impinges on the leading edge of a blunt body over a wide range of Mach numbers, from 2.65 to 19 (Newlander 1961, Siler & Deskins 1964, Francis 1965, Ray & Palko 1965, Hiers & Loubsky 1967). Edney carried out a thorough investigation of shock–shock interactions and presented the definitive treatise on the entire spectrum of interaction patterns. The details of his work were documented in the Flygtekniska Försöksanstalten (FFA) report (Edney 1968a), and a summary was published in Edney (1968b).

Shock–shock interaction's potential to cause catastrophic failure was fully realized in 1967 when a rocket-powered X-15 aircraft that carried a dummy ramjet engine suffered severe damage to the engine support pylon during a Mach 6.7 flight. The shock wave that propagated from the nose cone of the dummy engine impinged on the pylon, causing the interaction between the conical shock with the bow shock around the pylon. This interaction generated intense heat that burned through the pylon resulting in the loss of the dummy engine; it also caused severe damage to the aircraft's internal structure (Watts 1968).

Shock–shock interaction is of great importance in hypersonic flight as it can cause severe heating that can result in major structural damages, as shown in the case of the X-15 flight above. The interaction has great implications for hypersonic air-breathing engines as the leading-edge shock wave originating from the vehicle forebody is often designed to impinge on the inlet cowl lip. The interaction is also a concern on winged reentry vehicles, such as the space shuttle, as the bow shock originating from the nose can impinge on the wing leading edge. Another possible occurrence is during the flight of a hypersonic missile, the bow shock that originates from the nose cone can interact with the shocks around a booster or fuel tank attached to the missile. The effects of shock–shock interaction must therefore be quantified in ground experiments in order to validate hypersonic vehicle designs.

1.2.4 Hypersonic Trim Anomaly

The high-temperature chemically reacting nature of hypersonic flow can have a large impact on aerodynamic forces and moments. This fact was not fully recognized until the Space Shuttle's first flight (STS-1). During its first reentry, the Space Shuttle Columbia experienced a much larger nose-up pitching moment than the preflight prediction. The body flap had to be deflected more than twice the predicted amount (15° instead of 7.5°) – almost running out of available deflection – in order to achieve trimmed flight. This phenomenon has been referred to as hypersonic trim anomaly, which was also observed in Apollo flights (Crowder & Moot 1965, Park 1996a).

The predicted aerodynamic characteristics of the Space Shuttle were based on extensive wind-tunnel testing that spanned a wide range of Mach numbers ranging from subsonic to hypersonic, as documented in the preflight Aerodynamic Design Data Book (Anon. 1980). However, although real-gas effects were considered prior to flight, the wind-tunnel data did not include conditions that could simulate high-temperature hypersonic phenomena. Thus, real-gas effects were not adequately accounted for in the preflight predictions.

Investigations were conducted to resolve the observed aerodynamic anomaly. It was found that high-temperature real-gas effects can result in the pitching moment discrepancy (Maus et al. 1984), although other mechanisms, such as viscous–rarefaction effects (Koppenwallner 1987), could also be plausible explanations (see the discussion on pages 141–147 in Bertin [1994]). Nevertheless, there has been an increasing body of literature that supports real-gas effects as the most likely cause of the pitch-up anomaly (Brauckmann et al. 1995, Park 1995, Bertin & Cummings 2006). This realization is one of the driving forces for the development of large-scale shock-tunnel facilities in the world to replicate high-temperature real-gas effects at hypersonic flight conditions. The biggest facilities, including both the HEG in Germany and HIEST in Japan, were intended to assess high-temperature real-gas effects on aerodynamic characteristics.

1.2.5 Boundary-Layer Transition

Laminar–turbulent boundary-layer transition has great importance in hypersonic flight. For reentry vehicles, boundary-layer transition can drastically increase aeroheating, further aggregating the already severe problem. For vehicles powered by air-breathing propulsion systems, much of the hypersonic flight in the atmosphere takes place at conditions where the flow is transitional. This impacts not only the external aerodynamics and heating but also the flow characteristics in the engine and thus have ramifications for its performance. In the hypersonic regime, real-gas effects can influence the state of the boundary layer, as dissociation reactions can absorb the flow energy and have a stabilizing effect.

The prediction of boundary-layer transition is always a challenge, even for low speed, incompressible flows. The problem becomes more difficult at hypervelocity conditions. One of the difficulties in developing effective prediction tools is the lack of flight and ground testing data.

Flight experiments provide the most valuable data as they are taken at the actual flow conditions (see the survey by Schneider [1999]), but these are very expensive. Thus, only a handful of hypersonic flight experiments have been conducted. Two examples of such experiments are the Reentry-F in 1968, which was a slender cone boosted to Mach 20 (Wright & Zoby 1977) and has become a benchmark for ground experiments, and the Space Shuttle BLT experiment in 2006, which measured the effects of protuberances on the onset of transition (Berry et al. 2011). The most recent flight experiment that was specifically designed to study hypersonic boundary-layer transition was the NASA Hypersonic Boundary Layer Transition (HyBoLT) in 2008 (Chen & Berry 2010). The test model was mounted on top of an ATK rocket that was intended to collect data at Mach 5 and 7. The model is a double-sided symmetric wedge with a smooth flat plate on one side to study natural transition and a surface with roughness on the other side to study roughness transition. The flight took place on August 22, 2008, but was terminated prematurely as the rocket booster deviated off-course after lift-off (Berry et al. 2011).

On the ground, it remains a challenge to replicate the extreme flow conditions while avoiding large facility noise that can cause interference in the transition process. As will be seen in Chapter 4, significant progress has been made in using high enthalpy shock tunnels to study transition.

1.3 Simulation Requirements for Hypersonic Flow

In order to simulate high-temperature real-gas effects in ground testing facilities, a number of dimensionless parameters need to be replicated. As in other wind-tunnel simulations, the usual nondimensional parameters that need to be matched include Mach number, Reynolds number, ratio of specific heats, Prandtl number, and Lewis number. These parameters remain adequate for "cold" hypersonics where real-gas effects and chemical reactions do not dominate. Because of the high temperature and chemically reacting nature of hypersonic conditions, proper simulation requires additional parameters to be considered.

Since chemical reactions are highly sensitive to the gas composition, temperature, and pressure, it is hopeless to satisfy similarity unless the gas medium used in ground testing is the same as that in actual flight. As chemical reactions are driven by the energy content in the flow, duplication of the total enthalpy, and hence the velocity, is required. Furthermore, in order to simulate the thermochemical state of the flow over the vehicle, the duplication of Damköhler's first number, Da_1, is also required. Damköhler's first number is defined as the ratio of characteristic flow time to the chemical reaction time:

$$Da_1 = \frac{\text{Characteristic flow time}}{\text{Chemical reaction time}}. \tag{1.7}$$

Damköhler's first number is often rewritten in terms of length, as the freestream velocity can be used to construct both length scales:

$$Da_1 = \frac{\text{Characteristic flow length}}{\text{Chemical reaction length}}. \tag{1.8}$$

Thus, when Da_1 is duplicated, the same chemical reactive states occur at the same locations on a small-sized wind-tunnel model as those on the flight vehicle.

For chemical reactions dominated by two-body molecular collisions (e.g., dissociative reactions), Damköhler's first number leads to an important scaling parameter, the binary scale, which is the product of the flow density ρ and the characteristic length L:

$$\text{Binary scale} = \rho \times L. \tag{1.9}$$

The binary scale parameter can be derived for two-body chemical reactions and can be found in Anderson (2006).

If the temperature is also duplicated – as chemical reactions are very sensitive to temperature – the static pressure scales with the density according to the perfect gas equation:

$$p = \rho RT \sim \rho. \tag{1.10}$$

Thus, binary scaling can also be stated as

$$\text{Binary scale} = p \times L. \tag{1.11}$$

Furthermore, when the velocity and temperature are also duplicated, duplicating ρL implies that the Reynolds number, Re, is automatically duplicated:

$$\begin{aligned} Re &= \frac{\rho UL}{\mu}, \quad \mu = \mu(T)\,\text{duplicated}, \\ Re &\sim \rho L \end{aligned} \tag{1.12}$$

The duplication of the binary scale parameter implies that to test a small-scale model in the wind tunnel, the density (and therefore the pressure) needs to be raised. Since the test flow (freestream) pressure p_∞ scales with the reservoir pressure p_0, the facility total pressure needs to be increased accordingly.

1.3.1 Atmospheric Entry

For hypervelocity blunt-body flows in which Mach number independence applies, the significance of the Mach number is overtaken by other considerations. As pointed out in Section 1.2.1, instead of the ratio of kinetic energy to the thermal energy of the gas (i.e., Mach number), the ratio of the kinetic energy to a more characteristic energy is better suited to describe the flow. For reentry into Earth's atmosphere, the dominant energy is the dissociation energy of nitrogen, which has a value of approximately 34 MJ/kg. Therefore, the total enthalpy, which is essentially the kinetic energy, or the ratio of total enthalpy to nitrogen dissociation energy can be used as a measure of a wind tunnel's capability of generating such a condition. When the atmospheric gas composition is different (such as CO_2 for Mars entry, which has a dissociation energy of approximately 12.2 MJ/kg), the characteristic dissociation energy would be different as well.

As the test flow Mach number need not match that at flight, the test flow can be expanded to a Mach number that is lower than the flight value to achieve the required freestream pressure or density to achieve binary scaling. The advantage of this is that by expanding to a lower Mach number, the required total pressure decreases drastically, thereby reducing the facility pressure requirements.

On the other hand, the resulting lower test flow Mach number means that the test flow temperature is higher than that in flight, thus upsetting the matching of the freestream Reynolds number and other parameters. Fortunately, at the extreme high enthalpies encountered in atmospheric entry conditions, the thermal energy downstream of the bow shock wave is very closely approximated by the freestream kinetic energy:

$$h_t \approx \frac{U_\infty^2}{2} \approx c_{p_{\text{stag}}} T_{\text{stag}}, \tag{1.13}$$

where c_p is the specific heat at constant pressure and the subscript "stag" denotes the stagnation condition. Thus, the resulting temperature on the model, particularly in the shock-heated gas in the stagnation region, is still correct.

In summary, to simulate hypervelocity atmospheric entry, the primary matching parameters are the total enthalpy and the binary scale, as the same gas composition as in real flight is used. Total enthalpy is related to the energy, or power, that can be delivered by the facility, while binary scaling is related to the pressure level that can be achieved in the facility. The extremely large total enthalpies demanded in hypervelocity testing pose the biggest difficulty to overcome, although the fulfillment of adequate pressure is also a challenge in the design of such test facilities.

1.3.2 Air-breathing Propulsion

Ground testing of air-breathing engines requires a different set of simulation parameters than those for external flows over blunt bodies. First of all, the total enthalpy (i.e., velocity) must still be duplicated. But since the body shape for vehicles that utilize air-breathing engines for propulsion is not blunt, Mach number independence does not apply. Indeed, the simulation of the forebody and engine inlet shock system requires replicating the Mach number.

Proper simulation of the combusting flow in hypersonic air-breathing engines requires matching of the first and second Damköhler numbers. The timescale of the ignition process, which is a two-body chemical reaction, is usually used in the first Damköhler number, which then results in the binary scaling parameter. However, the subsequent heat release and equilibration in the combustion process are more complex and require different scaling laws. The second Damköhler number, Da_2, is defined as the ratio of the heat release by chemical reactions to the total enthalpy (Anderson et al. 1990):

$$Da_2 = \frac{\text{Heat release by chemical reactions}}{\text{Total enthalpy}}. \tag{1.14}$$

This number is relevant to propulsion as it pertains to the heat release in burning the fuel in the engine. It may be thought of as similar to the reciprocal of the ratio of total enthalpy to dissociation energy for atmospheric entry simulations.

In addition, the engine thrust (F) is proportional to the mass flow rate ($\dot{m}_{\text{air}}$) entering the engine, as seen in the thrust equation for an air-breathing engine:

$$F = \dot{m}_{\text{air}}\left(U_{\text{exhaust}} - U_\infty\right) = \rho_\infty U_\infty A_{\text{capture}}\left(U_{\text{exhaust}} - U_\infty\right), \tag{1.15}$$

where U_{exhaust} and U_∞ are the velocities at the exhaust and in the freestream, and A_{capture} is the inlet capture area. As the engine is scaled down in ground testing, the mass flow rate per unit area $(\rho_\infty U_\infty)$ must be replicated to correctly simulate the thrust produced.

Note, however, that the requirements of replicating the binary scaling $(\rho_\infty L)$ and mass per unit area $(\rho_\infty U_\infty)$ are incompatible with each other. For a small engine model of length L, binary scaling means ρ_∞ must increase. But if ρ_∞ is increased, the mass flow per unit area cannot be matched. This means that the test engine would generate a higher thrust for its size than the flight engine. To avoid this problem, ground testing of air-breathing engines should be carried out at, or close to, the same scale as the flight vehicle.

In many practical combustors, the timescale of the combustion process is dominated by mixing rather than by chemical reactions. In this so-called mixing-controlled combustion, although chemical reactions still take finite time, the process of fuel–air mixing requires far longer time and becomes the dominating timescale. Thus, the duplication of the binary scale parameter is not required. Furthermore, if the combustor flow is fully turbulent, which is an idealization, the boundary-layer thickness δ and mixing effects scale with the one-fifth power of the Reynolds number:

$$\frac{\delta}{L} \sim \frac{1}{Re^{1/5}}, \quad Re = \frac{\rho_\infty U_\infty L}{\mu_\infty}, \tag{1.16}$$

so that the mixing layer scales with the four-fifths power of the length:

$$\delta \sim \frac{L}{L^{1/5}} = L^{4/5}. \tag{1.17}$$

Fuel–air mixing then scales approximately linearly with the model size. This helps greatly the ground simulation of hypersonic air-breathing propulsion when the above ideal conditions are met.

Note also that the matching of $\rho_\infty U_\infty$ implies that the freestream dynamic pressure $\left(\frac{1}{2}\rho_\infty U_\infty^2\right)$ is automatically duplicated. Furthermore, as the dynamic pressure q_∞ can be rewritten in terms of the static pressure q_∞ and the Mach number M_∞:

$$q_\infty = \frac{1}{2}\rho_\infty U_\infty^2 = \frac{1}{2}\gamma_\infty p_\infty M_\infty^2, \tag{1.18}$$

duplicating the dynamic pressure and the Mach number implies that the static pressure is also duplicated. This is also equivalent to the duplication of the flight total pressure.

The significance of the last point is that hypersonic air-breathing engines encounter extremely high flight total pressures. For example, at a flight Mach number of 15, the freestream total pressure is of the order of 10,000 atmospheres. This represents a major challenge to ground-test hypersonic air-breathing propulsion as such pressures would exceed the structural load limits of facility components (~3,000 atmospheres).

1.4 Hypersonic Ground Testing Facilities

1.4.1 Pulse Facilities

For ground testing of hypersonic vehicles, the necessity to duplicate the extremely high enthalpy or velocity poses a severe power requirement. Consider the testing of a reentry vehicle at an LEO reentry velocity of 7 km/s and a freestream density of 0.0001 kg/m^3. For a 1/100th scale model, binary scaling dictates that the density in the wind-tunnel test section would be 0.01 kg/m^3. Assuming a test flow area of 1 m^2 in the cross section, the power (given by the mass flow rate × area × velocity = $\rho U^3 A$) is then 3.4 GW, or an energy flux of 3.4 GW/m^2. To get an appreciation of the magnitude of this, let us compare it with the energy flux at the surface of the sun. The power output by the sun, or solar luminosity (L_{sun}), is known to be 3.9×10^{26} W. The solar power per unit area at a distance from the sun is given by the total power divided by the surface area of the sphere ($4\pi r^2$), with its radius being the distance from the sun's center. At the sun's surface, the radius is 7×10^5 km, yielding an energy flux of 0.064 GW/m^2. Compared to this value, the wind-tunnel power is about 53 times that of the sun's surface. Clearly, this power cannot be sustained for a long time. In addition, the stagnation temperature of the test gas for the reentry velocity reaches about 10,000 K, or approximately twice the surface temperature of the sun (~5,800 K), which is beyond the melting temperature of all known materials. Therefore, the only practical way to achieve hypersonic test conditions is to use short-duration facilities and to rely on thermal inertia to avoid melting the facility.

The use of shock waves to rapidly heat the test gas is currently the main ground testing technique for generating high enthalpy flows. Shock-heated test facilities are a class of wind tunnels that allow the duplication of flight velocities and high-temperature real-gas effects. Two types of such facilities are the reflected-shock tunnel (RST) and the shock-expansion tube or tunnel (SET). Both will be the main subjects of this book. These facilities, whether they are RSTs or SETs, are derived from extremely powerful shock tubes. These are short-duration facilities and are therefore known variously as pulse, impulse, or simply shock-tunnel facilities.

Because of the short-duration nature of the hypersonic test process, an important issue is whether there is sufficient test time. A requirement is that the pulse facilities must have sufficient time to fully establish steady-state flow over the test model. The most pertinent measure of test time is the ratio of the test "slug length" to the characteristic length of the flow process. As shown in Figure 1.2, the slug length ($L_{\text{slug}} = U \times t$) is the distance traveled by the volume of test gas during the period of steady test time, where U is the flow velocity and t is the test time. The characteristic length L may be represented by the model length if the flow is attached, or it includes the length of the separated zone for separated flow. Different estimates of test time requirements are used in the hypersonic testing community, depending on the nature of the flow process and the shape and orientation of the test model. Some reasonable rules of thumb for testing relatively slender external flow models are as follows:

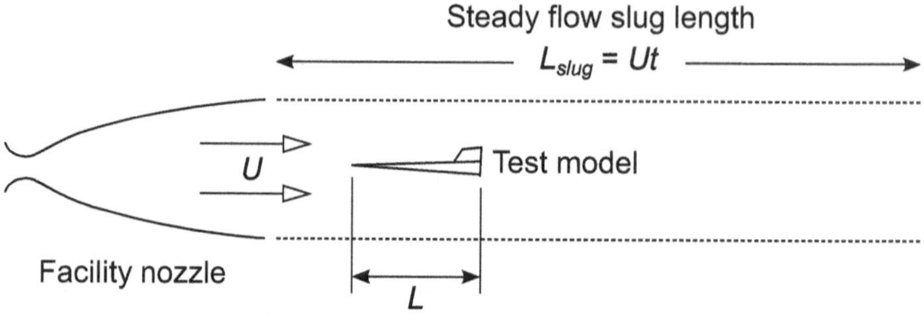

Figure 1.2 Slug length and characteristic length in the wind-tunnel test section.

$$\frac{Ut}{L} > 2, \quad \text{attached turbulent flow,} \tag{1.19}$$

$$\frac{Ut}{L} > 3, \quad \text{attached laminar flow,} \tag{1.20}$$

$$\frac{Ut}{L} > 10, \quad \text{separated flow.} \tag{1.21}$$

For testing streamline-shaped scramjet engines, it is a common practice to require a test slug of at least three model lengths in pulse facilities. Consider the testing of a 1-m long model. At Mach 10, the velocity is approximately 2,970 m/s (assuming a flight ambient temperature of 220 K). A 3-m gas slug would correspond to a test duration of 1 ms, which is readily achievable in RSTs. Similarly, at Mach 25, the velocity is 7,425 m/s. A 3-m slug length would correspond to a test period of 0.4 ms. The enthalpy level and the test time can be reached in SETs.

There are other considerations associated with RSTs and SETs, such as test gas contamination, which may be due to test gas dissociation in the test facilities, and the contamination of the test gas by the driver gas. Test gas dissociation is a consequence of the extremely high temperature encountered in the shock-heated gas in the facilities, which is especially a problem in the reflected-shock region in RSTs. This may also be an issue in SETs, but to a lesser degree. Driver gas contamination is related to the premature leakage of the gas medium used to drive the pulse facility which only affects RSTs.

1.4.2 Arc-jets

Because of the extremely short-duration nature of shock-tunnel facilities, intermittent flow facilities have also been used for hypersonic testing. At the lower boundary of hypersonic flow, such as at about Mach 7, these facilities can meet the requirements of most of the simulation parameters. However, at higher Mach numbers, they generally cannot meet the key simulation parameters, such as binary scaling.

The most common intermittent flow test facilities are arc-jets that have much longer run times (typically tens of minutes) than are available in pulse facilities. There are different types of arc-jet wind tunnels, but they all utilize electric discharge for heating the test medium to extremely high temperatures prior to expanding it in a nozzle to the test section (Smith et al. 2002). Arc-heaters can generate test flow total enthalpies of more than 30 MJ/kg or temperatures of the order of 10,000 K. On the other hand, the pressure in the arc-settling chamber, and thus the test flow stagnation pressure, is limited to about 100 atm. There is also a limit on the product of pressure and enthalpy an electric arc can achieve so that the maximum enthalpy is obtained at low pressures (Park 1995, 1996b). In contrast, RSTs can generate test flow stagnation pressures of about 1,000 atm. and SETs can reach even higher. Another metric to consider is the power level attained in the test facility. Arc-jets consume less than 150 MW of electrical power. Compared to the GW level delivered in large pulse facilities (as discussed in Section 1.4.1), arc-jets cannot be expected to be as capable as shock tunnels in providing hypersonic test conditions.

Arc-jets also suffer from metallic erosion from the electrodes. And since very high temperatures are generated in the gas ionized by the arc, significant amount of test-medium dissociation are formed in the plenum, just as in the case of RSTs, making the issues of chemical dissociation and contamination more severe. And although arc-jets provide a longer test duration, the test flows can be unstable and nonuniform. Because of the above reasons, arc-jets are not suitable for accurate aerothermodynamic and combustion experiments in the hypervelocity regime. However, arc-jets have indeed played a significant role in testing air-breathing propulsion in the lower hypersonic range, as demonstrated by the testing of the Hyper-X engine at the Mach 7 condition in NASA Langley's Arc-Heated Scramjet Test Facility (Guy et al. 1996, Voland et al. 1998).

Due to its relatively long test duration, arc-jets have been used extensively for testing thermal protection systems to assess materials' response and structural survival in reentry vehicles. For this reason, arc-jet wind tunnels are also referred to as aerothermal test facilities. But since arc-jets have lower total pressure capabilities than pulse facilities, they generally cannot satisfy binary scaling and Reynolds number requirements at the hypervelocity regime. Because of this, they will not be included in this book.

There are, indeed, other types of ground testing facilities to generate hypersonic flow conditions, such as those discussed in Chinitz et al. (1994) and in the book edited by Lu and Marren (2002). These facilities have different capabilities, advantages, and disadvantages, and are in different stages of readiness. Currently, shock-heated pulse facilities are the best option and the most widely used techniques for testing in the hypersonic regime. Therefore, they are selected as the focus of this book.

1.5 Outline of the Book

The purpose of the book is to present and discuss the theories and techniques for hypervelocity ground testing facilities. As the attention here is on shock-heated high-enthalpy pulse facilities, the fundamental theory of shock waves and its application to generate high-enthalpy flows in shock tunnels are presented in Chapter 2.

Shock-wave-heated facilities can be driven in a number of ways. The book then examines the different driver techniques that have been developed. These include the heated light-gas, free-piston, and detonation drivers, presented in Chapters 3–5. The theory of operation of these drivers and their applications in hypersonic testing are also discussed.

The development of long-duration shock tunnels is of particular interest to the hypersonic community as it can overcome the stringent limitation of test time. The concepts involved and the accomplishments in actual facilities are presented in Chapter 6.

The facility nozzle is a critical component in high-enthalpy shock tunnels, as it controls the quality of the flow delivered to the test section. Chapter 7 examines various methods for designing hypersonic flow nozzles, including conventional method of characteristics as well as the state-of-the-art CFD-based design approaches.

The book ends by discussing the phenomena associated with high-enthalpy flows, how they are manifested in real flight, and how they are studied in ground test facilities. Chapter 8 includes the experimental methods for measuring aerodynamic forces and moments, aerothermal heating, boundary-layer transition, as well as a discussion of supersonic combustion and scramjet testing.

Nomenclature

a	sound speed
A_{capture}	inlet capture area
c_p	specific heat capacity at constant pressure
Da_1	Damköhler's first number
Da_2	Damköhler's second number
E	total energy emitted
F	thrust
h	enthalpy
L	length
L_{slug}	slug length
$\dot{m}_{\text{air}}$	mass flow rate of air
M	Mach number
p	pressure
q_{conv}	convective heat transfer rate
q_{rad}	radiative heat transfer rate
R	gas constant
R_N	nose radius
Re	Reynolds number
t	time
T	temperature
U	velocity
U_{exhaust}	engine exhaust velocity

β	local shock-wave angle
δ	boundary-layer thickness
γ	ratio of specific heats

Subscripts:

s	post-shock condition
t	total or stagnation condition
∞	freestream condition

References

Abbany, Z. (2021). NASA Rover Attempting Most Difficult Martian Touchdown. *Deutsche Welle*, February 18. https://bit.ly/3rM214C (Accessed: February 28, 2021).

Aharonson, O., Russell, C. T., Head, J. W. III, et al. (2020). The Science Mission of SpaceIL's Beresheet Lander: Plans and Results. *51st Lunar and Planetary Science Conference*, Woodlands, Texas, 1267.

Allen, H. J. and Eggers, A. J. (1953). A Study of the Motion and Aerodynamic Heating of Ballistic Missiles Entering the Earth's Atmosphere at High Supersonic Speeds. *NACA-RM-A53D28*. https://ntrs.nasa.gov/citations/20050019430 (Accessed: May 2, 2023).

Allen, H. J. and Eggers, A. J. (1958). A Study of the Motion and Aerodynamic Heating of Ballistic Missiles Entering the Earth's Atmosphere at High Supersonic Speeds. *NACA-TR-1381*. (Supersedes NACA-TN-4017 by the same authors, 1957). https://history.arc.nasa.gov/hist_pdfs/awards/hjaa_1970a.pdf (Accessed: May 2, 2023).

Anderson, G., Kumar, A., and Erdos, J. (1990). Progress in Combustion Technology with Computation and Experiment. *AIAA Paper 90-5254*.

Anderson, J. D. Jr. (2006). *Hypersonic and High-Temperature Gas Dynamics*, 2nd ed. Reston, VA: AIAA.

Anon. (1980). *Aerodynamic Design Data Book – Orbiter Vehicle*, Vol. 1, Rockwell International, Space Division. *SD72-SH-0060-1M*.

Arney, D. (2016). Reusable Vehicles Come to Forefront. *Aerospace America*, AIAA, December, **54**(11), 70.

Barbosa, R. C. (2013). China's Chang'e-3 and Jade Rabbit Duo Land on the Moon. *Nasaspaceflight.com*, December 14. https://bit.ly/3lenwbO (Accessed: February 27, 2021).

Berry, S. A., Kimmel, R., and Reshotko, E. (2011). Recommendations for Hypersonic Boundary Layer Transition Flight Testing. *AIAA Paper 2011-3415*.

Bertin, J. J. (1994). *Hypersonic Aerothermodynamics*. Washington, DC: AIAA, pp. 231–276.

Bertin, J. J. and Cummings, R. M. (2006). Critical Hypersonic Aerothermodynamic Phenomena. *Annual Review of Fluid Mechanics*, **38**, 129–157.

Brauckmann, G. J., Paulson, J. W. Jr., and Weilmuenster, K. J. (1995). Experimental and Computational Analysis of Shuttle Orbiter Hypersonic Trim Anomaly. *Journal of Spacecraft and Rockets*, **32**(5), 758–764.

Button, K. (2018). Hypersonic Weapons Race. *Aerospace America*, AIAA, June, **56**(6), 20–25.

Charters, A. C. and Thomas, R. N. (1945). The Aerodynamic Performance of Small Spheres from Subsonic to High Supersonic Velocities. *Journal of the Aeronautical Sciences*, **12**(4), 468–476.

Chen, F.-J. and Berry, S. A. (2010). HyBoLT Flight Experiment. *NASA Technical Memorandum*, NASA/TM-2010-216725.

Chinitz, W., Erdos, J. I., Rizkalia, O., Anderson, G. Y., and Bushnell, D. M. (1994). Facility Opportunities and Associated Stream Chemistry Considerations for Hypersonic Air-Breathing Propulsion. *Journal of Propulsion and Power*, **10**(1), 6–17.

Costa, M., Pérez, M., Almeida, M., et al. (2016). Rosetta: Rapid Science Operations for a Dynamic Comet. *AIAA Paper* 2016–2538.

Cox, R. N. and Crabtree, L. F. (1965). *Elements of Hypersonic Aerodynamics*. New York: Academic Press.

Crowder, R. S. and Moot, J. D. (1965). Apollo Entry Aerodynamics, *Journal of Spacecraft and Rockets*, **6**(3), 302–307.

Desai, P. N., Lyons, D. T., Tooley, J., and Kangas, J. (2006). Entry, Descent, and Landing Operations Analysis for the Stardust Re-Entry Capsule. *AIAA Paper* 2006–6410.

Edney, B. E. (1968a). Anomalous Heat Transfer and Pressure Distributions on Bunt Bodies at Hypersonic Speeds in the Presence of an Impinging Shock. *FFA Report* 115, Flygtekniska Försöksanstalten (The Aeronautical Research Institute of Sweden), Stockholm.

Edney, B. E. (1968b). Effects of Shock Impingement on the Heat Transfer around Blunt Bodies. *AIAA Journal*, **6**(1),15–21.

eoPortal Chang'e-5. (n.d.). Chang'e-5 (China's Lunar Sample Return Mission)/CE-5. https://bit.ly/3lesd5g (Accessed: March 2, 2021).

eoPortal Hayabusa-2. (n.d.). Hayabusa-2, Japan's Second Asteroid Sample Return Mission. https://bit.ly/3lcw1nE (Accessed: March 2, 2021).

Fay, J. A. and Riddell, F. R. (1958). Theory of Stagnation Point Heat Transfer in Dissociated Air. *Journal of Aeronautical Sciences*, **25**(2), 73–85.

Francis, W. L. (1965). Experimental Heat-Transfer Study of Shock Impingement on Fins in Hypersonic Flow. *Journal of Spacecraft and Rockets*, **2**(4), 630–632.

Gibney, E. (2019). China Plans Mission to Earth's Pet Asteroid. *Nature*, April 30. www.nature.com/articles/d41586-019-01390-5 (Accessed: February 28, 2021).

Grinstead, J. H., Jenniskens, R., Cassell, A. M., Albers, J., and Winter, M. W. (2011). Airborne Observation of the Hayabusa Sample Return Capsule Re-entry. *AIAA Paper* 2011–3329.

Guy, R. W., Rogers, R. C., Puster, R. L., Rock, K. E., and Diskin, G. S. (1996). The NASA – Langley Scramjet Test Complex. *AIAA Paper* 96–3243.

Hadhazy, A. (2017). Space Station Experiment Targets Parkinson's. *Aerospace America*, AIAA, September, **55**(8), 9.

Hadhazy, A. (2019). Lunar Far Side Comes into Focus. *Aerospace America*, AIAA, April, **57**(4), 31–34.

Hiers, R. S. and Loubsky, W. J. (1967). Effects of Shock-Wave Impingement on the Heat Transfer on a Cylindrical Leading Edge. *NASA TN D-3859*.

Hodges, A. J. (1957). The Drag Coefficient of Very High Velocity Spheres, *Journal of the Aeronautical Sciences*, **24**(10), 755–758.

Hofactor, C. (2020a). How to Make a Megaconstellation. *Aerospace America*, AIAA, March, **58**(3), 16–23

Hofactor, C. (2020b). One of These Companies Could Deliver American Astronauts to the Moon in 2024. *Aerospace America*, AIAA, April 30. https://bit.ly/3qGqAyF (Accessed: February 27, 2021).

JAXA. (2019). Shooting bullets into Ryugu! *JAXA*, February 18. https://bit.ly/3bJVaDa (Accessed: February 27, 2021).

Justus, C. G. and Braun, R. D. (2007). Atmospheric Environments for Entry, Descent and Landing (EDL). In *5th International Planetary Probes Workshop and Short Course*, 23–29 June, Bordeaux, France.

Klotz, I. (2017). Recycling Rockets. *Aerospace America*, AIAA, September, **55**(8), 32–39.

Koppenwallner, G. (1987). Low Reynolds Number Influence on Aerodynamic Performance of Hypersonic Lifting Vehicles, Paper 11. *AGARD Conference Proceedings No. 428*, Aerodynamics of Hypersonic Lifting Vehicles, November.

Lauretta, D. S., Balram-Knutson, S. S., Beshore, E., et al. (2017). OSIRIS-REx: Sample Return from Asteroid (101955) Bennu. *Space Science Reviews*, **212**, 925–984.

Lebreton, J.-P. and Matson, D. L. (2002). The Huygens Probe: Science, Payload and Mission Overview. *Space Science Reviews*, **104**, 59–100.

Lebreton, J.-P., Witasse, O., Sollazzo, C., et al. (2005). An Overview of the Descent and Landing of the Huygens Probe on Titan. *Nature*, **438**(8), 758–764.

Lu, F. K. and Marren, D. E. (eds.) (2002). *Advanced Hypersonic Test Facilities*. AIAA Progress in Astronautics and Aeronautics, Vol. 198. Reston, VA: AIAA.

Maise, G., Powell, J., Paniagua, J., et al. (2003). Application of the MITEE Nuclear Ramjet for Ultra Long Range Flyer Missions in the Atmospheres of Jupiter and Other Giant Planets. *Paper IAC-03-Q.4.09, 54th International Astronautical Congress*, Bremen, Germany, September–October.

Malik, T. (2012). Death of DARPA's Superfast Hypersonic Glider Explained. *Space.com*, April 23. https://bit.ly/3lgwepV (Accessed: March 2, 2021).

Martos, J. F. d. A., Rêgo, I. d. S., Laiton, S. N. P., et al. (2017). Experimental Investigation of Brazilian 14-X B Hypersonic Scramjet Aerospace Vehicle. *International Journal of Aerospace Engineering*, 2017. Article ID 5496527. https://doi.org/10.1155/2017/5496527 (Accessed: May 2, 2023).

Marvin, J. G. and Deiwert, G. S. (1965). Convective Heat Transfer in Planetary Gases, *NASA Technical Report TR R-224*.

Maus, J. R., Griffith, B. J., Szema, K. Y., and Best, J. T. (1984). Hypersonic Mach Number and Real Gas Effects on Space Shuttle Orbiter Aerodynamics, *Journal of Spacecraft and Rockets*, **21**(2), 136–141.

McBride, B. J. and Gordon, S. (1996). Computer Program for calculation of Complex Chemical Equilibrium Compositions and Applications. II. Users Manual and Program Description. *NASA Reference Publication RP-1311*.

McCurdy, H. E. (2005). Low-Cost Innovation in Spaceflight. The Near Earth Asteroid Rendezvous (NEAR) Shoemaker Mission. *Monographs in Aerospace History*, **36**, NASA SP-2005-4536.

NASA. (2009). NASA and ISRO Satellites Perform in Tandem to Search for Ice on the Moon, August 20. www.nasa.gov/mission_pages/Mini-RF/news/tandem_search.html (Accessed: February 27, 2021).

NASA. (2020). NASA's OSIRIS-REx Successfully Stows Sample of Asteroid Bennu. *NASA News Release*, October 29. https://go.nasa.gov/3cojZng (Accessed: February 28, 2021).

NASA Mars. (2021). *Mars 2020 Perseverance Rover*. https://mars.nasa.gov/mars2020/ (Accessed: February 28, 2021).

NASA Perseverance. (2020). Mars 2020/Perseverance. *NASA Fact Sheet*. https://mars.nasa.gov/files/mars2020/Mars2020_Fact_Sheet.pdf (Accessed: February 28, 2021).

Newlander, R. A. (1961). Effect of Shock Impingement on the Distributions of Heat-Transfer Coefficients on a Right Circular Cylinder at Mach Numbers of 2.65, 3.51, and 4.44. *NASA TN D-642.*

Norris, G. (2013). Skunk Works Reveals SR-71 Successor Plan. *Aviation Week and Space Technology*, November 1. https://bit.ly/40VCJSv (Accessed: May 1, 2023).

Odom, T. and Johnston, D. (2018). Exploring High-Speed Propulsion for Weapons, Passenger Aircraft. *Aerospace America*, AIAA, December, **56**(11), 49.

Oswatitsch, K. (1951). Ähnlichkeitsgesetz für Hyperschallströmung (Similarity Laws for Hypersonic Flow). *Zeitschrift für Angewandte Mathematik und Physik ZAMP (Journal of Applied Mathematics and Physics)*, **2**(4), 249–264.

Oswatitsch, K. (1980). Similarity Laws for Hypersonic Flow. In *Contributions to the Development of Gas Dynamics*, ed. W. Schneider and M. Platzer. Braunschweig: Vieweg Verlag, pp. 76–88.

Owen, T., Raulin, F., McKay, C., et al. (1997). The Relevance of Titan and Cassini/Huygens to Pre-biotic Chemistry and the Origin of Life on Earth. In *Huygens: Science, Payload and Mission, ESA SP 1177*, ed. A. Wilson. Noordwijk: ESA Publications Division, ESTEC, pp. 231–233.

Park, C. (1995). Experimental Simulation and Evaluation of Chemical Effects. In *Aerothermochemistry for Hypersonic Technology, Lecture Series 1995–04*, ed. G. S. R. Sarma. Belgium: Von Karman Institute for Fluid Dynamics.

Park, C. (1996a). Experimental Evaluation of Real-Gas Phenomena in High-Enthalpy Impulse Test Facilities. *AIAA Paper 96–2234.*

Park, C. (1996b). Experimental Evaluation of Real-Gas Phenomena in High-Temperature Aerothermal Test Facilities: A Review. *AIAA Paper 96–2207.*

Planetary Society. (n.d.). *Tianwen-1 and Zhurong, China's Mars Orbiter and Rover.* www.planetary.org/space-missions/tianwen-1 (Accessed: May 1, 2023).

Ray, A. D. and Palko, R. L. (1965). An Investigation of the Effects of Shock Impingement on a Blunt Leading Edge. *AEDC-TR-65-153.* U.S. Air Force.

Schneider, S. P. (1999). Flight Data for Boundary-Layer Transition at Hypersonic and Supersonic Speeds. *Journal of Spacecraft and Rockets*, **36**(1), 8–20.

Sforza, P. M. (2012). *Theory of Aerospace Propulsion.* Waltham, MA: Elsevier, pp. 536–539.

Siddiqi, A. A. (2018). Beyond Earth: A Chronicle of Deep Space Exploration. *NASA SP-2018-4041*, NASA History Division.

Siler, L. G. and Deskins, H. E. (1964). Effect of Shock Impingement on Heat-Transfer and Pressure Distributions on a Cylindrical-Leading-Edge Model at Mach Number 19. *AEDC-TDR-64-228*, U.S. Air Force, November.

Singh, L., Sharma, S. P., and Moore, C. (2019). Space Exploration Focuses on Asteroids. *Aerospace America*, AIAA, December, **56** (11), 72.

Smith, D. M., Felderman, E. J., Shope, F. L., and Balboni, J. A. (2002). *Arc-Heated Facilities. In Advanced Hypersonic Test Facilities*, AIAA Progress in Astronautics and Aeronautics, Vol. 198. Reston, VA: AIAA, pp. 279–314.

SpaceX Falcon. (2021) *Falcon Heavy.* www.spacex.com/vehicles/falcon-heavy/ (Accessed: February 27, 2021).

Stevens, V. I. (1950). Hypersonic Research Facilities at the Ames Aeronautical Laboratory, *Journal of Applied Physics*, **21**(11), 1150–1155.

Sutton, K. and Graves, R. A. Jr. (1971). A General Stagnation-Point Convective-Heating Equation for Arbitrary Gas Mixtures, *NASA Technical Report TR R-376.*

Tauber, M. E. (1989). A Review of High-Speed, Convective, Heat-Transfer Computational Methods, *NASA Technical Paper 2914*.

Tauber, M. E., Meneses, G. P., and Adelman, H. G. (1987). Aerothermodynamics of Transatmospheric Vehicles. *Journal of Aircraft*, **24**(9), 594–602.

The Hindu. (2009). Chandrayaan Confirms Moon Was once Completely Molten: Scientist. *The Hindu*. https://bit.ly/41OO7R3 (Accessed: May 1, 2023).

Urzay, J. (2018). Supersonic Combustion in Air-Breathing Propulsion Systems for Hypersonic Flight. *Annual Review of Fluid Mechanics*, **50**, 593–627.

Voland, R. T., Rock, K. E., Huebner, L. D., et al. (1998). Hyper-X Engine Design and Ground Test Program. *AIAA Paper 98–1532*.

Watts, J. D. (1968). Flight Experience with Shock Impingement and Interference Heating on the X-15-2 Research Airplane, *NASA Technical Memorandum*, NASA/TM X-1669.

Wood, P. and Cliff, R. (2020). *A Case Study of the PRC's Hypersonic Systems Development*. E-book. China Aerospace Studies Institute, Air University, September 20. https://bit.ly/38B1MSm (Accessed: March 2, 2021).

Wright, R. L. and Zoby, E. V. (1977). Flight Boundary Layer Transition Measurements on a Slender Cone at Mach 20. *AIAA Paper 77–719*.

2 Shock-Wave Relations and Aerothermodynamic States

In this chapter, we summarize the aerodynamic fundamentals for the working theories of shock tunnels. Moving waves, including expansion waves, shock waves, and contact surfaces, can be found as the key points of the unsteady one-dimensional flows in any textbook on aerodynamics. As the theories of unsteady one-dimensional moving waves are also critical for the design and operation of shock tunnels, the important parts of the theories are selected and summarized in this chapter, instead of a full description.

2.1 Fundamentals of Moving Waves

Unsteady features of the fluid motion of a moving expansion wave or shock wave play a dominant role in the case of shock tunnels. Therefore, we shall concern ourselves with basic theories of unsteady one-dimensional motions of waves hereinafter. We shall start with continuous waves, such as expansion waves, and then move to discontinuous waves, such as shock waves and contact surfaces, in the following section.

2.1.1 Unsteady One-Dimensional Continuous Waves

Consider the one-dimensional unsteady gas flow in a duct with a constant cross-sectional area through the control surface as depicted in Figure 2.1. The flow is restricted to shock-free regions and friction is assumed negligible. Therefore, the entropy of the fluid remains constant anywhere and anytime, that is, $\frac{\partial s}{\partial t} = \frac{\partial s}{\partial x} = 0$. In addition, the gas is supposed to be a calorically thermal perfect gas.

The governing equations of the above flow through the control surface in a partial differential form have been provided in the following. These equations include the conservation of the mass or continuity equation, dynamic equation, and isentropic relation.

Conservation of mass or continuity equation:

$$\frac{\partial \rho}{\partial t} + u\frac{\partial \rho}{\partial x} + \rho\frac{\partial u}{\partial x} = 0. \tag{2.1}$$

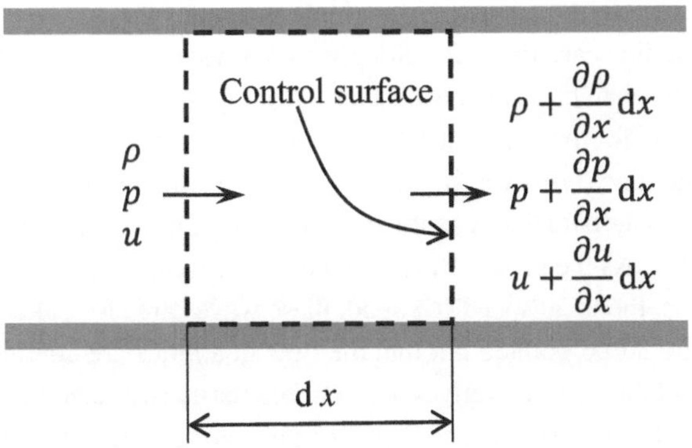

Figure 2.1 Flow through the control surface.

Dynamic equation or Euler's equation:

$$\frac{\partial u}{\partial t}+u\frac{\partial u}{\partial x}+\frac{1}{\rho}\frac{\partial p}{\partial x}=0. \tag{2.2}$$

Isentropic relation:
Under the homogenous entropy assumption, we have

$$\left(\frac{\partial p}{\partial \rho}\right)_s=\frac{\mathrm{d}p}{\mathrm{d}\rho}=c^2, \tag{2.3a}$$

so that

$$\frac{1}{\rho}\frac{\partial p}{\partial x}=\frac{1}{\rho}\frac{\mathrm{d}p}{\mathrm{d}\rho}\frac{\partial \rho}{\partial x}=\frac{c^2}{\rho}\frac{\partial \rho}{\partial x}. \tag{2.3b}$$

In the above equations, ρ, p, u, and c represent the flow variables of density, pressure, velocity, and speed of sound, respectively. From the theory of characteristics, the flow governed by the aforementioned equations, that is, Eqs. (2.1)–(2.3), have two families of characteristic curves in the x–t plane:

$$\text{Family I:}\begin{cases}\left(\dfrac{\mathrm{d}t}{\mathrm{d}x}\right)_{\mathrm{I}}=\dfrac{1}{u+c}, & \text{characteristic line} \quad (2.4a)\\[2ex] u+\dfrac{2}{\gamma+1}c=C_{\mathrm{I}}, & \text{compatiblity relation} \quad (2.4b)\end{cases}$$

$$\text{Family II:}\begin{cases}\left(\dfrac{\mathrm{d}t}{\mathrm{d}x}\right)_{\mathrm{II}}=\dfrac{1}{u-c}, & \text{characteristic line} \quad (2.5a)\\[2ex] u-\dfrac{2}{\gamma+1}c=C_{\mathrm{II}}, & \text{compatiblity relation} \quad (2.5b)\end{cases}$$

Here, C_{I} and C_{II} denote the Riemann invariants, respectively.

The characteristic lines are the paths along which a wave or disturbance propagates. For the family I of the characteristic lines, the propagating speed of a disturbance is $u + c$, while it becomes $u - c$ for the family II. The Riemann invariant maintains constant along a given characteristic line; however, it generally changes among different lines. Generally, the family I characteristic line is denoted as a right-running wave, while the family II is a left-running wave. As the propagating speed of the waves relative to the gas particle itself equals c, that is, the local speed of sound, these waves are also called Mach waves or Mach lines. It should be pointed out that the flow quantities are continuous across a characteristic line, while their derivatives are not necessarily restricted to be continuous. On the other hand, all flow quantities are kept constant along a characteristic line.

In an unsteady one-dimensional gas flow, if there are both left- and right-running waves, it is called a double-wave flow. Sometimes, the flow may be simple in which case there is only one family of waves, that is, either left-running or right-running waves, then it is called a simple-wave flow. In a simple-wave flow, since there is only one family of characteristic lines, the other family degenerates, which implies that the corresponding Riemann invariant remains constant within the whole simple-wave flow domain. Such degeneration can lead to great simplification to the solution of a flow problem. Furthermore, in a steady-state flow, both families of characteristics degenerate since all flow quantities remain constant, and hence there is no wave or disturbance in it.

Hereinafter, we have several examples to describe the features of simple-wave flows. As given in Figure 2.2(a), in an infinitely long duct, there is a piston that is originally at rest. At time transient $t = 0$, the piston suddenly starts moving rightward at an infinitesimal velocity of δu. There will be an expansion wave traveling to the left in the original static gas to the left of the piston as depicted in Figure 2.2(b), and a compression wave traveling to the right in the originally static gas to the right of the piston, as depicted in Figure 2.2(c). The former is a left-running expansion wave

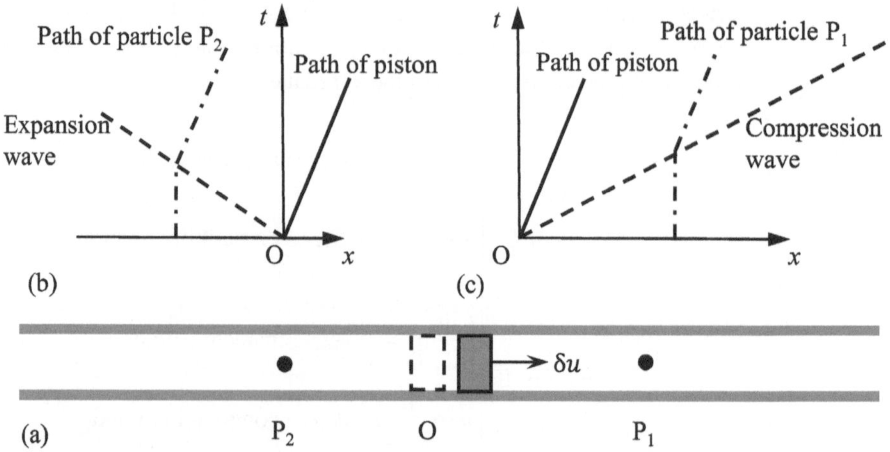

Figure 2.2 Continuous waves with small disturbance.

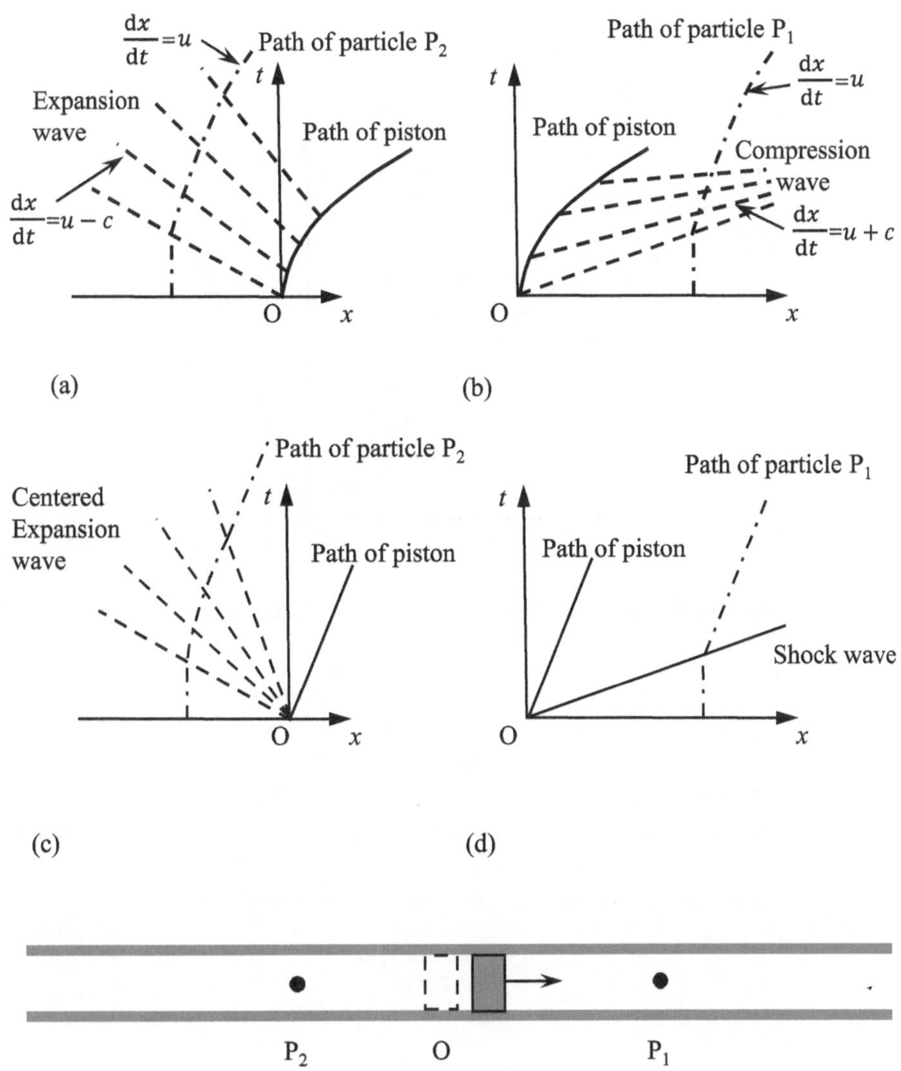

Figure 2.3 Continuous waves (except (d)) with a strong disturbance of finite amplitude.

(family I characteristics) with a speed of $\delta u - c \approx -c$, while the latter is a right-running compression wave (family II characteristics) with a speed of $\delta u + c \approx c$.

If the piston gradually accelerates rightward, a series of waves is generated that travels in both the directions, that is, left-running expansion waves (Figure 2.2(a)) and right-running compression waves (Figure 2.2(b)). If the acceleration occurs instantaneously, then the left-running expansion waves will change into a centered expansion wave as depicted in Figure 2.3(c). Such a flow pattern is analogical to a Prandtl–Meyer expansion fan in a two-dimensional supersonic corner flow. On the other hand, in the gas to the right of the piston, a shock wave will come into being and travel to the right. Such a flow structure is no longer continuous but a discontinuous wave, which will be discussed in Section 2.1.2.

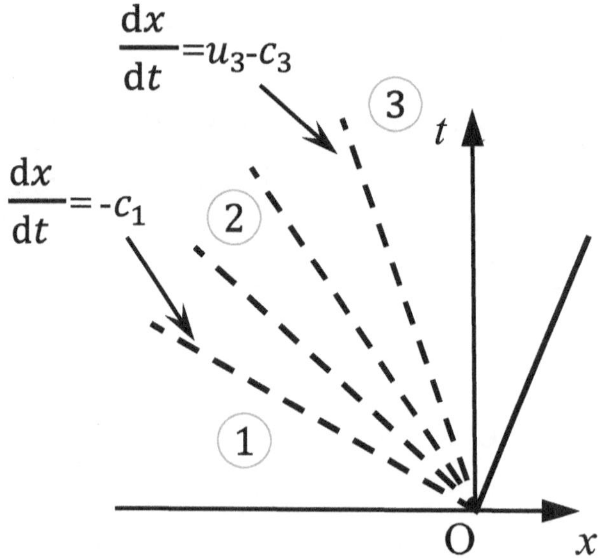

Figure 2.4 Solution of a left-running centered expansion wave.

For the flow pattern of a centered expansion wave, as depicted in Figure 2.4, the whole flowfield can be divided into three subzones, that is, the dead zone ①, the centered expansion wave ②, and the steady-state zone ③. The propagating speed of the wave head (the first character line) is $-c_1$ as $u_1 = 0$, while that for the wave tail, it is $u_3 - c_3$. Here, u_3 is the piston speed, a known quantity in the problem. As a left-running centered expansion wave is a simple wave where the right-running family of characteristics degenerates, the Riemann invariant $C_{\mathrm{I}} = \text{const}$ holds throughout the flow domain ②. Then we have

$$\frac{2}{\gamma-1}c_1 = u_3 + \frac{2}{\gamma-1}c_3, \tag{2.6a}$$

so that

$$\frac{c_3}{c_1} = 1 - \frac{\gamma-1}{2}\frac{u_3}{c_1}. \tag{2.6b}$$

From the definition of speed of sound, we can obtain the flow temperature in zone ③,

$$\frac{T_3}{T_1} = \left(\frac{c_3}{c_1}\right)^2 = \left(1 - \frac{\gamma-1}{2}\frac{u_3}{c_1}\right)^2. \tag{2.6c}$$

From the isentropic relations, we can obtain the other flow variables in zone ③,

$$\frac{p_3}{p_1} = \left(\frac{T_3}{T_1}\right)^{\frac{\gamma}{\gamma-1}}, \qquad \frac{\rho_3}{\rho_1} = \left(\frac{T_3}{T_1}\right)^{\frac{1}{\gamma-1}}. \tag{2.6d}$$

The solution as discussed above is for a left-running expansion wave. For a right-running centered expansion wave, just change the minus sign in Eq. (2.6b) into a positive sign, that is, $\frac{c_3}{c_1} = 1 + \frac{\gamma-1}{2}\frac{u_3}{c_1}$, we can solve it analytically. The centered expansion wave is one of the typical flow patterns in a shock tunnel. Therefore, the solution as discussed in Eqs. (26) above is critical and useful in understanding the working theory of a shock tube or shock tunnel, which will be presented in Section 2.3.

2.1.2 Unsteady One-Dimensional Discontinuous Waves

In unsteady one-dimensional flows, moving shock waves and contact surfaces are typical discontinuities. Figure 2.3(d) explains the generation of a right-running shock wave. Figures 2.5(a) and (c) present a right-running and a left-running shock wave, respectively. The flow variables post the shock wave, that is, region ②, are ρ_2, p_2, T_2, and u_2, while the corresponding ones ahead of the shock wave, that is, region ①, are ρ_1, p_1, T_1, and u_1, respectively. The shock wave is moving at a velocity of w relative to the gas ahead of it. Here, the shock Mach number is defined as

$$M_s = \frac{w}{c_1} \tag{2.7}$$

For a right-running shock wave, if we fix the coordinate at the shock front as shown in Figure 2.5b, then the flow velocities ahead of and post the shock wave change into

$$u_{(1)} = w \tag{2.8a}$$

$$u_{(2)} = w - u_1 + u_2. \tag{2.8b}$$

For a left-running shock wave, we have

$$u_{(1)} = w \tag{2.9a}$$

$$u_{(2)} = w - u_1 + u_2. \tag{2.9b}$$

In the coordinate fixed on the shock front, the one-dimensional conservation laws for a steady normal shock wave still hold and read

$$\dot{m}_s = \rho_1 u_{(1)} = \rho_2 u_{(2)} \tag{2.10a}$$

$$p_2 - p_1 = \dot{m}_s \left(u_{(1)} - u_{(2)} \right) = m_s^2 \left(\frac{1}{\rho_1} - \frac{1}{\rho_2} \right), \tag{2.10b}$$

as well as the Rankine–Hugoniot relation

$$\frac{p_2}{p_1} = \frac{\dfrac{\gamma+1}{\gamma-1}\dfrac{\rho_2}{\rho_1} - 1}{\dfrac{\gamma+1}{\gamma-1} - \dfrac{\rho_2}{\rho_1}}. \tag{2.10c}$$

Therefore, if we replace the flow Mach number ahead of a steady normal shock wave with the shock Mach number as given in Eq. (2.7), we can set up and rewrite the relations of thermodynamic variables for a moving shock wave,

$$\frac{p_2}{p_1} = \frac{2\gamma}{\gamma+1} M_s^2 - \frac{\gamma-1}{\gamma+1}. \tag{2.11a}$$

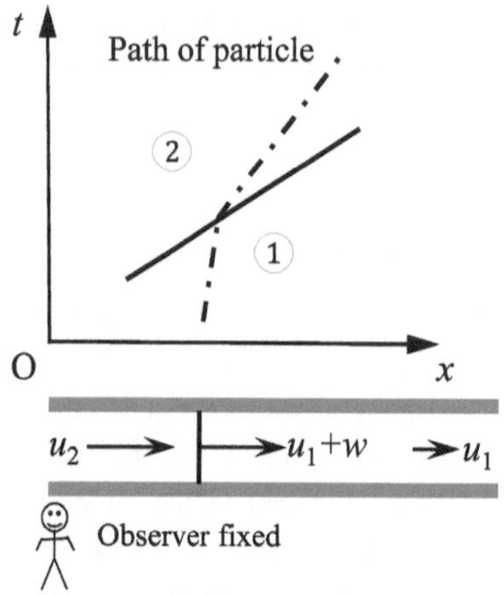

(a) right-running, observer fixed

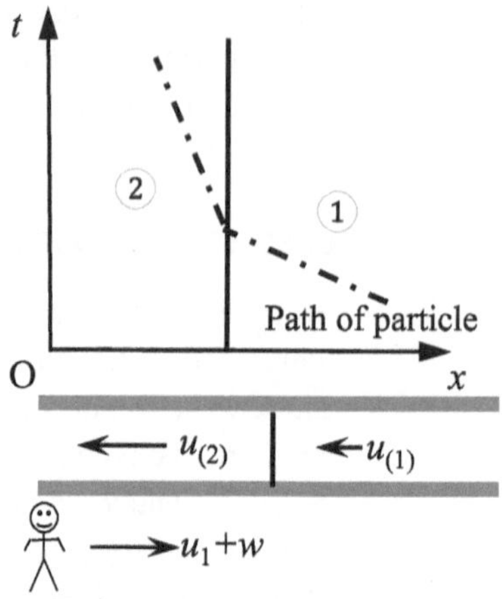

(b) right-running, shock fixed

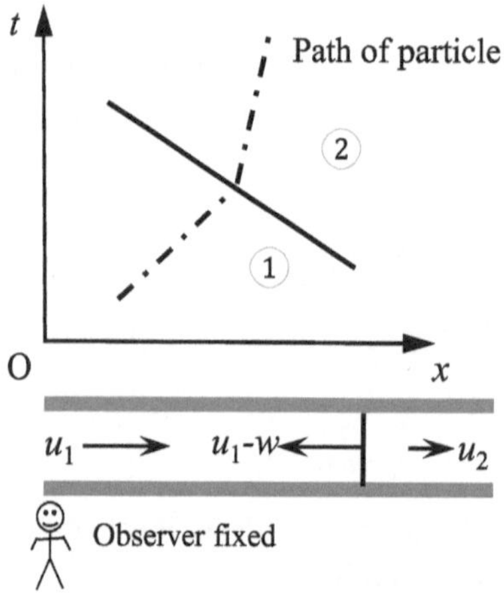

(c) left-running, observer fixed

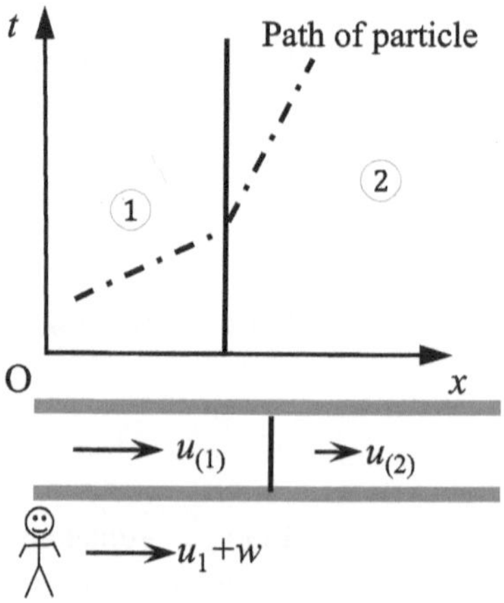

(d) left-running, shock fixed

Figure 2.5 Shock waves in different coordinates.

$$\frac{\rho_2}{\rho_1} = \frac{u_{(1)}}{u_{(2)}} = \frac{M_s^2}{1+\frac{\gamma-1}{\gamma+1}\left(M_s^2-1\right)}. \tag{2.11b}$$

$$\frac{T_2}{T_1} = \left(\frac{c_2}{c_1}\right)^2 = \frac{\left[2\gamma M_s^2-(\gamma-1)\right]\left[(\gamma-1)M_s^2+2\right]}{(\gamma+1)^2 M_s^2}. \tag{2.11c}$$

For the kinetic variables that are related to the coordinate where the flow motion is observed, we change Eq. (2.10a) into the original form in the laboratory coordinate and obtain

$$\rho_1 w = \rho_2\left[w \pm (u_1 - u_2)\right]. \tag{2.12a}$$

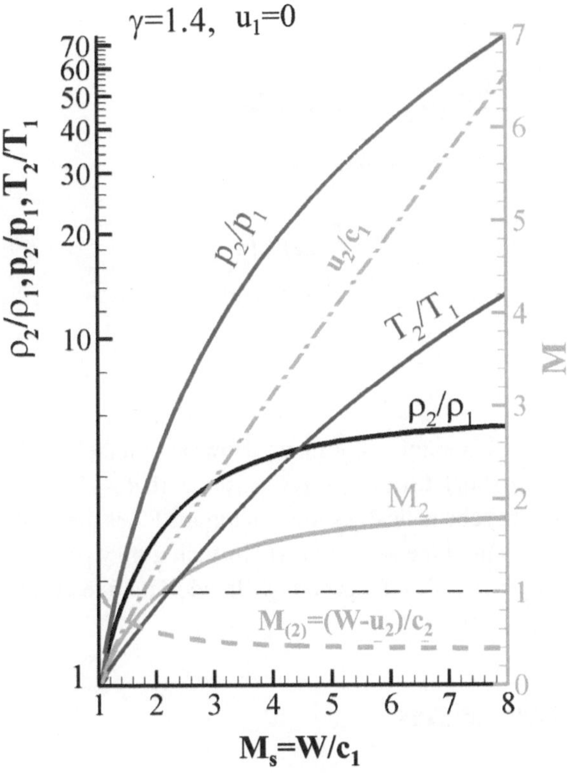

Figure 2.6 Diagram of relations for moving shock waves.

Substituting Eq. (2.11b) into (2.12a), we obtain

$$\frac{u_2 - u_1}{c_1} = \pm\left(\frac{w}{c_1} - \frac{w}{c_1}\frac{\rho_1}{\rho_2}\right) = \pm\frac{2\gamma}{\gamma+1}\left(M_s - \frac{1}{M_s}\right). \tag{2.12b}$$

Here, the operators "+" and "–" represent a right-running and left-running shock wave, respectively.

If the gas ahead of a right-running shock wave is stationary, that is, $u_1 = 0$, then (2.12b) can be rewritten as

$$\frac{u_2}{c_1} = \frac{2}{\gamma+1}\left(M_s - \frac{1}{M_s}\right), \tag{2.13a}$$

and

$$M_2 = \frac{u_2}{c_2} = \frac{2\left(M_s^2 - 1\right)}{\sqrt{\left[2\gamma M_s^2 - (\gamma-1)\right]\left[(\gamma-1)M_s^2 + 2\right]}}, \tag{2.13b}$$

$$M_{(2)} = \frac{w - u_2}{c_2} = \frac{w}{c_1}\frac{c_1}{c_2} - M_2 = \sqrt{\frac{(\gamma-1)M_s^2 + 2}{2\gamma M_s^2 - (\gamma-1)}}. \tag{2.13c}$$

The moving shock relations as illustrated in Eqs. (2.11a–c) and (2.13a–c) are combined in Figure 2.6. It can be seen clearly that the flow post a moving shock wave

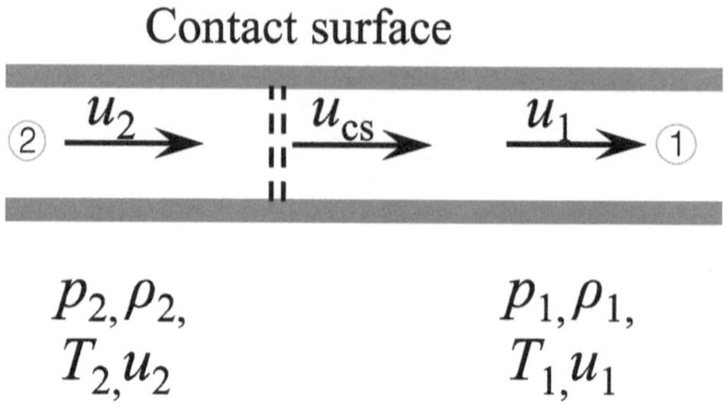

Figure 2.7 Contact surface.

may be supersonic in the laboratory coordinate. However, in the coordinate fixed on the shock front, the post-shock flow is always subsonic, that is, $M_{(2)} < 1$.

The contact surface, as schematically shown in Figure 2.7, is one of the typical discontinuities in unsteady one-dimensional flows. In such a flow pattern, $u_{cs} = u_2 = u_1$ and $p_2 = p_1$, while $\rho_2 \neq \rho_1$, $T_2 \neq T_1$ generally. In addition, the chemical composition on both sides may be different.

2.2 Interactions of Moving Waves

2.2.1 Reflection of an Expansion Wave from a Solid Surface

As depicted in Figure 2.8(a), when a piston moving to the right at an infinitesimal speed of δ_u in a duct, a left-running expansion wave will be generated. The particle originally in region ① will start moving at the speed of δ_u once swapped by the expansion wave and come into a new state ②. When the expansion wave reaches the left closed end of the duct, the wave will be reflected and changed into a right-moving expansion wave. To satisfy the boundary condition determined by the closed end, the particle of state ② should stop moving and reach a third state ③.

From the degenerated family I characteristics of the left-running expansion wave, we have

$$\frac{2}{\gamma-1}c_1 = \delta u + \frac{2}{\gamma-1}c_2, \tag{2.14a}$$

and then

$$\frac{c_2}{c_1} = 1 - \frac{\gamma-1}{2}\frac{\delta u}{c_1}. \tag{2.14b}$$

The isentropic relations between states ② and ① read

$$\frac{p_2}{p_1} = \left(\frac{T_2}{T_1}\right)^{\frac{\gamma}{\gamma-1}} = \left(\frac{c_2}{c_1}\right)^{\frac{2\gamma}{\gamma-1}} = \left(1 - \frac{\gamma-1}{2}\frac{\delta u}{c_1}\right)^{\frac{2\gamma}{\gamma-1}}. \tag{2.14c}$$

Analogously, for the right-running expansion wave, we can obtain

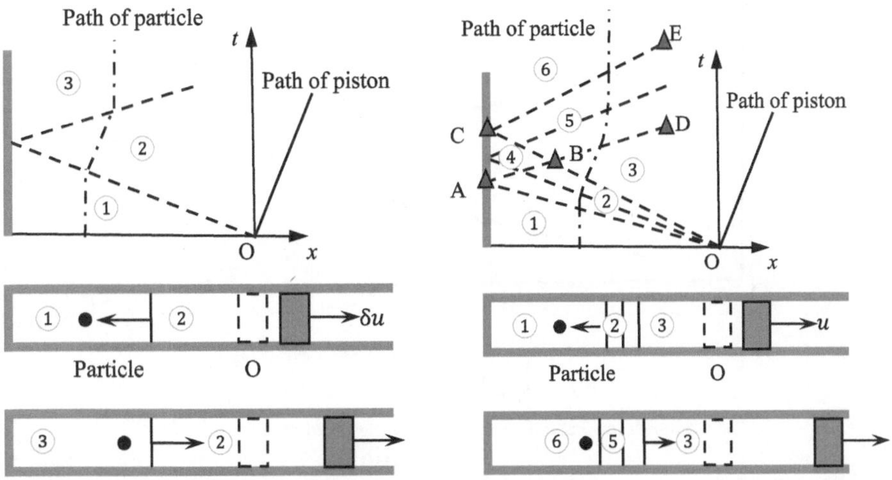

Figure 2.8 Reflection of an expansion wave or a centered expansion fan from a closed end.

$$\frac{c_3}{c_2} = 1 - \frac{\gamma - 1}{2}\frac{\delta u}{c_2}, \tag{2.14d}$$

$$\frac{p_3}{p_2} = \left(\frac{T_3}{T_2}\right)^{\frac{\gamma}{\gamma-1}} = \left(\frac{c_3}{c_2}\right)^{\frac{2\gamma}{\gamma-1}} = \left(1 - \frac{\gamma - 1}{2}\frac{\delta u}{c_2}\right)^{\frac{2\gamma}{\gamma-1}}. \tag{2.14e}$$

If the wave is a centered expansion wave as shown in Figure 2.8(b), the flow pattern will be more complex when it is reflected from the closed end. The whole flow state can be divided into six regions, that is, the steady-state regions ①, ③, and ⑥; the simple wave regions ② (OAB) and ⑤ (BCED); and the two-wave region ④ (ABC). The simple-wave and two-wave regions can be replaced with a finite number of expansion waves that can be individually solved as a single wave by Eqs. (2.14a)–(2.14e). One useful solution of the flow pattern is that of curve AB, that is, the reflected wave of the head of the incidental expansion wave that intersects with the wave tail at point B. The curve AB equation is

$$\frac{x}{L} = -\frac{\gamma + 1}{\gamma - 1}\left(\frac{c_1 t}{L}\right)\left[\left(\frac{c_1 t}{L}\right)^{-\frac{2(\gamma-1)}{\gamma+1}} - \frac{2}{\gamma + 1}\right]. \tag{2.15}$$

Here, L is the distance between the origin O and the closed end. The reflected wave head will change into a straight line, BD, in the simple wave region ⑤. This solution will be used to theoretically calculate the effective test time for a shock tube.

2.2.2 Reflection of a Shock Wave from a Solid Surface

As shown in Figure 2.9, when a right-running shock wave propagates at a velocity of w in an original static gas of region ① ($u_1 = 0$), and is reflected from the closed end, a left-running shock wave comes into being with a velocity of w_r. To satisfy the boundary condition, the particle in region ③ should be static, that is, $u_3 = 0$.

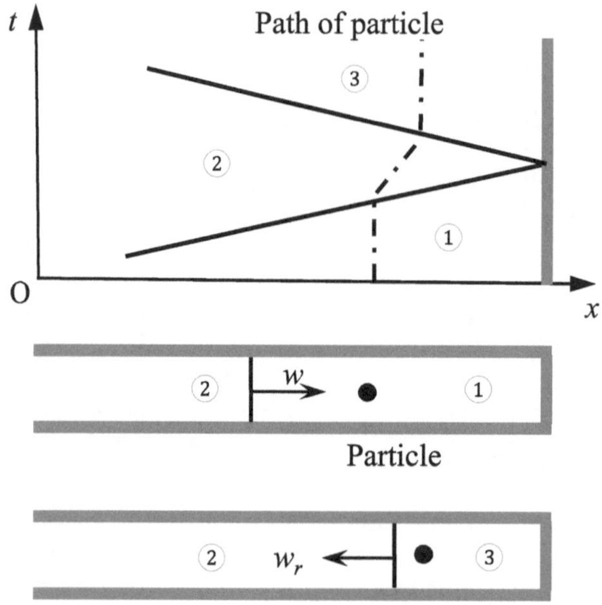

Figure 2.9 Reflection of a shock wave from a closed end.

From Eqs. (2.11a) and (2.12b), we can obtain

$$\frac{u_2-u_1}{c_1}=\pm\frac{1}{\gamma}\left(\frac{p_2}{p_1}-1\right)\left(\frac{\gamma+1}{2\gamma}\frac{p_2}{p_1}+\frac{\gamma-1}{2\gamma}\right)^{-0.5}, \tag{2.16a}$$

$$\left(\frac{c_2}{c_1}\right)^2=\frac{T_2}{T_1}=\frac{p_2}{p_1}\left(\frac{p_2}{p_1}+\frac{\gamma+1}{\gamma-1}\right)\left(\frac{\gamma+1}{\gamma-1}\frac{p_2}{p_1}+1\right)^{-1}, \tag{2.16b}$$

Applying Eq. (2.16a) for the incidental shock and the reflected shock, respectively,

$$\frac{u_2}{c_1}=\frac{1}{\gamma}\left(\frac{p_2}{p_1}-1\right)\left(\frac{\gamma+1}{2\gamma}\frac{p_2}{p_1}+\frac{\gamma-1}{2\gamma}\right)^{-0.5}, \tag{2.17a}$$

$$\frac{-u_2}{c_2}=-\frac{1}{\gamma}\left(\frac{p_3}{p_2}-1\right)\left(\frac{\gamma+1}{2\gamma}\frac{p_3}{p_2}+\frac{\gamma-1}{2\gamma}\right)^{-0.5}. \tag{2.17b}$$

Upon setting A = Eq. (2.16b), B = Eq. (2.17a), and C = Eq. (2.17b), and by noting that $\frac{B^2}{C^2}=A$, we have

$$\frac{p_3}{p_2}=\frac{\frac{3\gamma+1}{\gamma-1}\frac{p_2}{p_1}-1}{\frac{\gamma+1}{\gamma-1}+\frac{p_2}{p_1}}. \tag{2.18}$$

The relation for the reflected shock wave is

$$\frac{p_3}{p_2}=\frac{2\gamma}{\gamma+1}M_r^2-\frac{\gamma-1}{\gamma+1}. \tag{2.19}$$

From Eqs. (2.11a), (2.18), and (2.19), we can obtain the relation between M_r and M_s,

$$M_r^2 = \frac{2\gamma M_s^2 - \gamma + 1}{(\gamma - 1)M_s^2 + 2}. \tag{2.20}$$

With Eq. (2.20) and the moving shock relations Eqs. (2.11)–(2.13), the flow condition in region ③ can be solved:

$$\frac{p_3}{p_1} = \frac{\left[2\gamma M_s^2 - (\gamma - 1)\right]\left[(3\gamma - 1)M_s^2 - 2(\gamma - 1)\right]}{(\gamma + 1)\left[(\gamma - 1)M_s^2 + 2\right]}, \tag{2.21a}$$

$$\frac{\rho_3}{\rho_1} = \frac{\left[2\gamma M_s^2 - (\gamma - 1)\right]\left[(\gamma + 1)M_s^2\right]}{(\gamma + 1)\left[(\gamma - 1)M_s^2 + 2\right]}, \tag{2.21b}$$

$$\frac{T_3}{T_1} = \frac{\left[2(\gamma - 1)M_s^2 - (\gamma - 3)\right]\left[(3\gamma - 1)M_s^2 - 2(\gamma - 1)\right]}{(\gamma + 1)^2 M_s^2}. \tag{2.21c}$$

The above relations are very critical for the theoretical study of reflected shock tunnels (RSTs). The pressure profiles $\frac{p_3}{p_1}$, $\frac{p_3}{p_2}$, and $\frac{p_2}{p_1}$ with varying Mach number M_s of the incidental shock wave are presented in Figure 2.10.

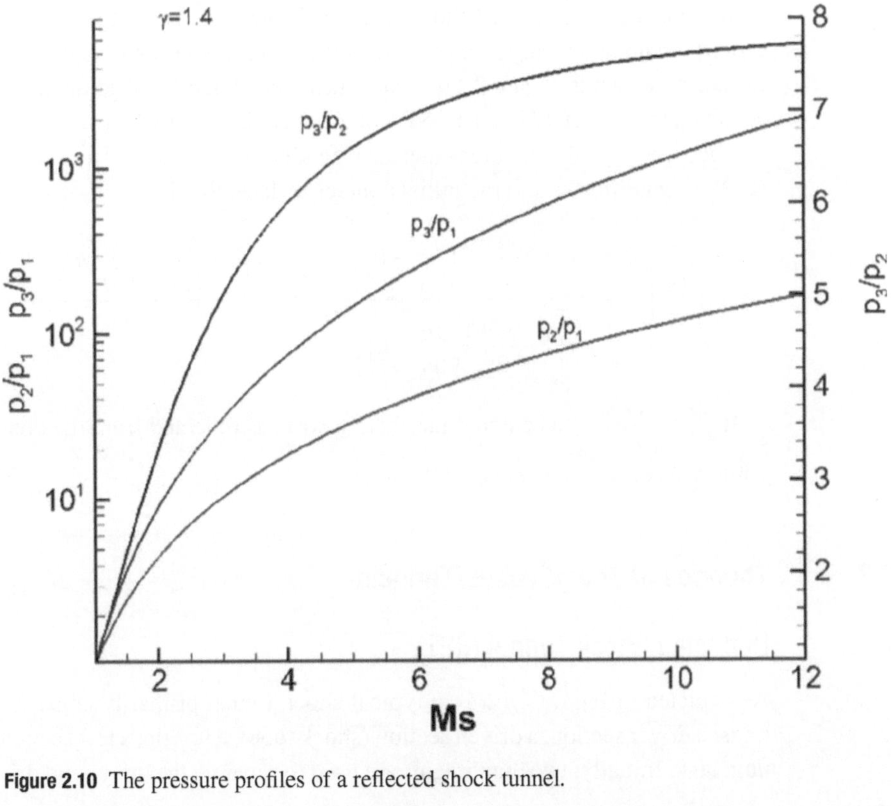

Figure 2.10 The pressure profiles of a reflected shock tunnel.

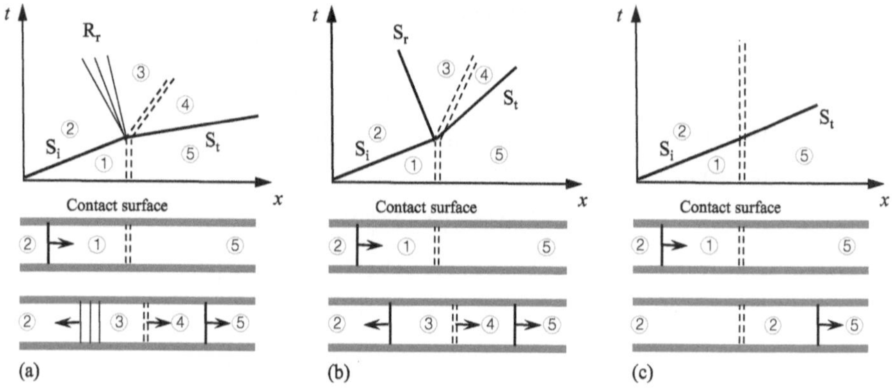

Figure 2.11 The interaction between a shock wave and a contact surface.

2.2.3 Interaction between a Shock Wave and a Contact Surface

When a shock wave interacts with a contact surface, there will be three kinds of flow patterns depending on the thermodynamic properties across the initial contact surface. A transmitted shock wave (S_t), which is weaker than the incidental shock wave (S_i), can be generated as depicted in Figure 2.11(a), while the reflected wave should be a centered rarefaction wave (R_r). In another case, as shown in Figure 2.11(b), a stronger transmitted shock wave followed by a reflected shock wave (S_r) is generated after the interaction. Under a certain condition, nothing changes when the incidental shock passes through the contact surface as shown in Figure 2.11(c), and such a condition is called the tailored-interface condition that is required for an RST to maintain a sufficiently long test time.

Theoretically, the tailored-interface condition can be satisfied under a special acoustic condition across the initial contact surface, that is,

$$\left(\frac{c_5}{c_1}\right)_{\text{TIC}} = \frac{\gamma_5\left[1+\dfrac{\gamma_5+1}{2\gamma_5}\left(\dfrac{p_2}{p_1}-1\right)\right]^{0.5}}{\gamma_1\left[1+\dfrac{\gamma_1+1}{2\gamma_1}\left(\dfrac{p_2}{p_1}-1\right)\right]^{0.5}} \overset{\gamma_2=\gamma_1}{\Rightarrow} \frac{\gamma_5\left[1+\dfrac{\gamma_1(\gamma_5+1)}{\gamma_5(\gamma_1+1)}\left(m_s^2-1\right)\right]^{0.5}}{\gamma_1 m_s}. \quad (2.22)$$

If $\dfrac{c_5}{c_1} > \left(\dfrac{c_5}{c_1}\right)_{\text{TIC}}$, a centered rarefaction wave is reflected from the contact surface, while a shock wave is reflected if $\dfrac{c_5}{c_1} < \left(\dfrac{c_5}{c_1}\right)_{\text{TIC}}$.

2.3 Theories of Shock Tubes/Tunnels

2.3.1 Reflected Shock Tunnel (RST)

As depicted in Figure 2.12(a), a typical shock tunnel primarily consists of five sections: a driver section, a driven section (shock tube), a nozzle, a test section, and a vacuum tank. Initially, two diaphragms are used to separate the driver and driven sections

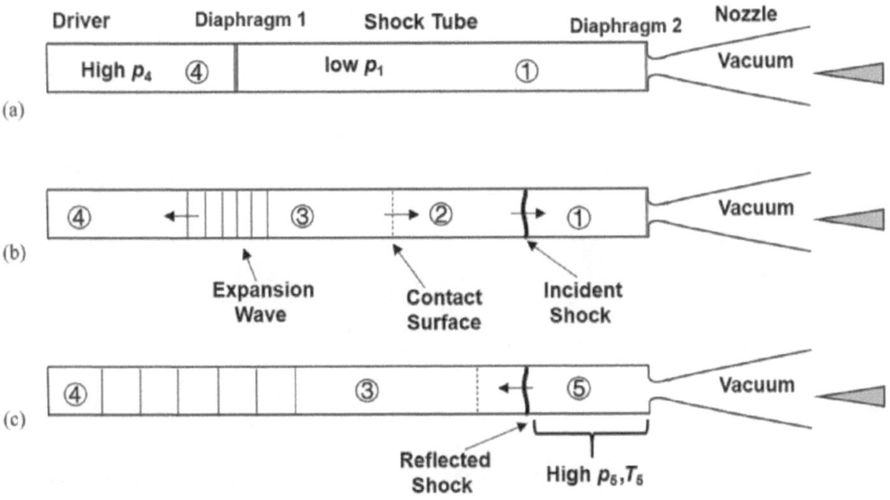

Figure 2.12 The wave dynamic in a reflected shock tube/tunnel.

and the driven section and the nozzle, respectively. Generally, the driver gas with a higher pressure p_4 is filled in the driver section, while the required test gas is filled in the driven section at a lower pressure, p_1. When the diaphragm separating the driver and driven sections is broken, a shock wave generates and propagates rightward in the test gas, followed by a contact surface. Simultaneously, a centered expansion wave comes into being and propagates leftward into the driver gas. Therefore, a left-running centered expansion wave, a right-running shock wave (incidental shock), and a contact surface are the key flow structures during this stage as shown in Figure 2.12(b). The flow domain is divided into four parts: the initial driver gas ④, the initial test gas ①, the uniform region after the shock wave ②, and the uniform region between the expansion wave and the contact surface ③. When the incidental shock wave reaches the right end of the driven section, it reflects and the reflected shock wave propagates leftward in the gas region ②. The flow region to the right of the reflected shock wave labeled ⑤ in Figure 2.12(c) corresponds to the stagnation condition for a shock tunnel working in a reflected mode. Later, the left-running reflected shock wave interacts with the contact surface, which will result in flow patterns discussed in Figure 2.11. As aforementioned, a tailored-interface condition is generally required to maintain the steady test condition with a sufficiently long test time.

For the initial driver gas ④ and the test gas ①, which are stationary, we have

$$u_4 = 0, u_1 = 0, \tag{2.23}$$

and other known parameters, $p_4, \rho_4, T_4, \gamma_4, c_4, p_1, \rho_1, T_1, \gamma_1$, and c_1. Between ④ and ③ is the left-running centered expansion wave. Here, the family I characteristics degenerate, and the corresponding compatible relation reads

$$u_4 + \frac{2}{\gamma_4 - 1} c_4 = u_3 + \frac{2}{\gamma_4 - 1} c_3. \tag{2.24a}$$

Substituting Eqs. (2.23) into (2.24a), we obtain

$$\frac{c_3}{c_4} = 1 - \frac{\gamma_4 - 1}{2}\frac{u_3}{c_4}. \tag{2.24b}$$

Note that the specific heat ratio in gas ③ is supposed to be the same as that in gas ④. Recalling the homogeneous entropy condition within the centered expansion wave, we have the isentropic relation

$$\frac{p_3}{p_4} = \left(\frac{T_3}{T_4}\right)^{\frac{\gamma_4}{\gamma_4 - 1}} = \left(\frac{c_3}{c_4}\right)^{\frac{2\gamma_4}{\gamma_4 - 1}}. \tag{2.24c}$$

Substituting Eq. (2.24b) into (2.24c), we obtain

$$\frac{p_3}{p_4} = \left(1 - \frac{\gamma_4 - 1}{2}\frac{u_3}{c_4}\right)^{\frac{2\gamma_4}{\gamma_4 - 1}}. \tag{2.24d}$$

For the contact surface between ② and ③, the compatible relation reads

$$u_2 = u_3, \; p_2 = p_3. \tag{2.25}$$

The relation for the right-running incidental shock wave between ② and ① reads

$$\frac{u_2}{c_1} = \frac{2}{\gamma_1 + 1}\left(M_s - \frac{1}{M_s}\right) = \frac{u_3}{c_1}. \tag{2.26a}$$

$$\frac{p_2}{p_1} = \frac{2\gamma_1}{\gamma_1 + 1}M_s^2 - \frac{\gamma_1 - 1}{\gamma_1 + 1} = \frac{p_3}{p_1}. \tag{2.26b}$$

Noting that the relation in Eq. (2.25) is already used in Eqs. (2.26a) and (2.26b). Substituting Eq. (2.26a) into (2.24d), we can obtain

$$\frac{p_3}{p_4} = \left(1 - \frac{\gamma_4 - 1}{2}\frac{u_3}{c_4}\right)^{\frac{2\gamma_4}{\gamma_4 - 1}} = \left[1 - \frac{\gamma_4 - 1}{\gamma_1 + 1}\left(M_s - \frac{1}{M_s}\right)\right]^{\frac{2\gamma_4}{\gamma_4 - 1}}. \tag{2.27}$$

Since

$$\frac{p_4}{p_1} = \frac{p_4}{p_3}\frac{p_3}{p_1}, \tag{2.28}$$

then substituting Eqs. (2.26b) and (2.27) into (2.28), we have

$$\frac{p_4}{p_1} = \left[1 - \frac{\gamma_4 - 1}{\gamma_1 + 1}\left(M_s - \frac{1}{M_s}\right)\right]^{-\frac{2\gamma_4}{\gamma_4 - 1}}\left(\frac{2\gamma_1}{\gamma_1 + 1}M_s^2 - \frac{\gamma_1 - 1}{\gamma_1 + 1}\right). \tag{2.29}$$

Such a relation indicates that with the initial driver and test gas pressures given, we can determine the Mach number of the incidental shock wave, which is a key parameter for the operation condition of a shock tube or tunnel. Figure 2.13 shows the capability of different kinds of driver gases based on Eq. (2.29). It indicates that a driver gas with a higher speed of sound (either higher temperature or smaller molecular weight) has a stronger driving capability.

The *x-t* wave diagram for the operation of a shock tube/tunnel is depicted in Figure 2.14. The figure corresponds to a tailored-interface condition in which the

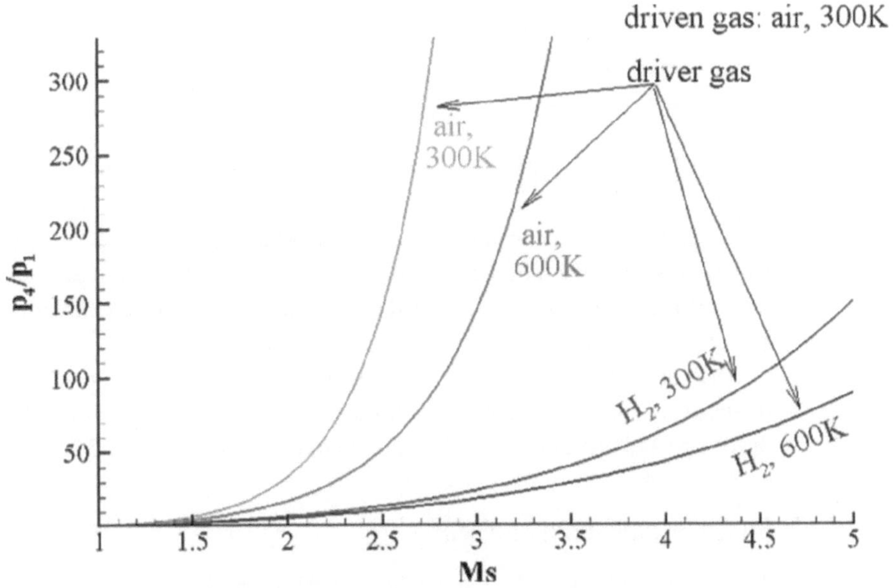

Figure 2.13 The driver capability of a shock tube (Jiang et al. 2020).

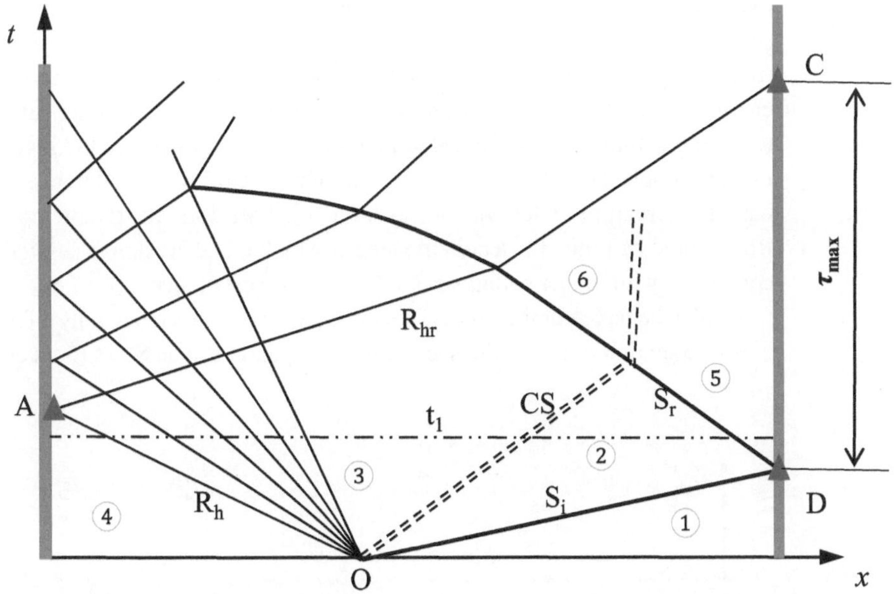

Figure 2.14 Wave diagram of a shock tube.

maximum test time $\tau_{\max}$ is theoretically determined by the arrivals of the incidental shock wave, S_i, and the head of the reflected expansion wave, R_{hr}. The tailored-interface condition is illustrated in Figure 2.11(c) and given in Eq. (2.22) for a stationary contact surface. Analogously for the shock tube, the tailored-interface condition reads

$$\frac{c_4}{c_1} = \frac{2}{\gamma_1 + 1}\left(M_s - \frac{1}{M_s}\right)$$
$$\left\{\left[\frac{(\gamma_1 - 1)M_s^2 + 2}{2\gamma_1\left(M_s^2 - 1\right)}\right]\left[\gamma_4^2 + \frac{\gamma_1\gamma_4(\gamma_4 + 1)\left(M_s^2 - 1\right)}{(\gamma_1 - 1)M_s^2 + 2}\right]^{0.5} + \frac{\gamma_4 - 1}{2}\right\}. \quad (2.30)$$

Equation (2.30) indicates that only one Mach number of the incidental shock wave can satisfy the tailored-interface condition once the driver gas and the test gas are given. For the operation of an RST, the tailored-interface condition is generally required to maintain the effective test time as long as possible.

RSTs are common and technically mature test facilities in the research community of shock dynamics in the world. For details on the basic theories and techniques on the design and operation of RSTs, one can refer to the literature (Ben-Dor et al. 2001).

2.3.2 Shock-expansion tunnel (SET)

As depicted in Figures 2.12 and 2.14, an RST works on the basis of the reflection of an incidental shock wave, which can generate a high-temperature and high-pressure test gas in the reservoir. Both the driver section and the reservoir must withstand extremely high pressures. The key parameter profiles correspond to the time transient t_1 (dash-dot-dot line) in the x-t diagram in Figure 2.14. The test gas in the reservoir (region ⑤) is stagnated as the kinetic energy in region ② is converted into internal energy. As a result, the total enthalpy, as well as the temperature, of the test gas reaches maximum simultaneously in the reservoir. This may lead to metal erosion or melting problems in the reflection region, which can further lead to test gas contamination or structural failure in the shock tunnel. Therefore, the total temperature of an RST should be limited to a certain value at which metal erosion is negligible. For test conditions with higher enthalpies, an SET is the best choice.

Unlike the large number of RST facilities, up to now, there are only a few SET facilities in operation in the world, such as the free-piston-driven SETs (Stalker et al. 1987,

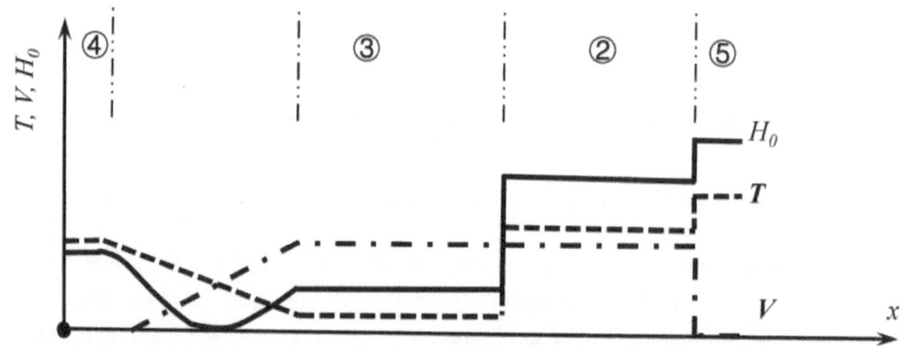

Figure 2.15 The transient parameter distributions in an RST (corresponding to time transient t_1 as shown in Figure 2.14).

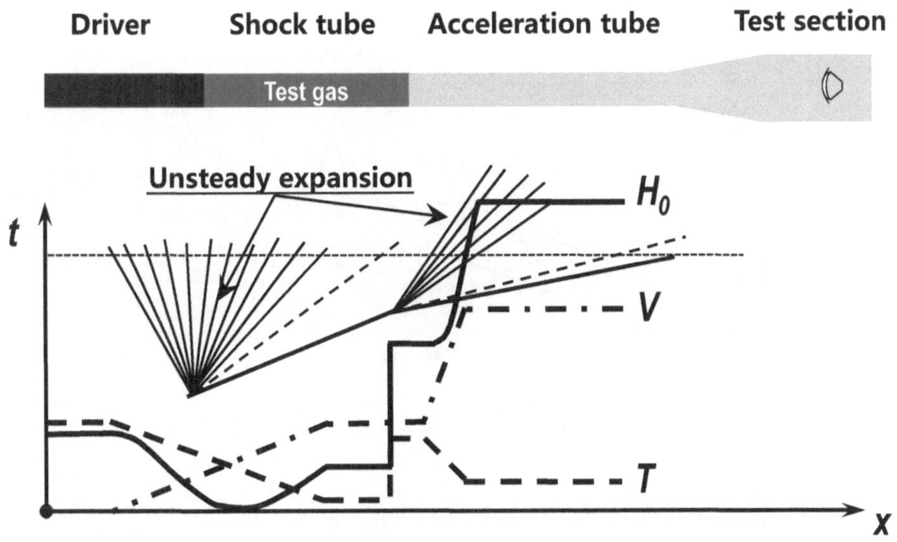

Figure 2.16 The transient flow properties in an SET, plotted with the *x*–*t* wave diagram.

Paull & Stalker 1989, 1992, Morgan 2000), the detonation-driven SETs (Chue et al. 2002, Jiang et al. 2015), and the light-gas-driven SETs (Holden et al. 2007). As shown in Figure 2.16, the reservoir with a reflection end wall upstream of the nozzle is replaced with a straight tube in an SET. The test flow in an SET is never stagnated as can be seen in the figure. Therefore, the aforementioned issues associated with an RST can be mitigated by using an SET to produce clean test flows of very high enthalpies or hypervelocities. The test gas is accelerated twice through two unsteady expansion before reaching the divergent nozzle, which can further accelerate the test gas through a steady expansion via the nozzle. The unsteady expansion in the acceleration tube can raise the total enthalpy of the test gas immensely while decreasing its static temperature. This is the substantial difference of the working basis of an SET from that of an RST.

2.4 Aerothermodynamic States for Hypersonic Flight

For a hypersonic vehicle, the aerothermodynamic design associated with the key fluid dynamic phenomena and the choice of flight trajectories depend on each other. Different from classical aircraft design, the aerodynamic performance of a hypersonic vehicle and its components must be closely tailored to its flight trajectory. Along with the guidance problem, the fulfillment of the basic mission is the primary objective of the trajectory design. Figure 2.17 shows typical flight trajectories for different missions of hypersonic vehicles. Generally, the trajectories of lifting entry, ballistic entry, and skimming entry are for the space shuttles (with wings), capsules (without wings), and return vehicles from other planets, respectively.

Once the flight trajectory is determined for a vehicle, the freestream flow conditions can be obtained, given that during reentering the atmosphere, the atmospheric properties are generally functions of altitude, as shown in Figure 2.18. Earth's atmosphere

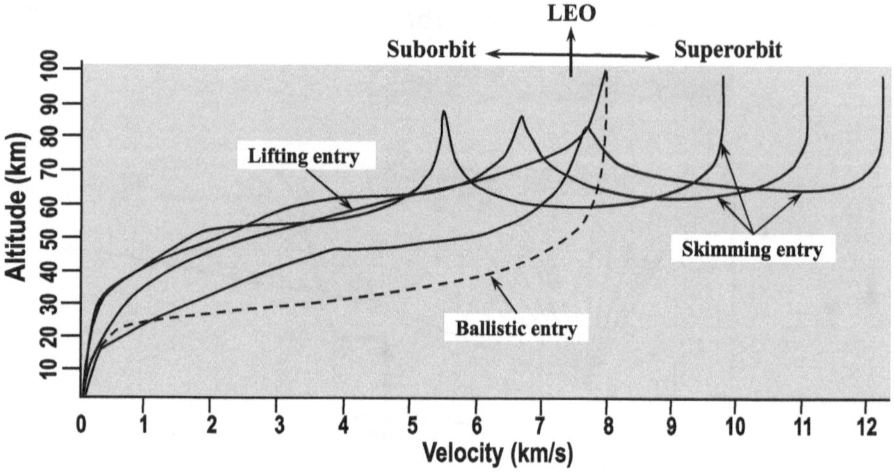

Figure 2.17 Typical flight trajectories of hypervelocity reentry.

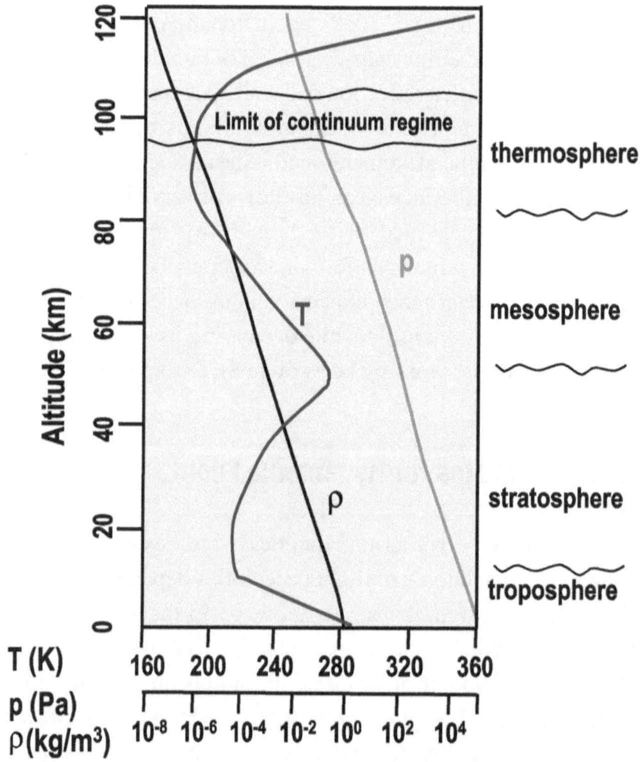

Figure 2.18 Thermodynamic properties of the atmosphere varying with altitude.

is approximately divided into four parts: troposphere, stratosphere, mesosphere, and thermosphere. The vertical distribution of temperature is nonmonotonic, which have different tendencies in different parts. On the other hand, the pressure p decreases rapidly and monotonously with increasing altitude and so does the density ρ. The dynamic viscosity μ is a function of temperature and is determined by the Sutherland formula (with acceptable accuracy at altitudes below 86 km). The composition of the atmosphere can be considered as constant in the domain where continuum flow regime maintains up to an altitude of approximately 100 km. Above the aforementioned altitude, the composition changes with altitude and the continuum flow regime stops.

At each point of the flight trajectory, the pressure p_∞, temperature T_∞, density ρ_∞, and the flight velocity V_∞ are known, and they provide the conditions of a definite solution for a given vehicle. Each parameter, Q, corresponding to the aerodynamic performance of the vehicle is the function of the abovementioned variables, that is, $Q = f\left(p_\infty, \rho_\infty, T_\infty, V_\infty\right)$.

During the design phase of a new hypersonic vehicle, wind-tunnel tests are necessary to verify the design, especially to verify features such as the aerodynamic forces and moments, the aerodynamic heating environment, stability, and maneuverability. Generally, the real flight conditions cannot be duplicated in a single ground-based test facility, especially for hypervelocity flights. Instead, the Reynolds number Re, the Mach number Ma, the ratio of total enthalpy to the characteristic enthalpy of chemical phenomena H_0/H_c, and others nondimensional parameters are selected to be simulated in the test based on similarity rules.

In an RST, as depicted in Figure 2.12, once the flight condition is specified at a given flight altitude, the required reservoir condition can be calculated as follows:

$$T_5 = T_\infty\left(1+\frac{\gamma-1}{2}M_\infty^2\right) \tag{2.31}$$

$$p_5 = p_\infty\left(1+\frac{\gamma-1}{2}M_\infty^2\right)^{\frac{\gamma}{\gamma-1}}. \tag{2.32}$$

The simulation of the above-mentioned parameters is the primary objective of the design or operation of a shock tunnel.

Nomenclature

c	speed of sound
C_I, C_{II}	Riemann invariant
H	enthalpy
L	distance between the origin O and the closed end
$\dot{m}_s$	mass flow rate of shock
Ma	Mach number
M_r	Mach number of the reflected shock
M_s	shock Mach number

p	pressure
Q	parameter
Re	Reynolds number
R	rarefaction wave
S	shock wave
s	entropy
t	time
T	temperature
w	moving velocity of shock wave
w_r	moving velocity of reflected shock wave
x	coordinate of streamwise
ρ	density
u	velocity
u_3	piston speed
u_{cs}	velocity of contact surface
V	flight velocity; flow velocity (Figures 2.15, 2.16)
γ	ratio of specific heats
δu	infinitesimal velocity

Subscripts:

c	chemical
t	transmitted
i	incidental
r	reflected
s	incident shock
TIC	tailored-interface condition
0	total or stagnation condition
1	initial state in the driven gas
2	state behind incident or primary shock
3	state between the expansion wave and the contact surface
4	initial state in the driver gas
5	state behind reflected shock in the test gas
(1)	state upstream of shock, shock-fixed coordinate
(2)	state downstream of shock, shock-fixed coordinate
∞	freestream condition

References

Ben-Dor, G., Igra, O., and Elperin, T. (2001). *Handbook of Shock Waves*. New York: Elsevier, pp. 553–601.

Chue, R. S. M., Tsai, C.-Y., Bakos, R. J., Erdos, J. I., and Rogers, R. C. (2002). Chapter 3: NASA's HYPULSE Facility at GASL: A Dual Mode, Dual Driver Reflected-Shock/Expansion Tunnel. In *Advanced Hypersonic Test Facilities*, Progress in Astronautics and Aeronautics, vol. 198. Reston, VA: AIAA.

Holden, M. S., Wadhams, T. P., MacLean, M., and Mundy, E. (2007). Experimental Studies in the LENS I and X to Evaluate Real Gas Effects on Hypervelocity Vehicle Performance. In *45th AIAA Aerospace Sciences Meeting and Exhibit*. Reno, Nevada, *AIAA Paper 2007–204*.

Jiang, Z. L., Li, J. P., Hu, Z. M., Liu, Y. F., and Yu, H. R. (2020). On the Theory and Methods for Advanced Detonation-driven Hypervelocity Shock Tunnels. *National Science Review*, **7**(7), 1198–1207.

Jiang, Z. L., Wu, B., Gao, Y. L., Zhao, W.,· and Hu, Z. M. (2015). Developing the Detonation-driven Expansion Tube for Orbital Speed Experiments. *Science China Technological Sciences*, **58**(4), 695–700.

Morgan, R. G. (2000). Development of X3: A Superorbital Expansion Tube. In *38th Aerospace Sciences Meeting and Exhibit. AIAA Paper 2000–0558.*

Paull, A. and Stalker, R. J. (1989). *Experiments on an Expansion Tube with a Free Piston Driver – Phase 2*. Department of Mechanical Engineering Report, The University of Queensland.

Paull, A. and Stalker, R. J. (1992). Test Flow Disturbances in an Expansion Tube. *Journal of Fluid Mechanics*, **245**, 493–521.

Stalker, R. J., Paull, A., and Stringer, I. (1987). *Experiments on an Expansion Tube with a Free Piston Driver – Phase 1*. Department of Mechanical Engineering Report, The University of Queensland.

3 Heated Light-Gas High-Enthalpy Shock Tunnels

3.1 Introduction

The achievable total enthalpy and pressure level in a shock tunnel depend on the ability to drive strong shock waves in the facility. To produce high shock strength, the driver gas needs to be at a high pressure and have a high sound speed. There are various techniques to increase the driver gas sound speed, which are essentially different approaches to raise the driver gas temperature. In this chapter and in Chapters 4 and 5, we will focus our attention on three gas dynamic techniques, namely heated light gas, free-piston compression, and detonative combustion drivers.

The first technique to increase the driver gas sound speed is to use a light gas and heat it to a high temperature using a heater. This chapter will discuss the use of electrical heaters, which include the relatively simple electrical resistance heaters and also electric-arc heaters. Strictly speaking, electric-arc heating is not a gasdynamic technique, and it is not capable of completely duplicating the conditions for hypervelocity testing. However, it is included because it can generate extremely high total enthalpies and are useful in certain applications.

3.2 Fundamentals of Heated Light-Gas Drivers

Shock-tunnel facilities generate high total enthalpy conditions by using strong shock waves. In the case of reflected-shock tunnels, heating of the test gas medium is accomplished by the incident (or primary) and reflected shocks. In shock-expansion tunnels, the primary shock also plays an important role before unsteady expansion waves further add total enthalpy to the test flow.

Shock tunnels are derived from the concept of the classical shock tube. Shown in Figure 3.1 is a schematic of a constant-area reflected-shock tunnel driven by a heated light gas. On the left side of the tube is the driver section that is initially separated by the main diaphragm from the driven tube (or shock tube). At the end of the driven tube is the reflected-shock region that is initially closed off from the nozzle by a thin diaphragm. States 1 and 4 denote the initial conditions of the driven and driver gases, respectively (Figure 3.1(a)). After bursting the main diaphragm, an incident (or primary) shock wave is generated in the driven gas propagating to the right, with state 2 denoting the condition behind the incident shock wave (Figure 3.1(b)). After

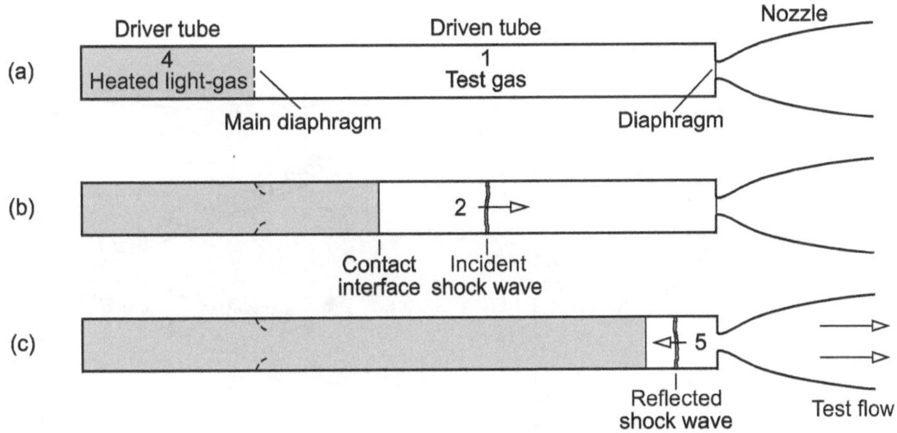

Figure 3.1 Heated light-gas-driven reflected-shock tunnel.

the incident shock has reflected off the driven tube end wall, state 5 is the condition behind the reflected shock (Figure 3.1(c)). After shock reflection, the diaphragm at the nozzle entrance is ruptured and the driven gas (i.e., test gas) expands through the convergent–divergent nozzle into the test section of the facility.

In essence, the facility depicted in Figure 3.1 is no different from a classical constant-area shock tube, except for the addition of a nozzle component. The driver tube has a constant volume and initially filled with a gas with uniform properties. The strength of the shock wave generated in a shock tube is a function of the initial pressure as well as the sound speed of the driver gas. The higher the driver gas pressure and sound speed, the stronger the shock wave. The initial pressure is controlled by the fill pressure in the driver section, which means that the driver tube must be constructed to withstand extremely high pressures.

For a perfect gas, the sound speed a is given by

$$a = \sqrt{\gamma RT}, \tag{3.1}$$

where γ is the ratio of specific heats, R is the gas constant, and T is the temperature. The sound speed of the driver gas is increased by using a gas that has a low molecular weight, such as helium and hydrogen, thus resulting in a large value of gas constant R. Further increase in the sound speed is obtained by heating the driver gas to a higher temperature. In a heated light-gas driver, a light gas is used as the driver gas, and it is heated using a relatively simple means. The main focus of this chapter will be on electrical-resistance-heated drivers as they are well suited for generating a wide range of test conditions in the hypersonic regime.

There are several advantages of using simple heated light-gas drivers. In such a driver, the driver gas state remains uniform and undisturbed until the arrival of the expansion wave subsequent to diaphragm rupture. Indeed, the wave dynamics during facility operation are essentially the same as in a classical shock tube, thus allowing the performance of the facility to be easily predictable. The wave processes for a

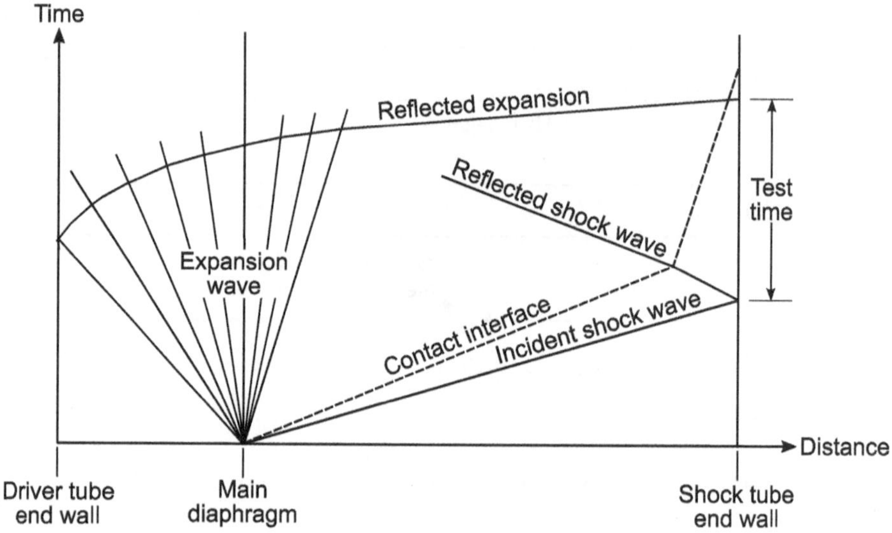

Figure 3.2 Wave diagram of a heated light-gas-driven reflected-shock tunnel.

heated light-gas-driven reflected-shock tunnel is illustrated in the wave diagram in Figure 3.2. After the main diaphragm ruptures, the expansion of the high-pressure driver gas generates an incident shock in the driven gas (i.e., the test gas). The incident shock eventually reflects off the driven tube end wall. The test gas that is processed twice by the shocks forms a stagnant high-pressure and high-temperature state that ruptures the thin diaphragm at the nozzle entrance. The reflected-shock region serves as the reservoir of the test flow, which expands into the test section of the facility. The testing period starts with the test gas in the reflected-shock region entering the nozzle and ends when all the test gas is depleted, provided that the reflected-shock state remains undisturbed during this period.

In the driver tube, the head of the expansion wave that originates from the diaphragm station propagates upstream toward the driver tube end wall. The expansion wave then reflects at the end wall and moves forward into the driven tube. If the reflected expansion wave reaches the reflected-shock region before all the test gas has been exhausted into the nozzle, it disturbs the test flow and terminates the useful test period. In heated light-gas drivers, the flow processes in the driver are well understood as they are the same as in classical shock tubes. The effects of the reflected expansion wave and the interference with the condition in the reflected-shock region can be delayed by using an appropriately long driver section. This results in high test flow quality, which is an important attribute in short-duration test facilities that are prone to noise and disturbances.

In contrast, in a free-piston-driven facility, the mechanical design, construction, and operation of the piston are much more complex. During operation, the piston-compressed driver gas undergoes a highly unsteady and complex process, with gas dynamic waves reverberating between the piston face and the diaphragm station. As the driver gas is compressed to a very small volume by the piston, the rapidly changing conditions in the driver gas can quickly propagate forward, which can then disturb the downstream test flow and shortens the useful test duration. Detonation drivers can also experience large acoustic wave disturbances because of the strong blast generated in the detonation process. On the other hand, the free-piston and detonation drivers can generate higher shock strength than that obtained from heating a light gas with conventional electrical heaters.

The use of light gases as drivers has long been recognized in classical shock-tube theory. Helium is the second lightest gas with a molecular weight of 4 kg/kmol and has been used in early shock tubes (Martin 1958) due to its combination of high sound speed and chemical inertness. Hydrogen, with a molecular weight of 2 kg/kmol, is the lightest gas possible and thus has the highest sound speed. However, there are disadvantages of using hydrogen as a driver gas. First, hydrogen is highly inflammable and detonable, so the safety of its handling is a major concern. Also, since the driver gas pressure must be high in order to generate strong shocks, it exacerbates the safety issue further. Problems such as leaks, inerting the post-run mixture, as well as the facility infrastructure for hydrogen usage must be carefully addressed to eliminate the chance of accidents. Furthermore, the cost of acquiring and storage of the light gas can be significant due to the large consumption in each test. Another problem with hydrogen is that it can react with the driven gas (i.e., air) at the contact interface, which then produces water vapor as a combustion product. This, in effect, reduces the amount of test gas available for testing, although this can be alleviated by lengthening the driven tube to compensate for the loss of test gas. Besides the issue of contamination of the test gas medium, the water vapor can also cause corrosion in the steel structure in the facility.

Heated hydrogen drivers have been used regularly for years at the Calspan-University at Buffalo Research Center (CUBRC), formerly known as the Cornell Aeronautical Laboratory (CAL). Although a severe accident caused by the use of high-temperature hydrogen did occur, heated-hydrogen drivers have largely been used successfully and safely. Issues of water vapor corrosion have also not been reported. To reduce the consumption of the driver gas, fast-acting valves have been installed in the driven tubes in all CUBRC tunnels to prevent the driver gas from entering the test section upon the completion of the useful test period. The excellent quality of the test data obtained in those facilities demonstrates that the heated light gas is a highly capable driver method for high enthalpy testing.

In the following section, we will demonstrate how the strength of the primary shock wave and the total enthalpy achieved are related to the driver gas sound speed.

3.3 Theories for Heated Light-Gas Shock Tunnels

Consider the incident shock wave generated in the constant-area reflected-shock tunnel as depicted in Figure 3.1. The shock-wave relations across the incident shock can be found from the normal shock relation:

$$\frac{p_2}{p_1} = 1 + \frac{2\gamma_1}{\gamma_1 + 1}\left(M_{s1}^2 - 1\right), \tag{3.2}$$

where p is the static pressure, M_{s1} is the primary or incident shock Mach number, and γ is the ratio of specific heats. The subscripts refer to the respective states described in Section 3.2. Equation (3.2) is written with the assumptions of perfect gases, constant specific heats, and one-dimensional inviscid flow.

The relationship between the post-shock pressure p_2 and the initial driver gas pressure p_4 can be found to be

$$\frac{p_4}{p_1}=\frac{p_2}{p_1}\left[1-\frac{\gamma_4-1}{\sqrt{2\gamma_1}}\frac{a_1}{a_4}\frac{\frac{p_2}{p_1}-1}{\sqrt{(\gamma_1+1)\frac{p_2}{p_1}+(\gamma_1-1)}}\right]^{-\frac{2\gamma_4}{\gamma_4-1}}, \tag{3.3}$$

where the ratio of the sound speeds is given by

$$\frac{a_4}{a_1}=\sqrt{\frac{\gamma_4}{\gamma_1}\frac{T_4}{T_1}\frac{W_1}{W_4}}, \tag{3.4}$$

with W being the molecular weight.

Substituting Eq. (3.2) into Eq. (3.3), an expression relating the pressure ratio p_4/p_1 to the shock Mach number M_{s1} can be obtained:

$$\frac{p_4}{p_1}=\left[1+\frac{2\gamma_1}{\gamma_1+1}\left(M_{s1}^2-1\right)\right]\left[1-\frac{(\gamma_4-1)}{(\gamma_1+1)}\frac{a_1}{a_4}\frac{\left(M_{s1}^2-1\right)}{M_{s1}}\right]^{-\frac{2\gamma_4}{\gamma_4-1}}. \tag{3.5}$$

In the limiting case when p_4/p_1 approaches infinity, M_{s1} can be solved directly as

$$\lim_{p_4/p_1\to\infty} M_{s1}=\frac{\gamma_1+1}{2(\gamma_4-1)}\frac{a_4}{a_1}+\sqrt{\left[\frac{\gamma_1+1}{2(\gamma_4-1)}\frac{a_4}{a_1}\right]^2+1}. \tag{3.6}$$

For $a_4/a_1 \geq 1$, the limiting value of M_{s1} is

$$M_{s1}=\frac{\gamma_1+1}{\gamma_4-1}\frac{a_4}{a_1},\quad \frac{p_4}{p_1}\to\infty. \tag{3.7}$$

Alternatively, the limiting value of the incident or primary shock velocity U_{s1} is given by

$$U_{s1}=\frac{\gamma_1+1}{\gamma_4-1}a_4,\quad \frac{p_4}{p_1}\to\infty. \tag{3.8}$$

Equations (3.7) and (3.8) show that the primary shock Mach number and velocity are approximately proportional to the driver gas sound speed.

When the incident shock wave reflects off the end wall of the driven tube, the flow velocity in the reflected shock is assumed to vanish, as in the case of a closed-end tube. The pressure and temperature in the reflected-shock state are given by the following relations:

$$\frac{p_5}{p_1}=\frac{p_2}{p_1}\frac{(3\gamma_1-1)\frac{p_2}{p_1}-(\gamma_1-1)}{\gamma_1+1+(\gamma_1-1)\frac{p_2}{p_1}}, \tag{3.9}$$

$$\frac{T_5}{T_1} = \frac{1}{\gamma_1}\left[(3\gamma_1 - 1)\frac{p_2}{p_1} - (\gamma_1 - 1)\right]\frac{1 + (\gamma_1 - 1)\frac{p_2}{p_1}}{(\gamma_1 + 1)\frac{p_2}{p_1} + (\gamma_1 - \ )}. \quad (3.10)$$

For large incident shock Mach numbers, the reflected shock temperature T_5 is approximated by

$$\frac{T_5}{T_1} \approx \frac{2(3\gamma_1 - 1)(\gamma_1 - 1)}{(\gamma_1 + 1)^2} M_{s1}^2. \quad (3.11)$$

In the case of the reflected-shock tunnel, the reflected-shock state represents the stagnation condition of the test flow. The total enthalpy is then given by

$$h_t = h_5 = c_p T_5. \quad (3.12)$$

If chemical reactions are neglected, the heat capacity c_p is given by the initial state of the driven gas:

$$c_p = \frac{\gamma_1}{\gamma_1 - 1} R_1.$$

The total enthalpy is thus approximated by

$$h_t \approx \frac{2(3\gamma_1 - 1)}{(\gamma_1 + 1)^2} U_{s1}^2. \quad (3.13)$$

For the case of air as a driven gas, γ_1 equals 1.4 so that Eq. (3.13) becomes

$$h_t \approx 1.11\, U_{s1}^2. \quad (3.14)$$

Chemical reactions are indeed significant at high temperatures encountered at high enthalpy conditions so that the exact expression for total enthalpy is more complex than that shown in Eq. (3.14). Nevertheless, Eqs. (3.13) and (3.14) show that the total enthalpy delivered in a reflected-shock tunnel is directly related to the incident shock strength and is proportional to the square of the incident shock velocity. For reflected-shock tunnels, the total enthalpy is commonly approximated to be

$$h_t \approx U_{s1}^2. \quad (3.15)$$

For infinitely large p_4/p_1, the total enthalpy can be found using Eq. (3.8) to be proportional to the square of the driver gas sound speed:

$$h_t = \left(\frac{\gamma_1 + 1}{\gamma_4 - 1}\right)^2 a_4^2, \quad \frac{p_4}{p_1} \to \infty. \quad (3.16)$$

For shock-expansion tunnels, the total enthalpy generated is partly due to the primary shock and partly due to the unsteady expansion process that occurs when the primary shock propagates into the acceleration section of the driven tube. Therefore, to reach high total enthalpies, it is crucial to maximize the strength of the primary shock, for both reflected-shock and shock-expansion tunnels.

Equation (3.5) shows that the incident shock strength depends not only on the driver-to-driven gas pressure ratio p_4/p_1 but also on the sound speed ratio a_4/a_1.

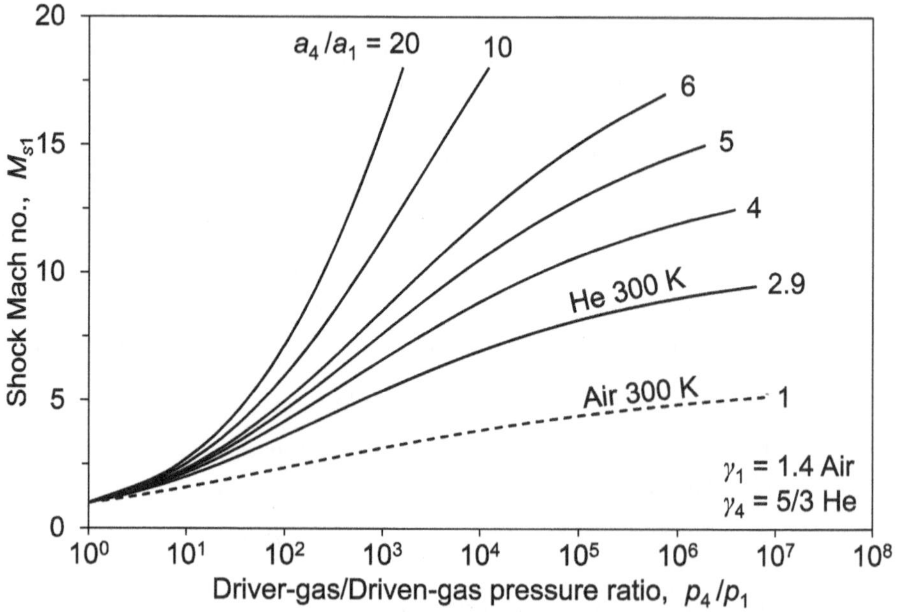

Figure 3.3 Incident shock Mach number versus driver gas pressure and sound speed. Except for the lowest curve, the driver gas is helium.

Shown in Figure 3.3 is the incident or primary shock Mach number M_{s1} plotted against p_4/p_1 for values of a_4/a_1 ranging from 1 to 20 for a constant-area shock tube. As a general trend, for a given driver gas sound speed (i.e., fixing a_4/a_1), the incident shock Mach number increases with increasing pressure ratio p_4/p_1 and eventually approaches the limiting value given by Eq. (3.7). By increasing the driver gas sound speed, the shock Mach number increases drastically. Note that the lowest curve in Figure 3.3 represents the case when both the driver and driven gases are air and initially at room temperature, so that a_4/a_1 is 1. It is observed that by replacing the driver gas with a light gas, helium, the sound speed ratio increases to 2.94 due to its low molecular weight, thereby yielding a significantly higher shock Mach number. The other curves in Figure 3.3 are obtained for the helium driver gas that is heated to yield larger values of a_4/a_1, which further increases the shock Mach number. It can be concluded that the incident shock strength is a strong function of the driver gas sound speed.

It is worth noting that the use of a high sound speed driver also permits the same shock velocity to be attained with a lower driver-to-driven fill pressure ratio p_4/p_1. A higher initial driven gas pressure p_1 translates into higher subsequent post-shock and test-section pressures at a given total enthalpy, which is especially desirable for satisfying total pressure, binary scaling, and Reynolds number. Indeed, most of the current techniques that enable impulse test facilities to achieve hypervelocity conditions are essentially different approaches to heat the driver gas and therefore to increase its sound speed.

3.4 Techniques for Heated Light-Gas Shock Tunnels

The main technique used to increase the temperature of driver gases is electrical heating. In this method, heat is generated by the internal electrical resistance of the heating element. Heat is transferred by conduction and convection from the heating element to the driver gas medium that is in a high-pressure quiescent state except for the convection induced by heating. A uniform volume of high-pressure and high-temperature gas that can satisfy the required test condition is obtained before the testing operation begins.

Another method is to use the discharge of an electric arc in the driver gas. In this method, the electric arc is rapidly discharged in the pressurized driver tube, and thus the testing operation immediately commences.

3.4.1 Electrical Resistance Heaters

Electrical heating can be applied either internally or externally to the driver tube. For the internal heater, the heating element is installed inside the tube so that the driver gas is in direct contact with the heater. However, the heating element can obstruct the flow in the driver tube. For the external heater, the heating element is placed outside the driver tube, and the heat is transferred through the tube wall to the driver gas. The schematics in Figure 3.4 illustrate the two types of resistance heaters: (a) the internal electrical heater and (b) the external electrical heater.

Electrical resistance heaters can increase the driver gas temperature up to 800 K. Similar to Figure 3.3, Figure 3.5 shows the variation of the incident shock Mach number with the driver-to-driven fill pressure ratio p_4/p_1. In this figure, the increase in the incident shock Mach number by using a light-gas driver and by heating it at a

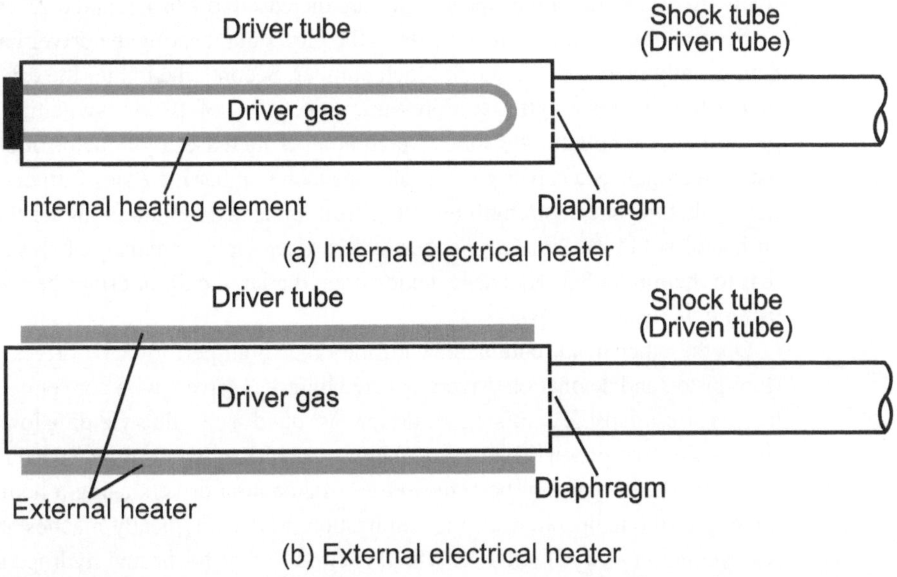

Figure 3.4 Electrical resistance heaters.

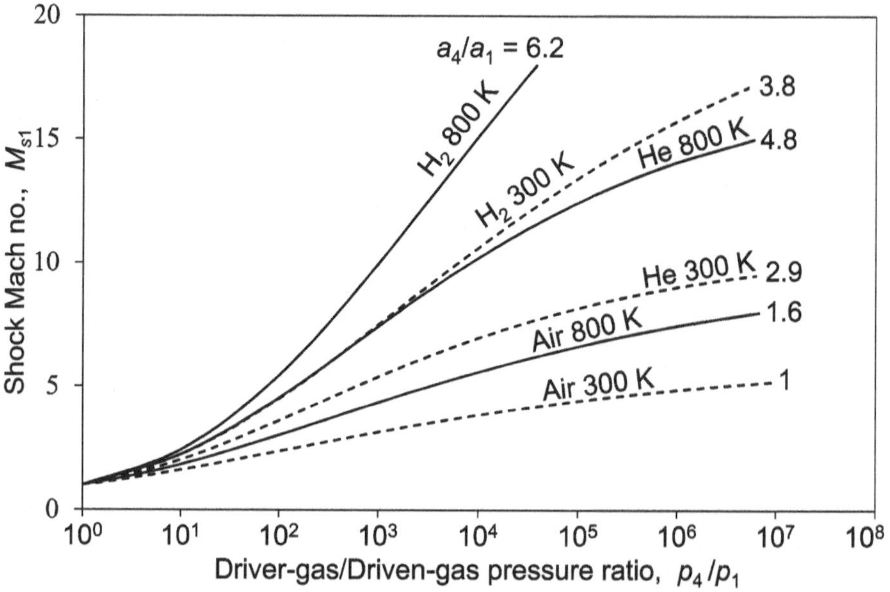

Figure 3.5 Incident shock Mach number versus driver gas pressure for heated hydrogen and helium as the driver gas. Air is also included as a comparison.

temperature of 800 K is shown. The curves are computed for three driver gases – air, helium, and hydrogen – which have the γ_4 values of 1.4, 5/3, and 1.4, respectively. The driven gas is assumed to be air. For each driver medium, the shock Mach number at driver temperatures of 300 K (unheated) and 800 K are plotted. At 300 K, the resulting sound speed ratios a_4/a_1 are 1.0, 2.9, and 3.8 for air, helium, and hydrogen, respectively. At 800 K, the sound speed ratios are increased to 1.6, 4.8, and 6.2, respectively.

Several observations can be made in the plot. First, heating the driver gas at 800 K significantly increases the shock Mach number. Second, the lighter the driver gas, the higher the shock strength. At a pressure ratio, p_4/p_1, of 10^4, by switching the driver gas from air to helium, the shock Mach number increases by a factor of about 2. By using hydrogen as a driver gas, the shock Mach number increases further. It is worth noting that the shock strength resulting from unheated helium is higher than that for air heated at 800 K. Similarly, unheated hydrogen yields a higher shock strength than that for helium at 800 K. These demonstrate the large gain in driver performance by using light, heated gases.

On the other hand, compared with the other high-performance driver techniques (free-piston and detonation drivers), heated light-gas drivers with conventional electric heaters is less capable of raising the driver gas sound speed, thus yielding lower strength shocks. For free-piston drivers using helium as a driver gas, a sound speed ratio, a_4/a_1, of about 10 can typically be achieved. For detonation drivers using a helium-diluted driver gas, the temperature of the combustion products typically reaches about 3,500 K, resulting in a_4/a_1 of about 6, which is close to that for heated hydrogen at 800 K. However, the smaller ratio of specific heats in the detonation products provides further gains in shock strength than for heated hydrogen. Nevertheless, the achievable sound speed ratio in heated hydrogen is more than adequate for reaching hypersonic conditions up to about Mach 15 in reflected-shock tunnels. If higher test flow Mach numbers or total enthalpies are required, heated helium or hydrogen drivers can be used to drive shock-expansion tunnels to reach those conditions. And because shock-expansion tunnels are more suitable for test conditions above Mach 15, as reflected-shock tunnels would suffer from significant test-gas dissociation, the use of heated light-gas drivers does not limit the total enthalpy capability of the test facility.

There are two major impulse facilities that have used electrical heaters in the driver: the NASA Langley Research Center 6-inch expansion tube and the various Large Energy National Shock (LENS) tunnel facilities at CUBRC.

For the Langley 6-inch expansion tube, the internal resistance heater was designed to heat hydrogen at 555 K at 4.8 MPa (Creel 1972), although the driver can use helium and hydrogen gases. To obtain the flexibility of operation, three different sized internal resistance heater elements were fabricated. The heating element is constructed of two parallel stainless-steel extrusions. The two members of the element are separated from each other by aluminum oxide insulators insulated from the driver tube wall by aluminum oxide cylinders. The heating element assembly is inserted into and runs along the driver tube. The heating element is connected at the upstream end to a 3-MW power supply by water-cooled beryllium-copper rods and at the downstream end by two stainless-steel crossover plates (Creel 1972, Moore 1975). A more detailed description and the performance of the heater is given in Creel (1972).

For the various LENS tunnel facilities at CUBRC, the driver sections can use helium, hydrogen, nitrogen, or a mixture of these gases. The drivers are heated externally by resistance heaters at temperatures between 600 K and 800 K (Bird 1962, Parker & Wakeman 2010). Although detailed information regarding the heater designs is not found in the literature, the rationale for selecting external rather than internal heating is clear. By applying heating on the exterior of the driver tube, the tube interior is devoid of heating element hardware that would interfere with the flow and wave processes during operation. The driver therefore works exactly as in a classical shock tube and free of flow complications.

3.4.2 Electric-Arc Heaters

Another method to increase the driver gas sound speed is to utilize the discharge of an electric arc to rapidly raise the driver gas temperature. The electric arc is generated by the release of electrical energy stored in a capacity bank charged at a high voltage. The arc can be discharged in various gases, including helium, hydrogen, and air, and therefore the technique is applicable for heating different driver gases in impulse facilities. The technique produces the highest velocities and temperatures among the shock tube and tunnel drivers.

On the other hand, electric-arc heating also has some notable disadvantages as well. An electric-arc discharge as a driver technique shares many of the same issues found in arc-jet blowdown wind tunnels discussed in Chapter 1. As in the case of the arc-jet, the electric-arc driver is suitable for relatively low operating pressures, thereby limiting the flow pressure and density achievable in the test section. Moreover, the flow produced by the driver is unsteady, which limits the test duration. Note also that the ability to produce extremely high shock velocities is not necessarily desirable for the generation of high-quality test flow. When an electric arc is used to drive a reflected-shock tunnel, the stagnated flow in the reflected-shock region would raise the already high post-shock gas temperature even higher, so that the issues of test gas dissociation and tube wall erosion will only worsen. If the technique is used to drive a shock-expansion tunnel, an overly strong primary shock would dissociate the test gas in which the shock propagates. This goes counter to the rationale for using unsteady expansion waves to yield high total enthalpies in an expansion tunnel in order to avoid test gas dissociation. Another disadvantage is cost. The capacitor bank is a highly complex system that is expensive to construct and operate. Because of this, the facility size is restricted to be small, with the consequence that the available test time is limited to the order of microseconds to tens of microseconds, which is insufficient for establishing a quasi-steady flow over reasonably-sized test models.

Electric-arc-driven shock tubes had been designed at the Jet Propulsion Laboratory (JPL) (Menard 1971) and at NASA Ames (Dannenberg 1972) for simulating planetary entry conditions into Venus, Jupiter, and Saturn. At NASA Langley, electric-arc drivers were used to drive the 6-inch shock tube (Nealy 1972) and also the 6-inch expansion tube (Miller 1976, Miller & Jones 1983), with the latter being the same facility as described in Section 3.4.1, which can accommodate both the electric arc and the electrical resistance heater.

The electric-arc driver in the Langley 6-inch expansion tube uses a parallel-rail design. Two electrodes (the rails) run parallel along the length of the driver tube on opposite sides of the circumference. During electric discharge, the current moves down one rail and back along the other rail. The energy to generate the electric-arc discharge is stored in a capacitor storage system rated at 12,000 V with a maximum energy of 5 MJ (Moore 1976). From the test runs conducted, it was noted that the arc-discharge process would induce a family of shock waves that were transmitted down the expansion tube and prove to be detrimental to the expansion tube test time and flow uniformity. There were a number of additional problems encountered, such as damage due to arcing between the rail and the driver tube wall, a long turnaround time required for facility cleanup after each run due to soot formation and deposits on parallel-rail components and on walls, poor run-to-run repeatability, and a relatively short life expectancy of the parallel-rail components. Because of these reasons, the electric-arc driver technique was abandoned in the Langley expansion tube (Miller 1976).

3.5 Development of Heated Light-Gas Shock Tunnels

3.5.1 Electrical Resistance Heaters

The development of the heated light-gas driver followed closely the early development of the reflected-shock tunnel to provide the quasi-steady test flow in the hypersonic regime. Early studies of hypersonic shock tunnels were carried out at the CAL in the 1950s. Various modes of shock-tube operation were explored, and the first shock tunnels were developed by adding a nozzle to the end of a nonreflecting shock tube. At the same time, different driver techniques to increase the capability of the test facilities were investigated (Hertzberg et al. 1961).

Deflagrative combustion was first examined as a means to increase the driver gas temperature and its sound speed, and it was found that such drivers can produce sufficient shock strength to meet the desired high total enthalpy conditions. However, a detailed examination of the measurements revealed that the flow generated by this type of combustion was of relatively poor quality when compared with those obtained using pure-gas drivers, such as helium or hydrogen. In particular, shock-wave attenuation was observed to be severe. Although the technique of deflagrative combustion could be improved, it was considered doubtful that its reliability could ever match that obtained using pure-gas drivers. The examination of available techniques at the time identified that high-pressure operation of hydrogen at 200 MPa was already commonplace in the chemical industry and that pressures up to 400 MPa were possible. With the ability to reach such pressures, the deflagrative combustion technique was abandoned in favor of high-pressure pure-gas drivers.

In order to drastically increase the test time, the nonreflecting shock tunnel was further modified into the reflected-shock tunnel (Hertzberg et al. 1961). The tailored mode of operating the reflected-shock tunnel was also developed so that when the incident shock was reflected at the end of the shock tube, the conditions across the driver–driven gas contact interface were matched to prevent the generation of additional waves from the interaction of the reflected shock and the interface. A high sound speed driver gas was critical in achieving the tailored-interface operation, which was perfected in 1957 (Bird 1962). The technique of using external electrical heaters was adopted, and several shock tunnels were constructed, including the 24-inch shock tunnel that began operation in 1959, which was later, in 1961, modified to the 48-inch shock tunnel and has been in operation since then. The tunnel can use either hydrogen or helium as the driver gas that can be heated up to 833 K (Bird 1962). It can simulate conditions from Mach 6 to 20 (Holden 2010).

In the 1980s, the development of a series of LENS tunnel facilities was initiated under a renamed research institute CUBRC. The first facility, LENS I, is a reflected-shock tunnel designed for high-altitude, high Mach number testing (MacLean et al. 2005). It is capable of full duplication of flight conditions from Mach 7 to 14 (Holden & Parker 2002). The second facility, LENS II, is also a reflected-shock tunnel that was designed for low-altitude, moderate Mach number conditions. LENS II is rated to produce conditions of Mach 4.5 at sea-level dynamic pressure in the freestream (MacLean et al. 2005). The size of the facility was unparalleled at the time it was built, with the total length of the driver and shock-tube components being more than 60 m and having a large uniform inner diameter of 60.95 cm. The size of LENS II allowed the test time to be as much as 100 ms, which is one to two orders of magnitude more than most other reflected-shock tunnels (MacLean et al. 2005). LENS II had also been configured into LENS X (Nompelis et al. 2004), which served as a prototype facility to validate

Table 3.1 Simulation capabilities and dimensions of heated light-gas facilities at CUBRC

			Driver tube		Driven tube		Nozzle
Facility name	Velocity (km/s)	Max. p_T (MPa)	L (m)	D (m)	L (m)	D (m)	Max. exit D (m)
LENS 48-inch	0.7–3	40	12.2	0.203	15.2	0.203	1.22
LENS I	1–4.5	200	7.62	0.279	18.3	0.203	1.83
LENS II	0.7–3	100	18.3	0.61	30.5	0.61	1.83
LENS XX	2.5–15	–	–	0.61	–	0.61	–

the design of a large shock-expansion tunnel that was to be used in a larger facility, LENS XX. The length of the LENS XX tube components is not known, but the end-to-end length of the facility is over 73 m (Dufrene & Holden 2011). It also has a large inner diameter of 60.95 cm (Dufrene et al. 2010). Its large size allows LENS XX to have a test time of over 4 ms for most conditions (Dufrene et al. 2010) and up to 8 ms for very low enthalpies (Dufrene et al. 2012), which is an order of magnitude higher than most other shock-expansion tunnels. LENS XX began operation in the late 2000s and was the largest shock-expansion tunnel at the time. It was designed to fully duplicate flight conditions from Mach 8 to 25 at altitudes covering sea level to 76 km (Holden et al. 2018).

The current family of shock-tunnel facilities at CUBRC includes LENS 48-inch, LENS I, LENS II, and LENS XX. LENS 48-inch is the original CAL 48-inch shock tunnel. It uses hydrogen or helium as the driver gas that can be heated up to 833 K (Bird 1962). LENS I uses hydrogen or helium as the driver gas that can be heated up to 700 K (Holden 1990). As LENS II has a lower enthalpy requirement, a mixture of helium and nitrogen rather than hydrogen is used as the driver gas. For LENS XX, the use of the hydrogen driver gas heated to 589 K has been reported (Parker & Wakeman 2010). The simulation capabilities and tube dimensions of the LENS facilities are provided in Table 3.1 (Holden 2010, Holden et al. 2018).

3.5.2 Electric-Arc Heaters

The electric arc as a shock-tube driver was developed in the late 1950s and in the 1960s. The shock tube at Avco-Everett Research Laboratory was a representative of such a facility (Camm & Rose 1963), which had an inner diameter of 15.24 cm and the driven tube lengths of 3.048 and 6.096 m. Electric-arc-driven shock tubes had also been designed and used at JPL, NASA Ames, and NASA Langley. The Langley facilities that used electric-arc drivers were the 6-inch shock tube and the 6-inch expansion tube as described in Section 3.4.2.

Currently, the Electric Arc Shock Tube (EAST) facility at NASA Ames is the only operational facility that utilizes the electric-arc discharge as a driver technique. The EAST can be operated with two types of electric-arc drivers: a conical driver and a cylindrical driver. The conical and cylindrical designs are arc chambers with geometries as described by their names. In each design, one electrode is placed at the upstream end of the chamber and the other at the downstream end. The arc chamber has a pressure rating of 500 atm. The energy of the driver is supplied by a 1.25-MJ capacitor storage system that can be charged up to 40,000 V (Sharma & Park 1990). The facility can operate as a shock tube or as a reflected-shock tunnel. The electrical discharge can heat hydrogen to a temperature of about 8,000 K, or helium to about 20,000 K, resulting in incident shock velocities of up to 15 km/s and 50 km/s for heated helium and hydrogen, respectively (Sharma & Park 1990).

3.6 Applications and Testing

3.6.1 Electrical Heaters

The NASA Langley Research Center 6-inch expansion tube was originally designed and constructed in the 1960s for radiative gas dynamics studies in support of the Apollo program (Trimpi 1962, 1966, Miller 1978, Miller & Jones 1983). It was also used for hypervelocity studies of aerothermal heating and aerodynamic performance of vehicles entering Earth and other planetary atmospheres (Miller 1976).

The Langley 6-inch expansion tube can be driven using unheated or heated helium, and two types of driver-gas heaters – a 3-MW internal resistance electric heater and a parallel-rail electric-arc heater – were incorporated in the design of the facility. The driven section has an inner diameter of 15.24 cm and a total length of 24.3 m, which includes the shock tube and the acceleration tube. The upstream section of the driver tube has a larger inner diameter of 35.56 cm and a length of 2.44 m in order to accommodate the resistance or electric-arc heater. To reduce the driver gas quantity when using unheated or resistance-heated operation, a 16.5-cm inner diameter liner can be installed (Miller 1976).

Test flow velocities up to 6.7 km/s were obtained with the helium driver gas heated to 510 K (Moore 1975). However, it was also found that when the helium driver is unheated, the performance is only modestly less than with the resistance-heated driver and yielded better run-to-run repeatability (Miller 1976). As mentioned in Section 3.4.2, the electric-arc heater was plagued with problems, and thus its use was discontinued.

Hypersonic real-gas research continued in the Langley expansion tube until it was decommissioned in 1983. One of the many studies includes the measurement of laminar heat transfer on biconics in hypervelocity flows (Miller et al. 1985). The biconics represent models of aeroassisted vehicles proposed for missions to a number of planets. In this study, unheated hydrogen was used as the driver gas, generating freestream velocities of 4.5 to 6.9 km/s. In 1989, the facility was relocated to the General Applied Science Laboratory (GASL) and recommissioned as the HyPulse (commonly written as HYPULSE) facility to support scramjet research for the NASP and later the Hyper-X programs. In the original installation of HyPulse at GASL, the internal resistance heater was included in the facility. However, as little performance gain was obtained with the heated helium, it was not put into use. Instead, the detonation-driver mode of operation was implemented.

Currently, the LENS facilities at CUBRC are the most notable ones in using electrically heated light-gas drivers. LENS is a family of large-size ground test facilities that include the LENS 48-inch, LENS I, and LENS II reflected-shock tunnels, and LENS XX shock-expansion tunnel (Holden & Parker 2002, MacLean et al. 2013). The 48-inch shock tunnel is the oldest and smallest among the facilities. The electrical resistance heaters are installed on the exterior of the driver tubes, thus avoiding any flow disturbances in the driver (Holden & Parker 2002).

The LENS I facility is a high-energy reflected-shock tunnel capable of fully duplicating flight conditions at Mach numbers from 7 to 14 (Holden & Parker 2002). It has been used to evaluate the aerothermal loads on hypersonic vehicles and reentry capsules, such as the X-38 Space Station recovery vehicle. It is also one of the facilities used by NASA for scramjet studies, including the ground testing of a full-scale X-43 scramjet (CUBRC LENS-I n.d.)

LENS II can generate high Reynolds number flows and can be operated as a reflected-shock or Ludwig tunnel. It can fully duplicate flight enthalpies from Mach 4 to 7 (Holden & Parker 2002). Because it is much larger than LENS I, it can accommodate much larger test particles, up to 9 m in length (CUBRC LENS-II n.d.) Like LENS I, LENS II has been used to study aerothermal loads and scramjet propulsion. Its large core-flow size allows it to conduct full-scale aerothermal testing of the Mach 6 HyFly model (MacLean et al. 2005) and the propulsion performance testing of the X-51 scramjet (CUBRC LENS-II n.d.)

LENS XX is a large-size shock-expansion tunnel and is the newest facility at CUBRC. Because of its size, it has overcome the limitation of short test time in other shock-expansion tunnels, and its test duration is over 4 ms. It can generate test flow velocities of greater than 13 km/s and shock velocities of up to 15 km/s (Dufrene et al. 2010). The high total enthalpy capability makes LENS XX an excellent facility to simulate the aerothermal loads and real-gas chemistry during entry into Earth and other planetary atmospheres (CUBRC LENS-XX n.d.) It has been used to study radiative emissions in atmospheric entry conditions (Dufrene & Holden 2011). In one study, the conditions that the NASA Crew Exploration Vehicle (CEV) capsule will experience during reentry from a Moon mission was replicated to measure shock-front emissions at velocities near 10 km/s (Parker & Wakeman 2010). The facility has also been used in fundamental research programs, such as the study of shock-wave/boundary-layer interactions in hypervelocity flows (MacLean et al. 2013).

The successful use of heated hydrogen over the years in the LENS facilities demonstrates that heated light-gas drivers are indeed very capable of generating the extreme flow conditions for hypervelocity ground testing.

3.6.2 Electric-Arc Heaters

The NASA Ames' EAST started operations in the 1960s and is the only pulse facility in operation that employs an electric-arc driver (Grinstead et al. 2010). Its driver can use either helium or hydrogen as the driver gas, increasing the driver gas temperature to about 20,000 K for helium or 8,000 K for hydrogen. Two driven tubes are available in the facility: one is 12 m in length with an inner diameter of 10 cm, and the other is 21 m in length with an inner diameter of 60 cm. Although the EAST can operate as either a shock tube or as a reflected-shock tunnel, the shock-tube mode is preferred for most of its applications.

The EAST facility has been used in fundamental and applied research. These include the studies of vibrational and rotational relaxation in expanding nitrogen flows (Sharma et al. 1993a, 1993b) and the investigation of radiative base heating of the Galileo probe during entry into the Jovian atmosphere (Shirai & Park 1979).

Because the electric-arc driver can generate extremely high shock velocities, the EAST has been used extensively to study bow shock radiation experienced during Earth and planetary entries. The conducted studies include shock radiation spectroscopic measurements and analyses for lunar return to Earth (Grinstead et al. 2008, Cruden et al. 2009), aerocapture for entry into Mars (Grinstead et al. 2009) and Titan (Bose et al. 2006) atmospheres. Because of its well-established achievements in this field, the radiative measurements obtained in the EAST often represent the benchmarks for others to compare to.

3.7 Summary

The capability of shock tunnels is seen to be enhanced by increasing the sound speed of the driver gas through the use of a low molecular weight gas medium. Further increase in sound speed is accomplished by heating the driver gas at a high temperature. From the basic shock-tube equations, it was demonstrated that the incident or primary shock Mach number produced by the driver scales approximately with the driver gas sound speed. By simply switching from unheated air to an unheated helium or hydrogen, the driver gas sound speed increases substantially. Indeed, using an unheated hydrogen driver gas would yield higher shock strength than that for heated helium at 800 K. Therefore, heated light-gas, and especially heated hydrogen, is a highly effective method for driving high enthalpy test facilities.

In this chapter, the relatively simple method of heating light-gas drivers using electrical heating has been discussed. The most common method used is electrical resistance heating, which has been adopted in the NASA Langley 6-inch expansion tube and in the various LENS facilities at CUBRC. The LENS facilities are the only ones currently using heated light-gas drivers, as electrical heating has been discontinued in the Langley expansion tube. For the LENS facilities, electrical heating is applied on the exterior of the driver tubes so that the flow processes are unobstructed, and the facilities operate exactly like a conventional shock tube. These facilities include both reflected-shock and shock expansion tunnels with test capabilities covering a wide range from low hypersonic to superorbital velocities. The use of heated hydrogen as a driver gas and their large size allow them to deliver high test flow qualities and long testing time. On the other hand, the safety of handling large volume of hydrogen remains an issue and must be dealt with continuously.

Although the electric-arc driver is not a gasdynamic technique to increase the driver gas sound speed, it is included in the chapter because of its ability to heat the driver gas to exceptionally high temperatures, thereby yielding extremely high shock velocities. However, the technique is not necessarily suitable for driving a hypersonic wind tunnel as an overly high shock strength can induce unwanted dissociation in the test gas. The electric-arc driver is also limited by the operating pressure, making it difficult to replicate the total pressure and Reynolds number encountered in hypersonic flight. The EAST facility at NASA Ames is currently the only electric-arc-driven shock tube/tunnel in operation. Its ability to generate extremely high shock velocities makes it an ideal tool to study radiative heating during Earth and planetary entries, and its measurements are considered as benchmarks in the community.

Nomenclature

a	sound speed
D	diameter
ht	specific total enthalpy
L	length
M_{s1}	incident or primary shock Mach number
p	pressure
R	gas constant
T	temperature
U_{s1}	incident or primary shock velocity
γ	ratio of specific heats

Subscripts:

t	total or stagnation condition
1	initial state in driven gas
2	state behind incident or primary shock
4	initial state in driver gas
5	state behind reflected shock in test gas

References

Bird, K. D. (1962). *48-Inch Hypersonic Shock Tunnel – Descriptions and Capabilities*. Buffalo, NY: Cornell Aeronautical Laboratory, Cornell University.

Bose, D., Wright, M., Bogdanoff, D., Raiche, G. A., and Allen, G. A. (2006). Modeling and Experimental Assessment of CN Radiation Behind a Strong Shock Wave. *Journal of Thermophysics and Heat Transfer*, **20**(2), 220–230.

Camm, J. C. and Rose, P. H. (1963). Electric Arc-driven Shock Tube. *Physics of Fluids*, **6**(5), 663–678.

Creel, T. R. Jr. (1972). Experimental Performance of an Internal Resistance Heater for Langley 6-inch Expansion Tube Driver. *NASA Technical Note TN D-7070*.

Cruden, B. A, Martinez, R., Grinstead, J. H., and Olejniczak, J. (2009). Simultaneous Vacuum Ultraviolet through Near IR Absolute Radiation Measurement with Spatiotemporal Resolution in an Electric Arc Shock Tube. *AIAA Paper 2009–4240.*

CUBRC LENS-I. (n.d.). *LENS I Mach 6 to 22 High Energy Shock Tunnel.* https://bit.ly/3eV9k6u (Accessed: May 3, 2023).

CUBRC LENS-II. (n.d.). *LENS II Mach 2 to 12 High Reynolds Number Shock/Ludwig Tunnel.* https://bit.ly/3s3emS2 (Accessed: May 3, 2023).

CUBRC LENS-XX. (n.d.). *LENS XX Expansion Tunnel for Velocities from 2,000 to 35,000 ft/sec.* https://bit.ly/30ZPw9X (Accessed: May 3, 2023).

Dannenberg, R. E. (1972). A Conical Arc Driver for High-Energy Test Facilities. *AIAA Journal,* **10**(12), 1692–1694.

Dufrene, A. and Holden, M. (2011). Experimental Characterization of the LENS Expansion Tunnel Facility Including Blunt Body Surface. *AIAA Paper 2011–626.*

Dufrene, A., MacLean, M., and Holden, M. (2012). High Enthalpy Studies of Capsule Heating in an Expansion Tunnel Facility. *AIAA Paper 2012-2998.*

Dufrene, A., MacLean, M., Parker, R. A., Wadhams, T., and Holden M. (2010). Characterization of the New LENS Expansion Tunnel Facility. *AIAA Paper 2010–1564.*

Grinstead, J. H., Wilder, M. C., Olejniczak, J., et al. (2008). Shock-Heated Air Radiation Measurements at Lunar Return Conditions. *AIAA Paper 2008-1244.*

Grinstead, J. H., Wilder, M. C., Reda, D. C., et al. (2010). Shock Tube and Ballistic Range Facilities at NASA Ames Research Center. In *RTO Educational Notes, Aerothermodynamic Design, Review on Ground Testing and CFD.* Paper NBR-1, RTO-EN-AVT-186.

Grinstead, J. H., Wright, M. J., Bogdanoff, D. W., and Allen, G. A. (2009). Shock Radiation Measurements for Mars Aerocapture Radiative Heating Analysis. *Journal of Thermophysics and Heat Transfer*, **23**(2), 249–255.

Hertzberg, A., Wittliff, C. E., and Hall, J. G. (1961). Summary of Shock Tunnel Development and Application to Hypersonic Research. *Report No. AD-1052-A-12*, Cornell Aeronautical Laboratory. *AFOSR TR-60-139.*

Holden, M. S. (1990). Large Energy National Shock Tunnel (LENS): Description and Capabilities. In *Calspan-University of Buffalo Research Center Brochure.* Washington, DC: Calspan-UB Research Center.

Holden, M. S. (2010). The LENS Facilities and Experimental Studies to Evaluate the Modeling of Boundary Layer Transition, Shock/Boundary Layer Interaction, Real Gas, Radiation and Plasma Phenomena in Contemporary CFD Codes. In *Aerothermodynamic Design, Review on Ground Testing and CFD*, ed. O. Chazot and K. Bensassi. RTO-EN-AVT-186. RTO/NATO, Paper 2. Belgium: von Karman Institute for Fluid Dynamics.

Holden, M. S. and Parker, R. A. (2002). LENS Hypervelocity Tunnels and Application to Vehicle Testing at Duplicated Flight Conditions. In *Advanced Hypersonic Test Facilities*, ed. F. K. Lu and D. E. Marren. Vol. 198 of AIAA Progress in Astronautics and Aeronautics. Reston, VA: AIAA, pp. 73–110.

Holden, M. S., Wadhams, T. P., and MacLean, M. (2018). Measurements in Regions of Shock Wave/Turbulent Boundary Layer Interaction From MACH 4 to 7 at Flight Duplicated Velocities to Evaluate and Improve the Models of Turbulence in CFD Codes. *AIAA Paper 2018–3706.*

MacLean, M., Candler, G., and Holden, M. (2005). Numerical Evaluation of Flow Conditions in the LENS Reflected Shock-Tunnel Facilities. *AIAA Paper 2005–903.*

MacLean, M., Holden, M. S., and Dufrene, A. (2013). Measurements of Real Gas Effects on Regions of Laminar Shock Wave/Boundary Layer Interaction in Hypervelocity Flows for "Blind" Code Validation Studies. *AIAA Paper 2013–2837.*

Martin, W. A. (1958). A Review of Shock Tubes and Shock Tunnels. *Report No. ZR-658–050*, Engineering Department, Convair, Division of General Dynamics Corporation, San Diego, September 10.

Menard, W. A. (1971). A Higher Performance Electric-Arc-Driven Shock Tube. *AIAA Journal*, **9**(10), 2096–2098.

Miller, C. G. (1976). Operational Experience in the Langley Expansion Tube with Various Test Gases and Preliminary Results in the Expansion Tunnel. Paper presented at the *AIAA 9th Aerodynamic Testing Conference*, Arlington, Texas, June 7–9. Full paper available from National Technical Information Service, Springfield, Virginia, 22151, as N78-12I05.

Miller, C. G. (1978). A Critical Examination of Expansion Tunnel Performance. Presented at the *AIAA 10th Aerodynamic Testing Conference*, San Diego, California, April 19–21, *AIAA Paper 78–768*.

Miller, C. G. and Jones, J. J. (1983). Development and Performance of the NASA Langley Research Center Expansion Tube/Tunnel, A Hypersonic-Hypervelocity Real-Gas Facility. Presented at the *14th International Symposium on Shock Tubes and Waves*, Sydney, Australia, August.

Miller, C. G. III, Micol, J. R., and Gnoffo, P. A. (1985). Laminar Heat-Transfer Distributions on Biconics at Incidence in Hypersonic-Hypervelocity Flows. *NASA Technical Paper 2213*.

Moore, J. A. (1975). Description and Initial Operating Performance of the Langley 6-inch Expansion Tube Using Heated Helium Driver Gas. *NASA Technical Memorandum TM X-3240*, September.

Moore, J. A. (1976). Description and Operating Performance of a Parallel-Rail Electric-Arc System with Helium Driver Gas for the Langley 6-inch Expansion Tube. *NASA Technical Memorandum TM X-3448*, December.

Nealy, J. E. (1972). Performance and Operating Characteristics of the Arc-Driven Langley 6-inch Shock Tube. *NASA TN D-6922*, August 1972.

Nompelis, I., Candler, G. V., Wadhams, T. P., and Holden, M. S. (2004). Numerical Simulation of High Enthalpy Experiments in the LENS-X Expansion Tube Facility. *AIAA Paper 2004-1000*, January.

Parker, R. and Wakeman, T. (2010). Shock Front Radiation Studies at CUBRC. *AIAA Paper 2010-1370*.

Sharma, S. P. and Park, C. (1990). Operating Characteristics of a 60- and 10-cm Electric Arc-Driven Shock Tube – Part I: The Driver. *Journal of Thermophysics and Heat Transfer*, **4**(3), 259–265.

Sharma, S. P., Ruffin, S. M., Gillespie, W. D., and Meyer, S. A. (1993a). Vibrational Relaxation Measurements in an Expanding Flow Using Spontaneous Raman Scattering. *Journal of Thermophysics and Heat Transfer*, **7**(4), 697–703.

Sharma, S. P., Ruffin, S. M., Meyer, S. A., Gillespie, W. D., and Yates, L. A. (1993b). Density Measurements in an Expanding Flow Using Holographic Interferometry. *Journal of Thermophysics and Heat Transfer*, **7**(2), 261–268.

Shirai, H. and Park, C. (1979). Experimental Studies of Radiative Base Heating of a Jovian Entry Model. *AIAA Paper 79-0038*, January.

Trimpi, R. L. (1962). A Preliminary Theoretical Study of the Expansion Tube, a New Device for Producing High-Enthalpy Short-Duration Hypersonic Gas Flows. *NASA Technical Report R-133*.

Trimpi, R. L. (1966). A Theoretical Investigation of Simulation in Expansion Tubes and Tunnels. *NASA Technical Report TR R-243*, June.

4 Free-Piston-Driven High-Enthalpy Shock Tunnels

4.1 Introduction

The free-piston driver is a powerful technique to increase the driver gas sound speed. It is highly capable of generating high-enthalpy conditions and offers the greatest performance among the various gasdynamic drivers. Indeed, it has been implemented in a number of major reflected-shock as well as shock-expansion tunnels (SETs) around the world. The method has the advantage that a high driver pressure is automatically generated in the process. Therefore, it can drive the strongest shocks and provides the highest total pressure capability amongst the other gasdynamic driver techniques (heated light-gas, combustion, and detonation drivers). On the other hand, the driver is far more complex mechanically and requires tuning in order to operate effectively.

In this chapter, the basic concepts of free-piston drivers are discussed. The analytical theory that describes the piston dynamics and the solution for tuned piston operation are presented. Examples of major free-piston-driven test facilities as well as their applications in hypersonic testing are also included.

4.2 Fundamentals of Free-Piston Drivers

A free-piston driver utilizes compressive heating to increase the temperature (and thus the sound speed) of the driver gas in order to drive a strong shock wave in shock tunnels. In a free-piston driver, high-pressure air is used to accelerate a heavy piston to high speed, which then adiabatically compresses the driver gas within the driver tube much like a bicycle pump, thus rapidly increasing the driver gas pressure and temperature. A diaphragm located at the compression tube exit area then bursts, and the driver gas enters into the shock tube (driven tube) to drive a primary shock for the subsequent RST or SET operation.

4.2.1 Piston Compression

The physical processes that occur in a free-piston driver are as follows. In the compression tube (also known as the pump tube), initially one side of the piston is filled with a light driver gas, typically at atmospheric pressure. The other side of the piston is connected to a reservoir (also known as the secondary reservoir) of compressed air at high pressure. When the piston is released, the high-pressure air expands and accelerates the heavy piston to a high velocity. The momentum of the heavy piston compresses the driver gas rapidly in front of the piston, thereby increasing its pressure and temperature. When the pressure reaches a certain value (burst pressure), the diaphragm at the downstream end of the compression tube bursts open, thus releasing the driver gas into the driven tube (or shock tube) connected to it. This then generates the primary shock wave in the shock tube and initiates the usual shock propagation in a RST or SET. A schematic of the free-piston driver is given in Figure 4.1.

Also shown on the top part of Figure 4.1 is the wave diagram of the processes during driver operation. As the diaphragm bursts open, the flow of the driver gas into the driven tube generates a forward-propagating shock wave and a contact interface behind it in the driven tube. At the same time, the expansion of the driver gas results in expansion waves that emanate from the diaphragm station. The expansion waves that propagate upstream into the driver tube will reflect off the piston's front face and propagate forward into the driven tube. Although not shown in Figure 4.1, the reflected expansion will eventually reach the end of the driven tube and will disturb the otherwise steady test flow condition in an RST or SET. When this happens, it signals the termination of the useful test duration of the facility.

In the compression tube, the driver gas can be considered to be compressed adiabatically during the piston motion. Usually, helium is used as the light driver gas, although it can also be blended with other gases such as argon. Because helium is inert even at high temperatures, there is no chemical reactivity so that the perfect gas isentropic relationships can be used to describe the compression process.

The volumetric compression ratio, or simply the compression ratio λ, defined as the ratio of the initial driver gas volume V_i to the final volume V_f at diaphragm rupture, is used to describe the amount of compression the driver gas undergoes. The final-to-initial driver gas pressure ratio is thus obtained from the compression ratio as

$$\frac{p_f}{p_i} = \left(\frac{V_i}{V_f}\right)^{\gamma} = \lambda^{\gamma}, \tag{4.1}$$

and the ratio of the final-to-initial driver gas temperature is

$$\frac{T_f}{T_i} = \left(\frac{p_f}{p_i}\right)^{\frac{\gamma-1}{\gamma}}, \tag{4.2}$$

where γ is the ratio of specific heats of the driver gas. The values p_f and T_f are closely related to the initial driver gas pressure and temperature in a conventional shock tube, respectively (i.e., p_4 and T_4 in conventional shock-tube terminology).

4.2.2 Advantages and Drawbacks of Free-Piston Driver

The free-piston technique has the advantage of delivering extremely high primary-shock strength in the driven tube (shock tube). By using a heavy piston, the large momentum of the piston motion becomes very efficient in compressing the driver gas to high pressure and temperature. For a large-size facility with a piston mass of the

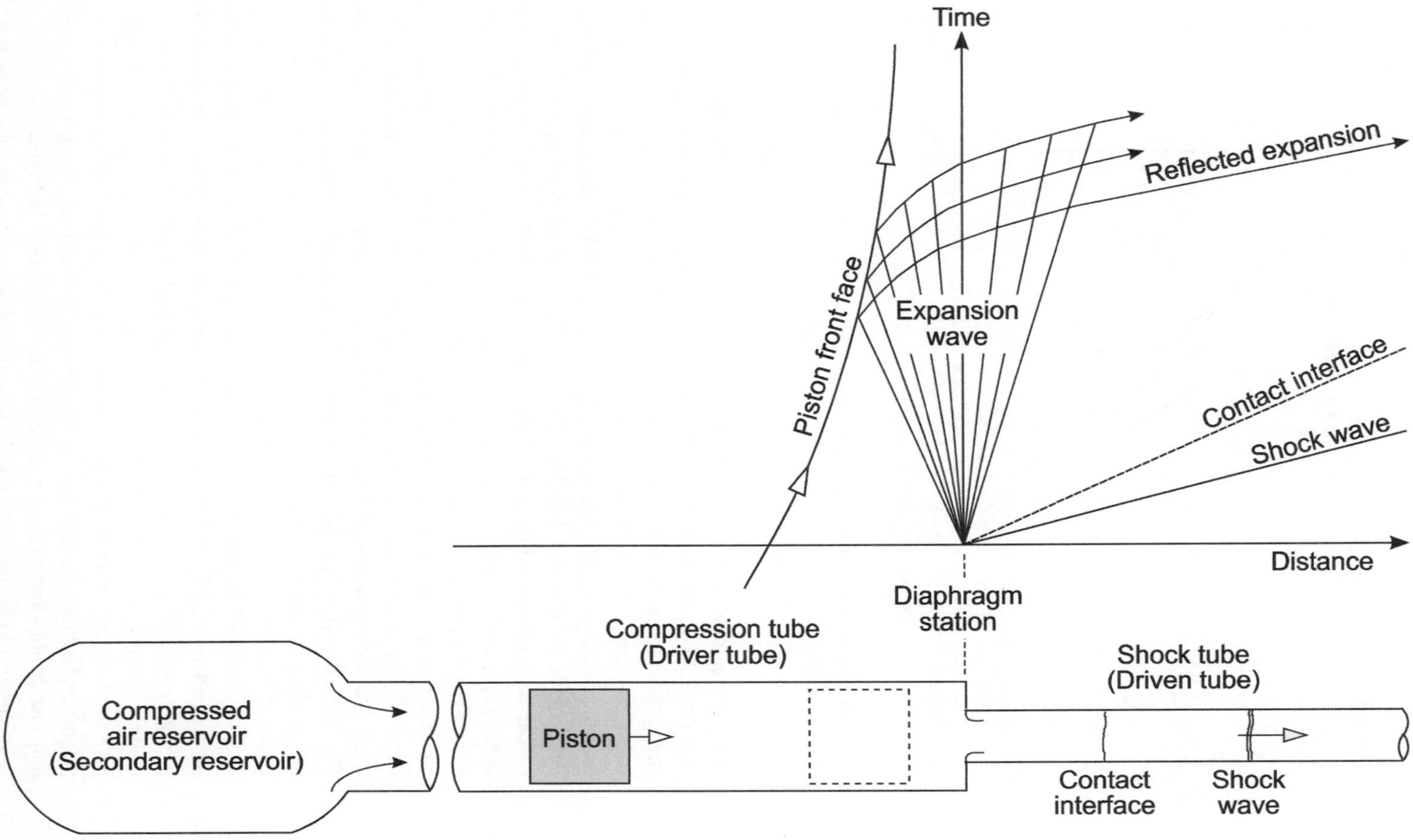

Figure 4.1 Schematic of a free-piston driver.

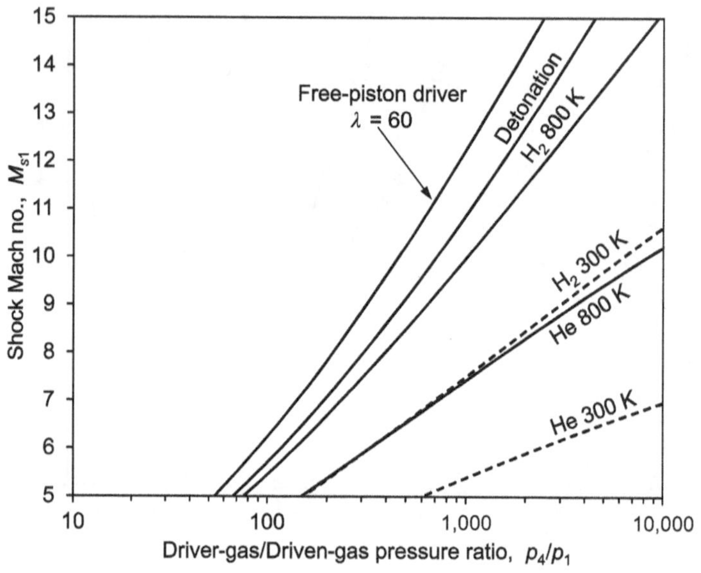

Figure 4.2 Comparison of the free-piston-driven primary shock Mach number with those driven by detonation and heated helium and hydrogen.

order of hundreds of kilograms, the free-piston driver can typically achieve a compression ratio (λ) of 50–60. A key parameter to assess the ability of a driver to generate strong shocks in a shock tube is the ratio of the initial driver gas sound speed a_4 to that of the pre–shocked-driven gas a_1. For the driver gas helium, with a γ of 5/3 and the driven gas being air, both initially at a temperature of 300 K, the compressed driver gas temperature for $\lambda = 60$ is 4,600 K, resulting in a sound speed ratio, a_4/a_1, close to 12.

The superior capability of the free-piston driver can be seen in a comparison with other gasdynamic drivers, namely the heated light-gas and detonative combustion drivers. In Figure 4.2, the primary shock Mach number M_{s1} produced by free-piston compression in a standard shock tube with an identical driver and driven tube diameters is compared with that obtained using the other driver techniques. For the free-piston driver, a compression ratio of 60 with helium as the driver gas is assumed. For the heated light-gas drivers, helium and hydrogen are assumed to be heated at a temperature of 800 K, yielding a_4/a_1 of about 5 and 6, respectively. The detonative combustion driver is assumed to have an initial driver gas composition of 30% H_2, 15% O_2, and 55% He (by volume), and the post-detonation driver gas temperature T_4 is assumed to be 3,600 K. This yields an approximate a_4/a_1 of about 6. As can be seen in Figure 4.2, the free-piston driver produces higher primary shock Mach numbers than the other driver techniques. The free-piston driver is more capable of generating high total enthalpies in both RST and SET, as the primary shock strength is a key parameter for facility capability. As a reference, the primary shock Mach numbers for unheated helium and hydrogen driver gas are also shown in Figure 4.2.

In contrast to heated light-gas drivers, the free-piston driver has the benefit of not needing to hold the high driver tube fill pressure and temperature, as the state of the driver gas is obtained by rapid adiabatic compression. This reduces the design requirements of the driver tube as the hardware used only needs to withstand the compressed gas pressure and temperature at the very last segment of the tube. Furthermore, there is also no need to handle dangerous combustible gases, as in the cases of heated hydrogen and combustion drivers.

High-enthalpy impulse facilities, be they RST or SET, utilize shock waves to generate high-velocity flows. Powerful, rapid, and complex flow events occur within the facility and are of the order of milliseconds. Such events are by nature highly transient and often lead to flow and acoustic disturbances that can interfere with the test flow quality. The facility operation starts with the rupturing of the main diaphragm at the junction between the driver and driven tubes. Under high total enthalpy test conditions, the diaphragm is a steel plate, and the burst pressure is of the order of 100 MPa. The rupturing event is a major source of mechanical and flow disturbances. In the case of a reflected-shock tunnel (RST), the reflection of the primary shock at the shock-tube end wall is another source. For SETs, the primary shock will hit and burst a secondary diaphragm located between the shock tube and the acceleration tube, which also generates disturbances.

Noise in high-enthalpy facilities is certainly not unique to the free-piston driver. However, the dynamics of the piston does introduce an additional source of mechanical and flow perturbations. Care is required to control the piston motion so that it does not result in large impact velocities and damage the compression tube end. Gas-dynamic waves can be generated by the piston, as it acts as a moving wall, and they can cause disturbances in the driven tube. In free-piston-driven RSTs, fluctuations in shock speed and shock pressure have been observed and attributed to wave processes originating from the piston driver. Furthermore, the high volumetric compression and the release of the driver gas from the resulting small compressed driver volume to the driven tube can lead to a rapid variation in driver gas conditions. The depressurization effects of the driver gas will be transmitted to the driven tube and subsequently arrive at the end of the driven tube to disrupt the test flow condition. In free-piston-driven facilities, this disturbance can reach the driven tube end earlier than those driven by long constant-volume driver tubes because of the short, compressed driver gas slug. However, the pressure decay in the driver tube can be delayed by tuning the piston motion to maintain an approximate constant-pressure and temperature condition after diaphragm rupture.

The motion of the piston, particularly if it impacts the compression tube end, will add to the mechanical vibrations and stress waves due to diaphragm rupture and shock reflection in a pulse facility. These disturbances can be transmitted across various tube sections in the facility. For a large-size piston-driven tunnel, such as the HEG, the entire facility recoils about 10–15 cm during an experiment. The design and operation of the piston also add to the complexity of the facility. The piston must be designed to withstand the pressure and impact loading during operation. The compression tube must be capable of withstanding the impact of the piston. This increases the complexity and cost of the facility. Furthermore, different test conditions will require different piston dynamics, and most facilities have a number of pistons with different mass at their disposal to meet the test requirements. It can be time-consuming to optimize the piston performance when changing test conditions.

4.3 Techniques for Free-Piston-Driven Shock Tunnels

There are basically two techniques to operate a free-piston driver – it can either be operated in a constant-volume or a constant-pressure mode. The difference between the two driving techniques is related to the difference in piston motion close to the end of the piston stroke (Jenkins 1974). In the constant-volume driver, the piston comes to rest momentarily just as the diaphragm ruptures (Figure 4.3(a)). This yields a constant volume of the compressed driver gas in the compression tube as in a conventional shock-tube driver. The compressed driver gas pressure p_f and temperature T_f in Eqs. (4.1) and (4.2) become the usual p_4 and T_4 in a conventional shock tube. From the moment of diaphragm rupture, the flow of the driver gas into the driven tube (shock tube) causes the pressure in the driver tube to decay, as in the case for a conventional shock-tube driver.

In the constant-pressure mode, the piston is allowed to travel at a high velocity after diaphragm rupture so that the driver gas pressure can be sustained at an approximate constant level until the piston decelerates and finally stops (Figure 4.3(b)). Thus, the constant-pressure mode of operation is capable of maintaining a steadier driver pressure than the constant-volume mode. The two modes of driver operation are described in more detail in Section 4.3.1.

4.3.1 Constant-Volume Driver

In the constant-volume driver, by allowing the piston to travel the maximum stroke length before the diaphragm opens, the driver gas is compressed to the smallest volume,

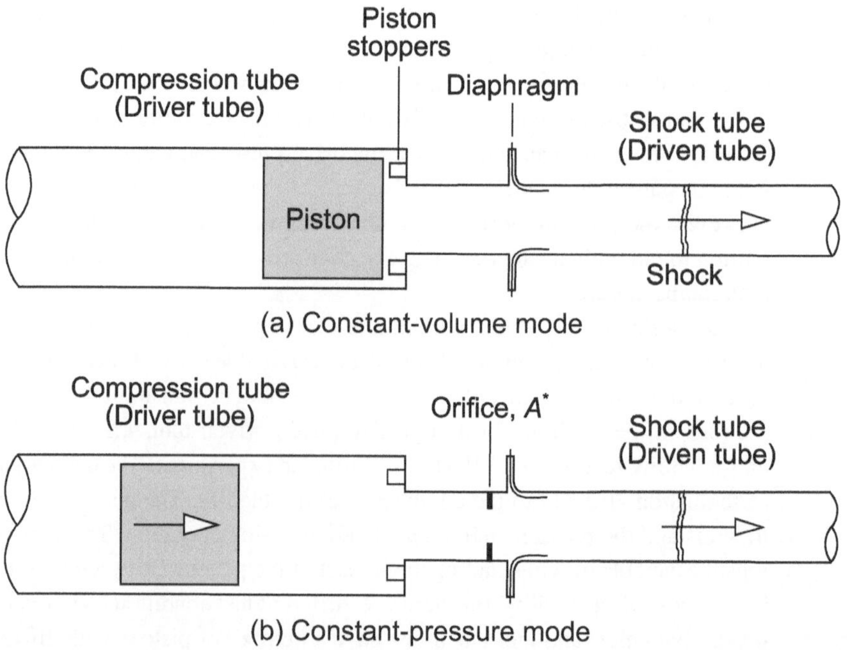

Figure 4.3 Free-piston driver operating modes.

yielding the highest possible driver pressure and temperature. However, the resulting small volume also means a very short slug of the driver gas between the piston face and the diaphragm station. Recall the wave diagram shown in Figure 4.1, the expansion waves – which result from the pressure decay in the compression tube as the driver gas flows into the driven tube after diaphragm rupture – will reach the piston and be reflected forward into the driven tube very quickly because of the small volume as well as the high sound speed in the hot driver gas. The reflected waves can thus reach the downstream end of the driven tube very early and interrupt the test flow condition before the experiment is completed. Because of this, the constant-volume mode of operation would result in very limited useful test time and is not a desirable technique in high-enthalpy testing when meeting test time requirement is already very challenging.

The constant-volume mode of operation is only an approximation as the piston remains motionless and the volume is unchanged only momentarily as the diaphragm ruptures. In practice, the piston will rebound because of the high-pressure buildup in front of the piston. The piston can reaccelerate forward and impact the compression tube end.

4.3.2 Constant-Pressure Driver

The constant-pressure driver technique was developed to obtain high compression ratios and high driver gas temperatures, but without the consequence of rapid variation in driver gas conditions (Stalker 1967). In this mode of operation, the diaphragm is allowed to burst before the piston reaches the end of its stroke, and the piston velocity is such that the piston displacement can compensate for the driver gas flowing into the driven tube. Thus, the driver gas pressure is approximately maintained at a constant level, which then suppresses the formation of expansion waves until the piston eventually decelerates and comes to rest at the compression tube end. All the major free-piston RSTs operate in the constant-pressure mode to prolong the duration of a useful test condition.

In the constant-pressure driver, it is important to control the amount of the driver gas mass flow into the driven tube as well as the piston velocity during and after diaphragm rupture. The mass flow from the driver tube to the driven tube depends on the area ratio of the two tubes. In shock-tube and shock-tunnel applications, there is usually an area reduction from the driver to the driven tube. To further limit the flow into the driven tube, an orifice that has an area smaller than the driven tube can be installed at the diaphragm station (see Figure 4.3(b)).

As will be seen in Section 4.5.1, an important parameter that characterizes piston motion is the reference velocity U_r, which is the piston velocity that would be required at the moment of diaphragm rupture to match the driver gas flow into the shock tube. From mass conservation, U_r is given by

$$U_r = \frac{\rho_r^* a_r^* A^*}{\rho_r A} = \frac{\rho_r^*}{\rho_r}\frac{a_r^*}{a_r}\frac{A^*}{A}a_r = \left(\frac{2}{\gamma+1}\right)^{\frac{\gamma+1}{2(\gamma-1)}}\frac{A^*}{A}a_r$$

$$U_r = \left(\frac{2}{\gamma+1}\right)^{\frac{\gamma+1}{2(\gamma-1)}} \frac{A^*}{A}\sqrt{\gamma R T_r}, \tag{4.3}$$

where the gas properties are for the driver gas, with ρ being the density, a the sound speed, T the temperature, γ the ratio of specific heats, and A the cross-sectional area of the compression tub. The subscript r refers to the condition at the moment of diaphragm rupture and the superscript * denotes the condition at the entrance to the driven tube or at an orifice near the diaphragm, which is assumed to be steady and sonic. The area of the orifice needs to be smaller than that of the pump tube. It can be as large as that of the driven tube, or it can be made smaller to restrict the flow.

In Figure 4.4, the reference piston velocity obtained using Eq. (4.3) is plotted against the reference driver gas temperature T_r for the helium driver gas for three different A^*/A area ratios. It can be seen in Figure 4.4 that if the orifice area is the same as that of the compression tube ($A^*/A = 1$), that is, if the driver gas flows directly into the shock tube with no area change, the reference piston velocity would become extremely high. Even at a modest reference driver gas temperature of 1,000 K, the reference velocity is at a high value of about 1,000 m/s. At a reference driver gas temperature of 3,000 K, the reference velocity becomes exceedingly high at 1,800 m/s. When A^*/A decreases to 0.25, the reference velocity is much lower, but is still impractical at over 400 m/s for $T_r = 3{,}000$ K. The lowest curve in Figure 4.4 represents a more typical shock tube-to-compression tube area ratio of 0.09 (taken from the HIEST facility, Itoh et al. 2002a), which yields far more reasonable reference velocities. Thus, by having a smaller shock-tube diameter than that of the driver, or by installing an orifice adjacent to the diaphragm, the piston velocity can be lowered to a reasonable range.

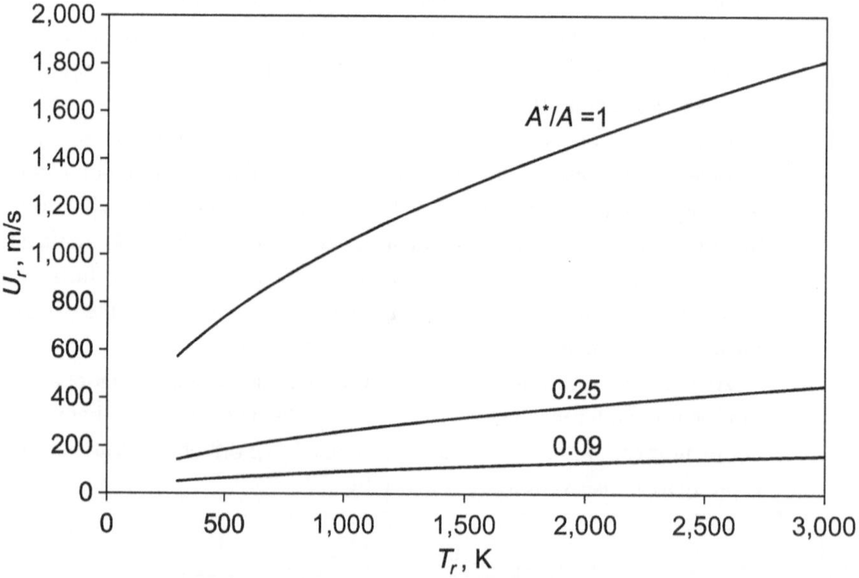

Figure 4.4 Reference velocity U_r versus driver gas temperature at diaphragm rupture.

From shock-tube theory (Glass & Hall 1959), it is known that an area reduction from the driver to the driven section yields a stronger driven shock wave for given diaphragm pressure and sound speed ratios than is obtainable in a constant-area tube. Essentially, with area reduction, the effective driver gas pressure becomes higher than the actual driver-gas fill pressure in a conventional shock tube. However, this pressure gain can be negated if an orifice is used, and its inner diameter is smaller than that of the driven tube. Therefore, to obtain the best performance for driving a shock tube, A^* should be equal to the driven tube area.

To achieve an approximately constant-pressure driver, the piston velocity at diaphragm rupture u_r should be somewhat greater than the reference velocity U_r so that the driver gas pressure can be made to increase slightly after diaphragm rupture before the pressure decreases as the piston decelerates and comes to rest. The ratio of u_r to U_r that will yield optimum performance will be determined later in Section 4.5.

4.4 Theory of Constant-Volume Free-Piston Driver

In the constant-volume mode of operation, the piston comes to rest just as the diaphragm ruptures. The volume bounded by the piston and the diaphragm thus remains constant momentarily after the diaphragm opens. This driver mode is essentially the same as for a conventional shock tube.

As the diaphragm is always closed when the piston is in motion, the piston can be analyzed, as though it is traveling in a closed tube. The analysis describes the flow processes during piston motion until the piston stops at the moment of diaphragm rupture. The constant-volume mode of operation is analyzed with the following assumptions:

1. The gases are perfect.
2. The expansion and compression of the gases are isentropic.
3. The gas properties behind and ahead of the piston are spatially uniform.
4. Friction forces on the piston are negligible compared with the pressure forces.
5. Gas leakage between the piston and the tube wall is negligible.
6. The frame of reference is assumed stationary with respect to the tube assembly.

Assumptions 2 and 3 are approximations that impose additional constraints on the piston. Assumption 2 assumes the piston motion to be sufficiently rapid so that the heat transfer to the walls is negligible. On the other hand, assumption 3 implies that the motion is sufficiently slow to avoid the generation of unsteady waves in the gases. The latter constraint is easily satisfied in the helium driver gas, as the sound speed of helium is high and the undergoing compression process further increases it. However, assumption 3 for the expanding air behind the piston may be poorer as the piston velocity can reach a substantial fraction of the air sound speed. Nevertheless, assumption 3 allows a simple closed-form solution to be easily constructed. Assumption 6 implies that the piston mass is much smaller than that of the tube assembly so that the acceleration of the latter can be ignored. The constant-volume driver analysis presented below is similar to that developed by Roffe (1963) and also used by Jenkins (1974). A more

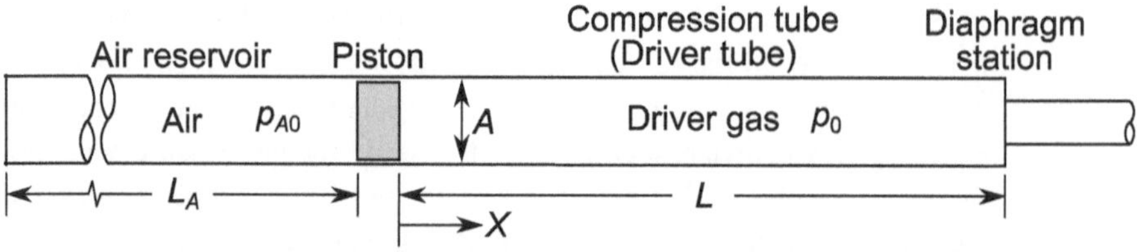

Figure 4.5 Schematic of a constant-volume driver with initial conditions before piston motion.

exact analysis without imposing assumption 3 for the reservoir gas can be found in Hornung (1988).

The relationship between the piston motion and the forces acting on it is given by the momentum equation

$$m\frac{d^2X}{dt^2} = A(p_A - p), \tag{4.4}$$

where m is the piston mass, X is the piston displacement, t is the time, A is the cross-sectional area of the compression tube, p is the pressure of the driver gas ahead of the piston, and p_A is the pressure behind the piston. Subscript A refers to the air behind the piston.

Let the initial pressure in the driver gas before piston motion be p_0, the initial distance between the piston and the diaphragm be L, and the displacement of the piston be X (see Figure 4.5). When the piston is set in motion, the driver gas pressure p is obtained from the ratio of the geometric volume or length:

$$p = p_0\left(\frac{L}{L-X}\right)^{\gamma}, \tag{4.5}$$

where γ is the ratio of specific heats of the driver gas. Equation (4.5) ceases to be valid as the diaphragm breaks and the driver gas is allowed to flow into the driven tube. From that moment onward, the pressure of the driver gas that remains in the compression tube will drop, and it cannot be described by Eq. (4.5).

For the gas behind the piston, an equation similar to Eq. (4.5) can be written for the expansion of the air in the reservoir. As a simplification, let's consider the air reservoir to be a tube with the same cross-sectional area as the compression tube and has an effective length L_A that would constitute the same initial volume before piston motion (see Figure 4.5). As the piston is displaced, the air-side pressure is given by

$$p_A = p_{A0}\left(\frac{L_A}{L_A+X}\right)^{\gamma_A}, \tag{4.6}$$

where p_{A0} denotes the initial pressure of the air and γ_A is the ratio of specific heats for air. The momentum equation then becomes

$$\frac{m}{A}\frac{d^2X}{dt^2} = p_{A0}\left(\frac{L_A}{L_A+X}\right)^{\gamma_A} - p_0\left(\frac{L}{L-X}\right)^{\gamma}. \tag{4.7}$$

Integrating with respect to time and applying the initial condition $X = 0$ and $u = 0$ at $t = 0$, the piston velocity u can be obtained as a function of piston displacement:

$$\frac{mu^2}{2A} = \frac{p_0 L}{(\gamma - 1)}\left[1 - \left(\frac{L}{L - X}\right)^{\gamma - 1}\right] - \frac{p_{A0} L_A}{\gamma_A - 1}\left[\left(\frac{L_A}{L_A + X}\right)^{\gamma_A - 1} - 1\right]. \tag{4.8}$$

The piston velocity u for a given displacement can be calculated if the initial gas pressures and the tube geometry are known.

The location at which the maximum piston velocity occurs can be obtained from Eq. (4.7) by setting its left-hand-side to 0. This corresponds to the condition when the pressures at the rear and front faces of the piston are identical. From that point onward, the piston decelerates continuously until it comes to rest at the maximum displacement. For a given geometry, the location of maximum velocity depends only on the initial pressure ratio p_{A0}/p_0 but not on the piston mass.

The maximum displacement (full stroke length) can be calculated by setting $u = 0$ in Eq. (4.8), which yields, upon rearranging:

$$\frac{p_{A0}}{p_0} = \left(\lambda^{\gamma - 1} - 1\right)\frac{L_0}{L_A}\frac{\gamma_A - 1}{\gamma - 1}\left[1 - \left(\frac{L_A}{L_A + X_f}\right)^{\gamma_A - 1}\right]^{-1}, \tag{4.9}$$

where X_f and λ are the piston displacement and compression ratio, respectively, at full stroke. The maximum displacement also depends only on the initial pressure ratio p_{A0}/p_0 for a given geometry. For the constant-volume driver, the required initial driver gas fill pressure p_0 to result in a desired diaphragm burst pressure can be found from Eq. (4.5) using the maximum displacement.

As an example, the piston velocity has been computed for the following assumed parameters:

- Initial length of the driver gas in the compression tube, $L_0 = 10$ m
- Initial length of air in the reservoir, $L_A = 5$ m
- Tube cross-sectional area, $A = 0.1$ m^2
- Initial pressure of air in the reservoir, $p_{A0} = 50$ atm.
- Initial air and the driver gas temperature, T_{A0}, $T_0 = 300$ K
- Helium driver gas: $\gamma = 5/3$; reservoir air: $\gamma_A = 1.4$

The piston velocity versus the displacement of three different initial driver gas pressures (p_0) that correspond to compression ratios at a full stroke of $\lambda = 40, 50, 60$ is shown in Figure 4.6. The piston motion is seen to undergo a strong initial acceleration and terminating with a much more rapid deceleration near the tube end. When the initial driver gas pressure is low, the reservoir air exerts a stronger push on the piston so that the piston is accelerated to a higher maximum velocity and undergoes a more rapid deceleration to bring it to rest. This also yields a longer piston stroke, and therefore a higher volumetric compression ratio and driver gas pressure.

Also plotted in Figure 4.6 is the driver gas pressure ratio p/p_0 versus piston displacement. The driver gas pressure changes very little for much of the piston displacement but undergoes a very steep rise as the piston is very near the full stroke. Since the

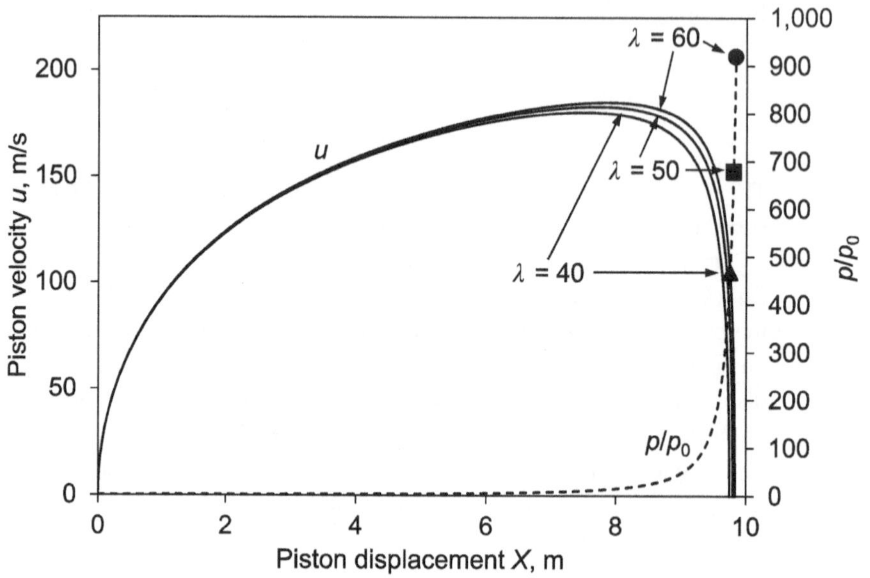

Figure 4.6 Piston velocity and driver gas pressure ratio as a function of displacement. The symbols represent the driver gas pressure ratios at maximum displacement.

pressure ratio, as given by Eq. (4.5), is only a function of piston displacement, there is only one curve shown in Figure 4.6 for all the three cases. The driver gas temperature follows the same behavior as the pressure, and it rises very rapidly close to the end of piston stroke. However, the final pressure or temperature reached at the maximum displacement does depend on the initial air-to-driver gas pressure p_{A0}/p_0. For the current example of fixed initial reservoir air pressure, a higher final driver gas pressure is obtained by lowering the initial driver gas pressure. The very rapid increase in pressure and temperature at the very end of the piston stroke allows the constant-volume driver to reach higher driver gas pressure and temperature than the constant-pressure driver. On the other hand, once the diaphragm opens, the pressure in the driver tube will drop and this information will be transmitted quickly by expansion waves to the driven tube.

The driver gas properties at the maximum piston displacement are shown in Table 4.1. By reducing the initial driver pressure from 1.37 to 1.03 atm., the compression ratio λ is increased from 40 to 60. Accordingly, both the final-to-initial pressure ratio p/p_0 and temperature ratio T/T_0 increase. For a typical compression ratio of 50, a driver gas temperature of over 4,000 K can be obtained. Included in Table 4.1 is the ratio of the compressed driver gas sound speed to that of pre-shocked air in the driven tube, a/a_1, which corresponds to the sound speed ratio a_4/a_1 in a conventional shock tube. The sound speed ratio is found to be at least 10 for the cases considered.

Note also that the final driver gas pressure is much higher than the initial air reservoir pressure p_{A0}. This effect of pressure amplification is one of the advantages of using piston compression to drive pulse facilities. The achieved driver gas pressure is well above the initial pressure anywhere in the driver system. However, the initial

Table 4.1 Driver gas properties at maximum piston displacement

p_0 (atm.)	λ	p/p_0	p/p_{A0}	p (atm.)	T/T_0	T (K)	a/a_1
1.37	40	468	12.8	640	11.7	3,510	10.0
1.17	50	679	15.8	792	13.6	4,070	10.8
1.03	60	920	18.9	943	15.3	4,600	11.5

driver gas pressure p_0 should be set with care so that it is not too low and result in a peak pressure that exceeds the pressure rating of the hardware components.

4.5 Theory of Constant-Pressure Free-Piston Driver

The piston motion in a constant-pressure free-piston driver can be separated into two phases: before and after diaphragm rupture. The processes that occur in the former phase are exactly those analyzed in the constant-volume driver above, with the exception that the diaphragm is allowed to open before the full piston stroke. The latter phase affects directly the final characteristics of the compressed state of the driver gas. This phase is also important in determining whether the piston would decelerate monotonically and come to a complete stop, or whether it would rebound or run into the tube end and damage the compression tube.

A theory that describes the so-called tuned operation of free-piston driver was thus developed to yield the optimum driver performance. The idea behind the theory is that the piston would continue to be in motion after diaphragm rupture such that the driver gas pressure remains approximately constant, and the piston would stop at the moment it reaches the compression tube end, thereby avoiding damage to the tube.

The theory of tuned piston operation was first proposed by Stalker (1967). It was refined by Itoh (Itoh et al. 1994, 1998, Itoh 1996) and was applied to design the HIEST facility at JAXA in Japan. The resulting pistons have mass that are lower than those for untuned operation, although for the size of HIEST, the pistons still weigh several hundreds of kilograms.

4.5.1 Governing Equations for Tuned Operation

Consider the piston motion shown in Figure 4.7. The piston moves from left to right. Behind the piston is pressurized air that is used to propel the piston. On the right side of the piston is the driver gas, typically helium, which is being compressed. After diaphragm rupture, the driver gas exhausts either directly into the shock tube or through an orifice at the diaphragm station.

The set of assumptions used in deriving the equations for the constant-volume driver in Section 4.4 will be used for the constant-pressure driver as well, except for the assumption regarding the gas behind the piston. The driver gas ahead of the piston is again assumed to be isentropic and spatially uniform. For the gas behind the

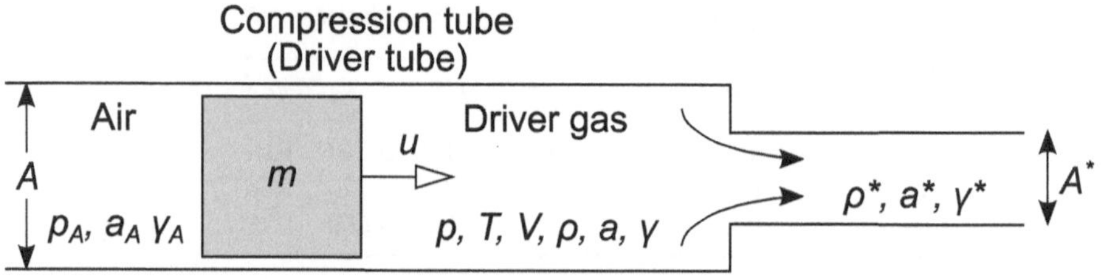

Figure 4.7 Piston motion in a compression tube during constant-pressure operation.

piston, the previous assumption of spatial uniformity is replaced by the assumption that one-dimensional isentropic expansion or compression waves are permitted. For the flow exhausting into the driven tube, it is assumed steady and sonic. The basic equations for the events at and after the moment of diaphragm rupture are as follows.

Mass conservation of the driver gas:

$$\frac{d(\rho V)}{dt} = -\rho^* a^* A^*. \tag{4.10}$$

Momentum equation of the piston, assuming frictionless motion:

$$m\frac{du}{dt} = A(p_A - p). \tag{4.11}$$

In the above equations, V is the driver gas volume ahead of the piston in the compression tube, ρ is the driver gas density, a is the driver gas sound speed, A is the cross-sectional area, m is the piston mass, u is the piston velocity, p is the pressure of the driver gas ahead of the piston, and p_A is the pressure behind the piston. Subscript A refers to the gas behind the piston, which is typically air. Equation (4.10) describes the depletion of driver gas mass through the exit of compression tube after diaphragm rupture. Before diaphragm bursts, the gas mass remains constant, and Eq. (4.10) is not applicable. Equation (4.11) is a force balance on the piston and is applicable before and subsequent to diaphragm rupture. The following analysis will focus on the portion of piston motion at and after diaphragm rupture. The initial conditions are those at the moment of diaphragm rupture, which include the driver gas volume ahead of the piston and the gas pressures on both sides of the piston.

The pressure ahead of the piston is given by the isentropic relation:

$$p = p_r\left(\frac{\rho}{\rho_r}\right)^{\gamma}, \tag{4.12}$$

where subscript r refers to the condition at the moment of diaphragm rupture.

The air-side pressure p_A in Eq. (4.11) is given by

$$p_A = p_{Ar}\left[1+\frac{\gamma_A - 1}{2}\frac{u_r}{a_{Ar}}\left(1-\frac{u}{u_r}\right)\right]^{\frac{2\gamma_A}{\gamma_A - 1}}. \tag{4.13}$$

Differential equations (4.10) and (4.11) can be rewritten in a form that can be more readily solved. This will be accomplished below in several steps.

4.5.2 Nondimensional Governing Equations

The following nondimensional variables are defined:

$$v \equiv \frac{V}{V_r}, \quad x \equiv \frac{\rho V}{\rho_r V_r}, \quad y \equiv \frac{u}{u_r}, \tag{4.14}$$

$$\beta \equiv \frac{u_r}{U_r}, \tag{4.15}$$

$$k \equiv \frac{p_r V_r}{m u_r^2}. \tag{4.16}$$

The parameter β is the ratio between the actual piston velocity at diaphragm rupture u_r and the reference piston velocity U_r that is required to match the driver gas mass flow into the driven tube at the moment of diaphragm rupture. As discussed in Section 4.3.2, the piston velocity should be higher than the reference velocity in order to achieve an approximately constant-pressure driver condition. Indeed, because the piston starts to decelerate before the diaphragm ruptures, the driver gas pressure decreases rapidly unless the piston velocity at the moment of diaphragm rupture is much larger than U_r. Thus, β is expected to be larger than 1.

Using the nondimensional variables in Eqs. (4.14)–(4.16), the mass conservation equation (4.10) can be written in nondimensional form as

$$\frac{dx}{dv} = \frac{\left(\frac{x}{v}\right)^{\frac{\gamma+1}{2}}}{\beta y}. \tag{4.17}$$

The piston momentum equation (4.11) becomes

$$y\frac{dy}{dv} = k\left(\frac{x}{v}\right)^{\gamma} - \epsilon k f, \tag{4.18}$$

where f is the ratio of p_A to p_{Ar} given by Eq. (4.13):

$$f(y) = \frac{p_A}{p_{Ar}} = \left[1 + \frac{\gamma_A - 1}{2}\frac{u_r}{a_{Ar}}(1-y)\right]^{\frac{2\gamma_A}{\gamma_A - 1}}, \tag{4.19}$$

and ϵ is the ratio of the pressure behind the piston to that ahead at the moment of diaphragm rupture:

$$\epsilon = \left(\frac{p_A}{p}\right)_r. \tag{4.20}$$

The nondimensional driver gas volume v can be eliminated by expressing it in terms of the nondimensional driver gas mass y and piston velocity x in the following steps. Equation (4.18) is first rewritten in terms of the derivative of x using Eq. (4.17) to yield

$$\frac{dy}{dx} = k\beta\left(\frac{x}{v}\right)^{\frac{\gamma-1}{2}} - \epsilon k \beta f \left(\frac{x}{v}\right)^{-\frac{\gamma+1}{2}}. \tag{4.21}$$

Taking the derivative of Eq. (4.21) with respect to x, a second-order differential equation is obtained:

$$\frac{d^2y}{dx^2} = k\beta\frac{\gamma-1}{2x}\left[\left(\frac{x}{v}\right)^{\frac{\gamma-1}{2}} - \left(\frac{x}{v}\right)^{\frac{\gamma+1}{2}}\frac{dv}{dx}\right] - \epsilon\ \beta\frac{d}{dx}\left[f\left(\frac{x}{v}\right)^{-\frac{\gamma+1}{2}}\right]. \tag{4.22}$$

From Eq. (4.17), the second term on the right-hand side of Eq. (4.22) is given by βy; Eq. (4.22) then becomes

$$\frac{d^2y}{dx^2} = k\beta\frac{\gamma-1}{2x}\left[\left(\frac{x}{v}\right)^{\frac{\gamma-1}{2}} - \beta y\right] - \epsilon k\beta\frac{d}{dx}\left[f\left(\frac{x}{v}\right)^{-\frac{\gamma+1}{2}}\right]. \tag{4.23}$$

Equation (4.21) can be used to rewrite the first term of Eq. (4.23). Thus, Eq. (4.23) becomes

$$\frac{d^2y}{dx^2} = \frac{\gamma-1}{2x}\left[\frac{dy}{dx} + \epsilon k\beta f\left(\frac{x}{v}\right)^{-\frac{\gamma+1}{2}}\right] - \frac{\gamma-1}{2x}k\beta^2 y - \ k\beta\frac{d}{dx}\left[f\left(\frac{x}{v}\right)^{-\frac{\gamma+1}{2}}\right]. \tag{4.24}$$

Upon rearranging,

$$\frac{d^2y}{dx^2} - \frac{\gamma-1}{2x}\frac{dy}{dx} + \frac{\gamma-1}{2x}k\beta^2 y = \epsilon k\beta\frac{\gamma-1}{2x}f\left(\frac{x}{v}\right)^{-\frac{\gamma+1}{2}} - \epsilon k\beta\frac{d}{d}\left[f\left(\frac{x}{v}\right)^{-\frac{\gamma+1}{2}}\right]. \tag{4.25}$$

Equation (4.17) can be used to eliminate x/v in the above equation as follows:

$$\frac{dx}{dv} = \frac{\left(\frac{x}{v}\right)^{\frac{\gamma+1}{2}}}{\beta y},$$

$$dv = \beta y\left(\frac{x}{v}\right)^{-n} dx,$$

where n is defined as

$$n \equiv \frac{\gamma+1}{2}. \tag{4.26}$$

Thus,

$$v^{-n}dv = \beta y x^{-n}dx.$$

Upon integrating,

$$\left.\frac{v^{1-n}}{1-n}\right|_{v_1}^{v_2} = \beta\int_{x_1}^{x_2}\frac{ydx}{x^n}. \tag{4.27}$$

State 1 in Eq. (4.27) represents the initial condition at diaphragm rupture, such that

$$v_1 = \left.\frac{V}{V_r}\right|_r = 1,$$

$x_1 = \left.\dfrac{\rho V}{\rho_r V_r}\right|_r = 1$. State 2 is an arbitrary state after diaphragm rupture.

Thus, Eq. (4.27) becomes

$$\frac{v^{1-n}}{1-n} - \frac{1}{1-n} = \beta \int_1^x \frac{y dx}{x^n},$$

$$\left(\frac{1}{v}\right)^n = \left[1-(n-1)\beta \int_1^x \frac{y dx}{x^n}\right]^{\frac{n}{n-1}}. \tag{4.28}$$

Define a new variable ξ as

$$\xi(x,y) \equiv f\left(\frac{x}{v}\right)^{-n}. \tag{4.29}$$

Using Eq. (4.28), ξ is

$$\xi(x,y) = \frac{f}{x^n\left[1-(n-1)\beta \int_1^x \frac{y dx}{x^n}\right]^{\frac{n}{n-1}}}. \tag{4.30}$$

The right-hand side of Eq. (4.25) can be rewritten using Eq. (4.30) to give

$$\sigma x \frac{d^2 y}{dx^2} - \frac{dy}{dx} + \theta y = \epsilon k \beta \sigma x h(x,y), \tag{4.31}$$

where

$$\sigma \equiv \frac{2}{\gamma-1} = \frac{1}{n-1}, \quad \theta \equiv k\beta^2,$$

and

$$h(x,y) = \frac{n-1}{x}\xi - \frac{d\xi}{dx}. \tag{4.32}$$

The second-order differential equation (4.31) is the final form of the governing equation.

4.5.3 Homogeneous Solution – Pressure Behind Piston Neglected

At the moment of diaphragm rupture, the pressure behind the piston can be assumed to be much smaller than that ahead of it as the driver gas is greatly compressed and the piston motion is largely supported by the piston momentum rather than by the pressure force behind it. Thus,

$$p_{Ar} \ll p_r$$

and the value of ϵ in Eq. (4.20) can be treated as small. It is expected that ϵ should roughly scale inversely with the compression ratio λ and ϵ can be neglected for $\lambda \gg 1$, which is typically the case for high-enthalpy test operations.

As a first approximation, ϵ is assumed to be negligible in the following derivation so that the right-hand side of the governing equation (4.31) is set to zero to yield the homogeneous differential equation:

$$\sigma x \frac{d^2 y}{dx^2} - \frac{dy}{dx} + \theta y = 0. \tag{4.33}$$

This implies that the pressure behind the piston is always neglected. The solution is then the homogeneous solution to the full governing equation of Eq. (4.31).

4.5.3.1 Transformation to the Bessel Equation

Note that the homogeneous equation can be recast into the Bessel differential equation by performing the following transformations:

$y \to z$:

$$y = x^r z. \tag{4.34}$$

$x \to \tau$:

$$x = \frac{\tau^s}{\alpha}. \tag{4.35}$$

Upon substituting Eqs. (4.34) and (4.35) into Eq. (4.33), it can be found that

$$r = \frac{\gamma+1}{2} = \frac{\sigma+1}{2\sigma} = \frac{n}{2}, \tag{4.36}$$

$$s = 2, \tag{4.37}$$

$$\alpha = \frac{4\theta}{\sigma} = 2(\gamma-1)\theta. \tag{4.38}$$

Therefore,

$$y = x^{n/2} z, \tag{4.39}$$

$$x = \frac{\tau^2}{2(\gamma-1)\theta}, \quad \tau = \omega\sqrt{x}, \tag{4.40}$$

$$\omega = \sqrt{2(\gamma-1)\theta}, \quad \theta = k\beta^2. \tag{4.41}$$

Finally, the homogeneous differential equation (4.33) is transformed into the standard Bessel differential equation:

$$\tau^2 \frac{d^2 z}{d\tau^2} + \tau \frac{dz}{d\tau} + \left(\tau^2 - n^2\right) z = 0. \tag{4.42}$$

The solutions to the homogeneous equation are

$$z_1 = J_n(\tau), z_2 = J_{-n}(\tau), \tag{4.43}$$

where $J_{\pm n}$ is the Bessel function of the first kind, given by

$$J_{\pm n}(\tau) = \left(\frac{\tau}{2}\right)^{\pm n} \sum_{i=0}^{\infty} \frac{(-1)^i}{i!\,\Gamma(i \pm n + 1)} \left(\frac{\tau}{2}\right)^{2i}. \tag{4.44}$$

The full homogeneous solution is the sum of the two Bessel solutions:

$$z = c_1 z_1 + c_2 z_2 = c_1 J_n(\tau) + c_2 J_{-n}(\tau). \tag{4.45}$$

The homogenous solution in terms of the original variables x and y is therefore

$$y = c_1 y_1 + c_2 y_2 = c_1 x^{\frac{n}{2}} J_n\left(\omega\sqrt{x}\right) + c_2 x^{\frac{n}{2}} J_{-n}\left(\omega\sqrt{x}\right). \tag{4.46}$$

Taking the derivative of y in Eq. (4.46) then yields

$$\begin{aligned}\frac{dy}{dx} &= c_1 \frac{n}{2} x^{\frac{n-2}{2}} J_n\left(\omega\sqrt{x}\right) + c_1 x^{\frac{n-1}{2}} \frac{dJ_n(\tau)}{d\tau}\frac{\omega}{2} \\ &+ c_2 \frac{n}{2} x^{\frac{n-2}{2}} J_{-n}\left(\omega\sqrt{x}\right) + c_2 x^{\frac{n-1}{2}} \frac{dJ_{-n}(\tau)}{d\tau}\frac{\omega}{2}.\end{aligned} \tag{4.47}$$

4.5.3.2 Boundary Conditions at Diaphragm Rupture

The derivatives of the Bessel functions are known to have the following properties:

$$\frac{dJ_n(\tau)}{d\tau} = J_{n-1} - \frac{n}{\tau} J_n, \tag{4.48}$$

$$\frac{dJ_{-n}(\tau)}{d\tau} = -J_{-n+1} - \frac{n}{\tau} J_{-n}. \tag{4.49}$$

Using the recurrence formulae of Eqs. (4.48) and (4.49), Eq. (4.47) becomes

$$\frac{dy}{dx} = \frac{\omega}{2} x^{\frac{n-1}{2}} \left[c_1 J_{n-1}\left(\omega\sqrt{x}\right) - c_2 J_{-n+1}\left(\omega\sqrt{x}\right)\right]. \tag{4.50}$$

The constants c_1 and c_2 can be determined using the boundary conditions at the moment of diaphragm rupture. At diaphragm rupture

$$x = 1, y(1) = 1. \tag{4.51}$$

Also, from Eq. (4.21), for the current homogenous solution with $\epsilon = 0$,

$$\left.\frac{dy}{dx}\right|_{x=1} = k\beta. \tag{4.52}$$

Applying the boundary conditions to Eqs. (4.46) and (4.50), we obtain

$$\begin{aligned}1 &= c_1 J_n(\omega) + c_2 J_{-n}(\omega), \\ k\beta &= \frac{\omega}{2}\left[c_1 J_{n-1}(\omega) - c_2 J_{-n+1}(\omega)\right].\end{aligned} \tag{4.53}$$

Solving for c_1 and c_2,

$$c_1 = \frac{J_{-n+1}(\omega) + \dfrac{\omega}{(\gamma-1)\beta} J_{-n}(\omega)}{D}, \tag{4.54}$$

$$c_2 = \frac{J_{n-1}(\omega) - \dfrac{\omega}{(\gamma-1)\beta} J_n(\omega)}{D}, \tag{4.55}$$

where the denominator D is given by

$$D = J_n(\omega) J_{-n+1}(\omega) + J_{n-1}(\omega) J_{-n}(\omega). \tag{4.56}$$

Equation (4.46) with conditions (4.54)–(4.56) is therefore the homogeneous solution for the driver condition when the pressure behind the piston is neglected (i.e., $\epsilon = 0$).

4.5.3.3 Soft-Landing Condition

For the operation such that the piston is tuned to stop precisely at the moment it arrives at the compression tube end (*soft-landing* condition), two further boundary conditions are required. First, the piston velocity vanishes when the driver gas volume is zero, which is

$$y(x = 0) = 0. \tag{4.57}$$

Second, the piston acceleration also vanishes at $x = 0$, so that there is no subsequent reacceleration. This is achieved when $du/dt = 0$ and when the piston reaches the tube end. From the original momentum equation (4.11), this implies that the forces on the piston are exactly balanced, that is, $p_A = p$. Initially, at the moment of diaphragm rupture, the pressure p ahead of the piston is much larger than that behind the piston p_A, so that the piston is rapidly decelerated. As the driver gas flows into the driven tube, the driver gas pressure decreases. In contrast, the deceleration of the piston generates compression waves behind it as the gas that travels there close to the sound speed is decelerated.

However, by setting $\epsilon = 0$, the contribution of the pressure behind the piston (p_A in Eq. (4.13) and $f(y)$ in Eq. (4.19)) is always neglected so that the condition of zero piston acceleration at $x = 0$ cannot be imposed for the homogenous solution.

For the condition of $y(0) = 0$, the Bessel function $J_n\left(\omega\sqrt{x}\right)$ in y_1 is expanded as

$$y_1(x) = x^{\frac{n}{2}} J_n\left(\omega\sqrt{x}\right) = x^{\frac{n}{2}} \left(\frac{\omega\sqrt{x}}{2}\right)^n \sum_{i=0}^{\infty} \frac{(-1)^i}{i!\,\Gamma(i+n+1)} \left(\frac{\omega\sqrt{x}}{2}\right)^{2i}.$$

Since n is a positive real number, y_1 automatically satisfies the boundary condition in Eq. (4.57):

$$y_1(0) = 0.$$

For y_2, $J_{-n}\left(\omega\sqrt{x}\right)$ is expanded and rearranged as

$$y_2(x) = x^{\frac{n}{2}} J_{-n}\left(\omega\sqrt{x}\right) = x^{\frac{n}{2}} \left(\frac{\omega\sqrt{x}}{2}\right)^{-n} \sum_{i=0}^{\infty} \frac{(-1)^i}{i!\,\Gamma(i-n+1)} \left(\frac{\omega\sqrt{x}}{2}\right)^{2i}$$

$$y_2(x) = \left(\frac{\omega}{2}\right)^{-n} \left[\frac{1}{(-n+1)} + \sum_{i=1}^{\infty} \frac{(-1)^i}{i!\,\Gamma(i-n+1)} \left(\frac{\omega\sqrt{x}}{2}\right)^{2i}\right]. \tag{4.58}$$

It can be seen that the first term in the square bracket remains finite even at $x = 0$. Therefore, y_2 can only satisfy the boundary condition at $x = 0$ if c_2 is identically zero. The soft-landing condition therefore implies that Eq. (4.55) for c_2 is reduced to

$$J_{n-1}(\omega)-\frac{\omega}{(\gamma-1)\beta}J_n(\omega)=0. \tag{4.59}$$

This yields a unique relationship between β and ω,

$$\beta=\frac{\omega}{(\gamma-1)}\frac{J_n(\omega)}{J_{n-1}(\omega)}. \tag{4.60}$$

Alternatively, the above relation can be expanded to relate β to k as follows:

$$\beta=\sqrt{\frac{2k\beta^2}{(\gamma-1)}}\frac{J_{\frac{\gamma+1}{2}}\left(\sqrt{2(\gamma-1)k\beta^2}\right)}{J_{\frac{\gamma-1}{2}}\left(\sqrt{2(\gamma-1)k\beta^2}\right)}. \tag{4.61}$$

Recall from Eqs. (4.15) and (4.16) that β is the nondimensional piston velocity at diaphragm rupture, while k involves the conditions at diaphragm rupture as well as the piston mass. Equation (4.60) or (4.61) thus describes the required piston mass and the diaphragm rupture condition in order to achieve soft landing in a tuned free-piston driver, assuming that the pressure behind the piston at diaphragm rupture is negligible.

After imposing $c_2 = 0$, c_1 can simply be found using Eq. (4.53) to be

$$c_1=\frac{1}{J_n(\omega)}. \tag{4.62}$$

Finally, the homogeneous solution y in Eq. (4.46) becomes

$$y=x^{\frac{n}{2}}\frac{J_n\left(\omega\sqrt{x}\right)}{J_n(\omega)}. \tag{4.63}$$

For finite values of ϵ (i.e., finite p_A/p at diaphragm rupture), an approximate general solution has also been obtained by Itoh et al. (1998) by linearization of the driving force term in Eq. (4.31). Interested readers can refer to that reference.

4.5.4 Tuned Operation Results

Equation (4.60) is computed for helium as the driver gas, and the variation of β with ω for tuned operation is shown in Figure 4.8. For the current case of $\epsilon = 0$ (i.e., infinite compression ratio), the resulting homogenous solution is such that β and ω are of order 1 and that β increases with ω. While the solution exists for β smaller or greater than 1.0, but since the piston velocity at diaphragm rupture should be greater than U_r in order that the driver gas pressure can continue to rise slightly to achieve an approximately constant-pressure driver, β should be larger than unity. This also results in a relatively large value of ω.

For finite ϵ, the full solution depends on the compression ratio and lies approximately parallel to the homogenous solution, but the general trend that β increases with ω remains unchanged. Also plotted in Figure 4.8 are the theoretical values of β calculated for tuned operation experiments in the HIEST facility, taken from Itoh et al. (2002a). In the experiments, the compression ratio ranges from 40.9 to 49.4,

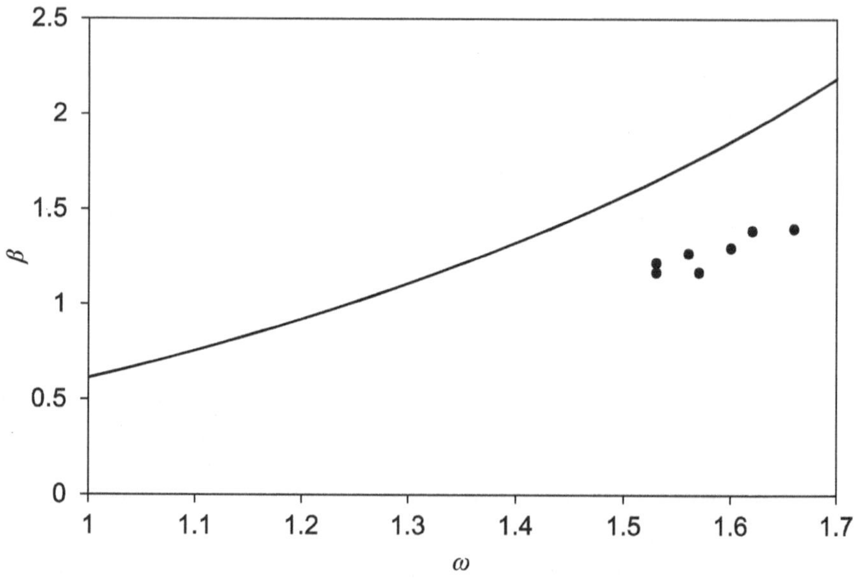

Figure 4.8 β versus ω for tuned operation. Pressure behind piston is neglected (homogenous solution). The solid points are theoretical values calculated for experiments in HIEST.

and ω is varied by changing the piston mass, driver gas composition (pure helium or blended with argon), and the initial driver gas pressure. Lower values of β than the homogeneous solution can be seen.

It is shown in Section 4.5.3 that ω is a key parameter to characterize the piston dynamics. Equation (4.41) can be rewritten more explicitly as

$$\omega = \sqrt{2(\gamma-1)\left(\frac{\gamma+1}{2}\right)^{\frac{\gamma+1}{\gamma-1}} \frac{p_0 V_0}{m\left(\frac{A^*}{A}\right)^2 a_0^2}}, \tag{4.64}$$

where the parameters are as defined in the previous section and p_0 and V_0 are the initial driver gas pressure and volume before piston motion. Equation (4.64) shows that ω is a function of the driver gas fill condition, the piston mass, and the area ratio at the compression tube exit. The parameter ω can be considered as a deceleration factor. When ω is too large, such as the case if the piston mass is too small, the piston is decelerated too much and rebounds at the tube exit. If ω is too small, the piston is not decelerated sufficiently, and it impacts the tube exit. An optimum ω corresponds to the condition when the piston decelerates monotonically and comes to rest near the tube exit. From the analysis performed in Section 4.5.3 for the case of an infinite compression ratio, this condition is satisfied if β follows the relationship of Eq. (4.60). However, the condition for zero piston reacceleration at the tube exit requires the solution for finite ϵ, which would slightly modify the relationship between β and ω.

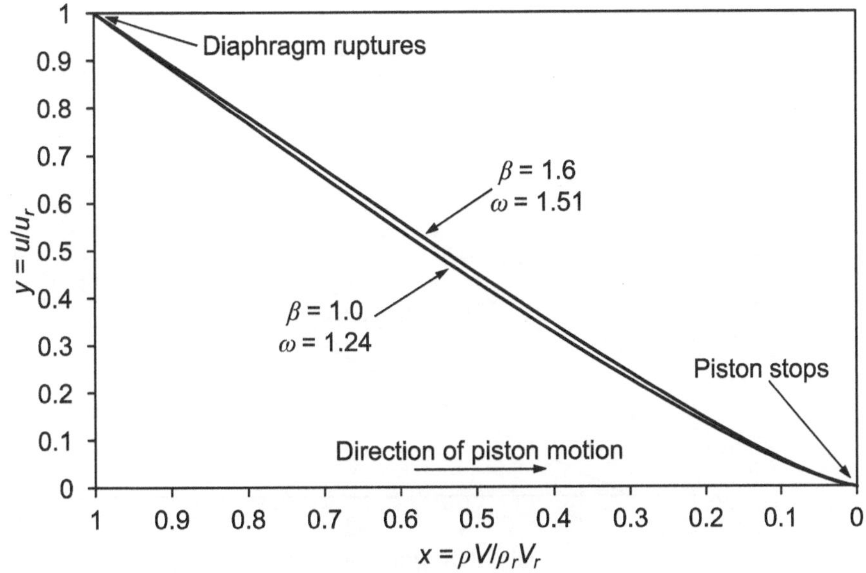

Figure 4.9 Homogeneous solution from diaphragm rupture to piston stoppage.

For a given facility that needs to attain high-enthalpy conditions, there are several options to reach a sufficiently high value of ω in order to achieve tuned piston operation. By inspection of Eq. (4.64), these include the use of a lower sound speed driver gas or a smaller compression tube exit area ratio A^*/A. However, lowering the driver gas sound speed or using an A^* that is smaller than the driven tube area would lead to performance degradation in the shock tube. An alternative is to reduce the piston mass. This approach was adopted in the design of HIEST to achieve a tuned piston driver (Itoh 1996). The implication of using a light piston is that it is more difficult for the piston to have sufficient strength to withstand high pressures so that its mechanical design becomes more critical.

The homogenous solution for the nondimensional piston velocity y in Eq. (4.63) is numerically calculated and a graph of y versus the nondimensional driver gas mass x is shown in Figure 4.9. The calculation starts from the moment of diaphragm rupture (upper-left corner in the plot) and ends when the piston comes to rest (lower-right corner). The solutions for $\beta = 1.0$ and 1.6, corresponding to $\omega = 1.24$ and 1.51, respectively, are shown, which are fairly close to each other.

Figure 4.10 shows the piston trajectory and velocity obtained from an experiment in HIEST reported in Itoh et al. (2002a). For this experiment, the characteristic piston parameter ω is 1.53. At the moment of diaphragm rupture, the compression ratio is 46.6, the piston speed u_r is 146.9 m/s, and the theoretical and measured β are 1.17 and 1.15, respectively. The trajectory corresponds to a soft-landing operation as the piston can be seen to decelerate monotonically and come to rest at the location of the piston stopper, which is located at 0.460 m from the end of the compression tube. If tuned operation is not achieved, the piston would either rebound or hit the piston stopper at a

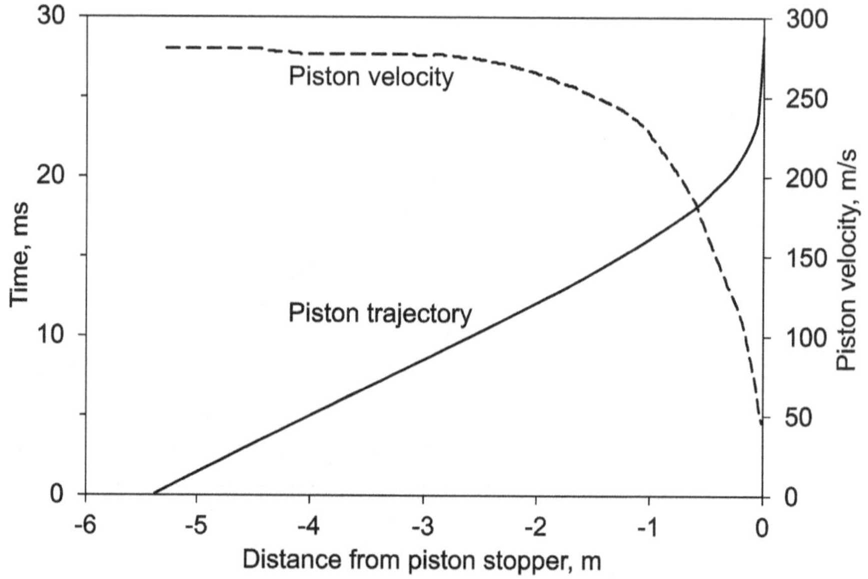

Figure 4.10 Piston trajectory and velocity near the compression tube end in tuned operation.

finite velocity. The final piston motion can be more clearly seen in the piston velocity curve, which shows that the velocity gradually approaches to a low value at the piston stopper.

4.6 Development of Free-Piston-Driven High-Enthalpy Shock Tunnels

The two-stage light-gas gun, the gun tunnel, and the free-piston RST belong to a family of hypersonic test techniques that utilize a rapidly moving piston to compress the driver gas. In a gas gun, a moving piston compresses the driver gas, which then propels a projectile to a high velocity. In a gun tunnel, the piston-compressed driver gas expands directly to the test section, and the facility operates as a blowdown wind tunnel.

By the 1950s, gas guns were already being used extensively for the launching of free-flight models to simulate the atmospheric entry of long-range ballistic missiles (e.g., Eggers 1955, Charters et al. 1957), and also for studying the mechanisms involved in hypervelocity impact damage (e.g., Atkins & Halperson 1960). In the late 1950s, hypersonic gun tunnels started to be in operation for low enthalpy aerodynamic testing (e.g., Bray et al. 1959).

The RST, although not free-piston driven, was developed at the CAL in the 1950s (Hertzberg et al. 1961) and tailored-interface shock-tunnel operation was perfected in 1957 to substantially increase run time (Wittliff et al. 1959, Bird 1962). Several RSTs were constructed and operated at CAL by the late 1950s, one of which was the 24-inch Hypersonic Shock Tunnel, which has a driver tube length of 12.19 m and a driven tube length of 15.24 m, both tubes have 203.2 mm inner diameters (Hertzberg et al. 1961,

Bird 1962). The 24-inch tunnel started operation in 1959 and was upgraded in 1961 to be the 48-inch Hypersonic Shock Tunnel (24-inch and 48-inch refer to the facility nozzle exit diameters). The CAL RSTs were originally driven by constant-volume combustion and later by heated light gas. To further increase performance, Hertzberg proposed employing the inertia of a rapidly moving piston to compress the driver gas, a technique used in light-gas guns (Hertzberg et al. 1961). The proposed driver technique was indeed the free-piston driver.

Besides the Cornell facilities, the GE Research Laboratory Hypersonic Shock Tunnel was another RST that started operation in 1957 (Nagamatsu et al. 1959). The GE tunnel has a driver tube length of 6.1 m and a driven tube length of 31.4 m. The inner diameters of the driver and driven tubes are 152.4 mm and 101.6 mm, respectively. The driver gas is heated using deflagrative combustion (Nagamatsu et al. 1961).

The present-day free-piston shock tunnel was originally developed by R. J. Stalker; free-piston-driven RSTs are also known as Stalker tubes. The first free-piston driver was developed in 1959 at the National Research Council in Ottawa, Canada (Stalker 1961). Two aircraft hydraulic accumulators, which were surpluses from the Avro CF-105 Arrow supersonic fighter, were joined end-to-end to produce a compression tube 73 mm in diameter and 1.61 m long to drive a shock tube that is 25 mm in diameter and 1.7 m long. At about the same period, theoretical studies of free-piston-driven shock tubes were also conducted at the Jet Propulsion Laboratory by G. A. Roffe (1963).

4.6.1 Current Major Free-Piston-Driven High-Enthalpy Facilities

The success of Stalker's free-piston concept led to several "waves" of development of free-piston-driven high-enthalpy test facilities in the world. Below are some representative examples of major free-piston-driven test facilities:

- FD-21, CAAA
- RHYFL, NASA (not completed)
- HIEST, JAXA
- HEG, DLR
- X3, University of Queensland
- T5, Caltech
- T4, University of Queensland

Until 2017, the HIEST facility at JAXA (Itoh 2002, Tanno et al. 2005) and HEG at the German Aerospace Center (DLR) (Hannemann et al. 2018) were the largest free-piston-driven facilities in the world. The FD-21 facility at China Academy of Aerospace Aerodynamics (CAAA) started operation in 2017 and is currently the largest free-piston-driven RST (Bi et al. 2018).

The Rocketdyne Hypersonic Flow Laboratory (RHYFL) (Morrison et al. 1989) included in the examples was initiated by the former Rocketdyne Division of Rockwell International during the National AeroSpace Plane program. However, RHYFL was never completed, although most of the hardware components were built and placed in storage (Chinitz et al. 1994).

Table 4.2 Simulation capabilities and dimensions of major free-piston test facilities

Facility name	Velocity (km/s)	Max. p_T (MPa)	Compression tube L (m)	Compression tube D (m)	Driven tube L (m)	Driven tube D (m)	Nozzle Max. exit D (m)
FD21	3–7	–	75	0.668	34	0.29	2
RHYFL	4–9	200	46.6	0.61	31.7	0.203	–
HIEST	3–7	150	42	0.6	17	0.18	1.2
HEG	1.7–6.2	200	33	0.55	17	0.15	0.88
T5	1.5–6.5	100	30	0.3	12	0.09	0.33
T4	2.2–5.5	90	26	0.228	10	0.075	0.270
X3	3–14	~10^3	14.5	0.5	48.6	0.2, 0.18, 0.183	0.573

T4, T5, and X3 are research facilities at universities (Stalker 2006, Hornung 2010, Toniato 2019). They have been used extensively for fundamental studies of hypersonic flow and propulsion.

Figure 4.11 shows the layout of the various facilities. The schematics are constructed using information from the cited references and are only approximate. However, the relative lengths of the compression and driven tubes are preserved to show a size comparison. Except for X3, all the facilities shown in Fig. 4.11 are free-piston RSTs. X3 is an SET, although it can also be configured to operate as an RST. The SET configuration of X3 is shown in Figure 4.11.

The simulation total enthalpies, maximum total pressure p_T, and the dimensions of the compression and driven tubes of the example free-piston-driven facilities are provided in Table 4.2. The details shown for X3 are for the SET configuration, and the driven tube length includes the secondary driver (12 m), shock tube (12.6 m), and the acceleration tube (24 m), and their respective diameters are 0.2 m, 0.18 m, and 0.1826 m.

4.6.2 Two-Stage Piston Driver

To further enhance the performance of the free-piston driver, a two-stage driver concept was developed at the University of Queensland. The X2 shock-expansion tube was constructed as a pilot facility to validate principles and designs that were to be used for a much larger facility, X3 (Steward 2004, Scott 2006). X2 and X3 remain the only shock-tunnel facilities that have utilized this driver technique. The two-stage piston (also known as a compound piston) driver attempts to improve the efficiency of the free-piston driver that still utilizes adiabatic compression to heat the driver gas. A schematic of the two-stage piston driver is shown in Figure 4.12.

The compound piston consists of two components that are held together initially, with the primary (outer) piston being light in weight and acts as a carrier sabot for the secondary (inner) piston. After the compound piston is accelerated and traversed in the launch tube (primary compression tube), the outer piston hits a buffer and is

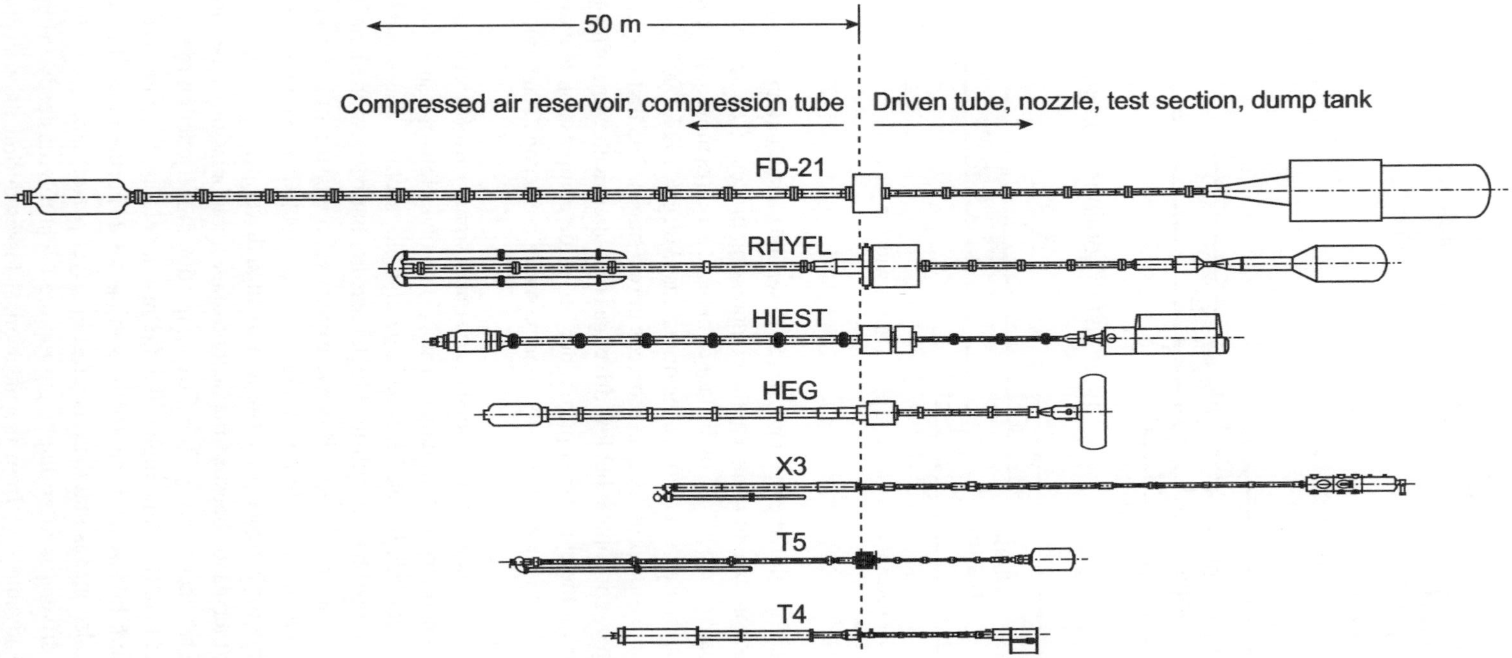

Figure 4.11 Selected major free-piston-driven high-enthalpy test facilities. The layouts are aligned so that the primary diaphragms are at approximately the same axial location for all facilities.

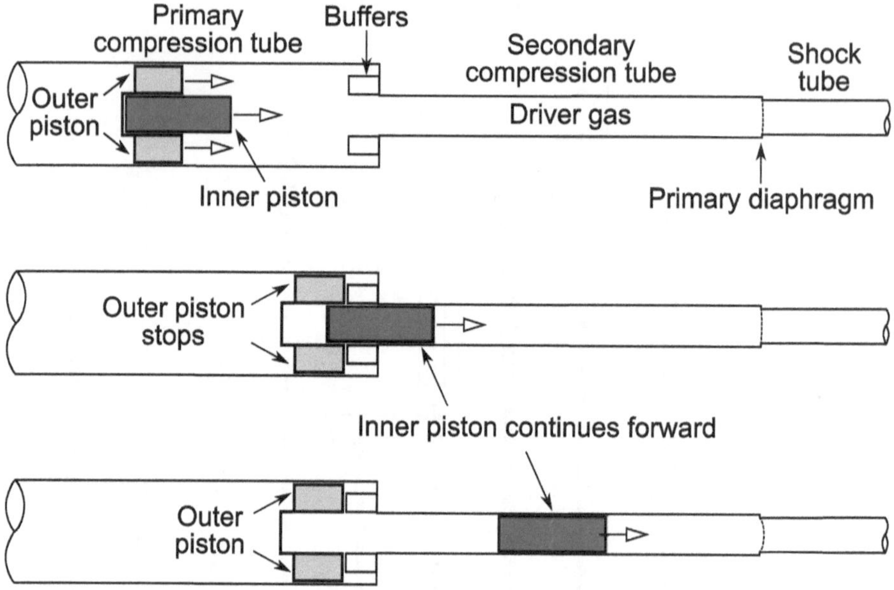

Figure 4.12 Schematic of the two-stage piston driver with the piston at different stages of the compression process.

stopped, thereby allowing the inner piston to separate and continue down the secondary compression tube to complete the compression process. The secondary piston is heavy and carries about 75% of the launch mass. It acquires most of the kinetic energy and momentum used to compress the driver gas (Morgan 1997, 2000).

The compression process thus proceeds in two stages. The first stage is characterized by high swept volume but low driver gas pressures, with work being done by the air reservoir on the compound piston, but very little compression work done on the driver gas. The driver gas is mostly compressed in the second stage but with a small swept volume.

The main advantage of the two-stage piston design is cost (Morgan 2000). As the first-stage launch tube (0.5 m diameter in X3) involves low-pressure compression, a thinner wall tube can be used for the large diameter section. The second-stage tube (compression tube) has to withstand high pressure, but because the inner diameter is small (0.2 m diameter in X3), it is less expensive to construct the thicker tube. The design is especially suitable for smaller facilities, such as those in academic institutions with modest budgets but desire to have high driver performance.

On the other hand, because the second-stage (compression) tube already has a small diameter, there is very little area contraction at the junction to the driven tube (in X3, A^*/A is 0.81). This implies that the piston velocity becomes very high when the diaphragm bursts, as discussed in Section 4.3.2. Furthermore, the lack of area contraction also means that there is a lost in driver performance because the expansion of the driver gas after diaphragm rupture is entirely unsteady (Morgan 1997). Another disadvantage is from the mechanical losses during the transition process between the stages. Realistic performance may therefore be drastically reduced in a large-scale facility.

The X3 shock-expansion tube was originally designed to have a two-stage piston driver. The theoretical capability of X3 was expected to generate superorbital velocities up to about 16 km/s (Abdel-Jawad et al. 2001). However, because of difficulties operating the two-stage piston driver and the reduction in performance, X3 was later converted to have a single-piston driver with a driver tube of uniform diameter (0.5 m) (Dann et al. 2012). The current capability of X3 is around 14 km/s or 100 MJ/kg (UQ X3 n.d.).

4.7 Applications and Testing

Free-piston-driven facilities are well suited for generating high total pressure as well as high total temperature flow. An advantage of the free-piston driver is that it has a high-pressure recovery because the driver gas pressure is obtained by an essentially isentropic compression. In an RST, a high-pressure recovery means that the test-section total pressure can be quite close to the maximum pressure experienced in the facility, which is the driver gas pressure (p_4 in conventional shock-tube terminology). For example, the pressure recovery in HIEST is close to 1.0. A high-pressure recovery allows a facility to generate large binary scaling parameters (ρL) for reentry-type experiments and high stagnation pressures for air-breathing propulsion studies.

As the free-piston driver is currently the most powerful gasdynamic technique to generate a shock wave, it is a favored choice in the hypervelocity testing community. The majority of the above-mentioned facilities (FD21, HIEST, HEG, T5, and T4) have adopted the free-piston approach to drive RSTs, whereas X3 uses the technique for both SET and RST operations.

Below are some examples of the applications and testing performed in selected facilities.

4.7.1 Aeroheating Measurements

The ability to duplicate high-enthalpy conditions makes free-piston-driven tunnels very suitable for simulating reentry flight conditions. HEG was constructed to support the development of a shuttle-like vehicle, Hermes, while the initial application of HIEST was for testing the H-II Orbiting Plane (HOPE), a Japanese reusable manned winged spacecraft.

Aeroheating is of great interest for the survival of earth reentry and planetary entry vehicles, and free-piston-driven tunnels have been used extensively for aeroheating measurements. As an example, heat flux measurements on the Apollo command module (CM) have been conducted in HIEST. The test campaign examined the effects of both fully laminar and fully turbulent boundary layers (with a boundary-layer trip) on heating. The total enthalpy tested ranged from 3.5 to 21 MJ/kg, and total pressure between 14 and 60 MPa (Tanno et al. 2014). Figure 4.13 shows an Apollo CM test model installed in the HIEST test section (Tanno 2013).

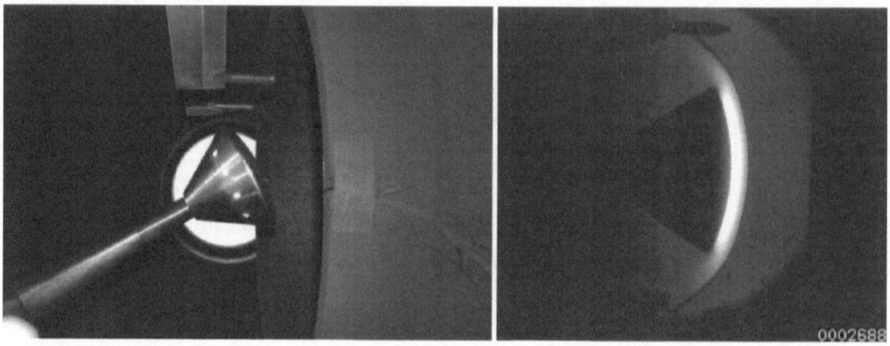

Figure 4.13 Left: Photograph of a 6.4% scaled (250-mm diameter) Apollo CM reentry capsule model in the HIEST test section. Right: Schlieren image of the Apollo CM model in HIEST. (Photos courtesy of H. Tanno, JAXA.)

At extremely high flight speeds, radiation becomes a significant component of aerodynamic heat transfer. When the medium is air, as for earth reentry flights, as the velocity increases beyond about 9.5 km/s, radiative heating becomes comparable with and eventually exceeds convective heating. For other gas media, the velocity at which radiative heating becomes significant can be much lower. For example, Titan's atmosphere consists mainly of nitrogen with a small amount of methane and other trace gases. When a spacecraft enters Titan's atmosphere, the methane dissociates in the bow shock layer and combines with atmospheric nitrogen to form CN, which produces intense radiation. This leads to higher levels of radiation at relatively low entry velocities when compared with earth reentry because carbon-containing species radiate strongly.

Free-piston-driven facilities and, in particular, free-piston-driven SETs are excellent tools for studying radiative heating because of the high-enthalpy capability. The X2 SET has been used to study radiative heat transfer at Titan entry conditions. A 25-mm diameter cylindrical model was used in the experiment, and the freestream velocity was 6.5 km/s (Porat et al. 2012). The test gas mixture consists of 95% of N_2 and 5% of CH_4, representative of Titan's atmosphere. Figure 4.14 shows a frame from the high-speed video taken in the experiment. The radiating shock layer in front of the cylindrical model is clearly visible. The entry condition at 8.5 km/s was also investigated.

4.7.2 Force and Moment Measurements

Force measurement is a challenge in pulse facilities because of the extremely short test duration and also because of severe vibrations induced by the passage of the starting shocks over the test model. However, force measurements have been successfully performed in high-enthalpy free-piston facilities. Force measurements in short-duration hypersonic facilities have been in development since the 1960s, and early works have been discussed in Bernstein (1975). The sting force balance that is widely used in

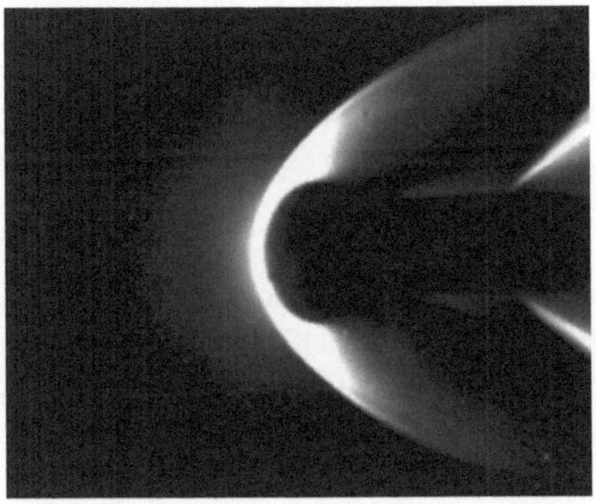

Figure 4.14 Flow over a cylindrical model simulating entry into Titan atmosphere at 6.5 km/s. (Photo courtesy of H. Porat and UQ Centre of Hypersonics.)

long-duration wind tunnels has been adapted to meet the fast response requirements in shock tunnels. A six-compound sting balance that uses semiconductor strain gauges to measure the stress waves in the balance and signal processing to yield the aerodynamic loads was designed in the Shock Wave Laboratory at Aachen University (Jessen & Grönig 1989). The balance was employed to measure forces and moments of the winged reentry vehicle Hopper in the detonation-driven RST TH2-D in the Shock Wave Laboratory. The test model, which has a fuselage length of 0.24 m and a mass of approximately 250 g (Glössner & Olivier 2005), was tested at velocities of 3.95 and 4.84 km/s, corresponding to total enthalpies of 9.0 and 14.2 MJ/kg, respectively.

A different stress-wave measurement technique was developed at the University of Queensland and has been used in the free-piston-driven RST T4 and the shock-expansion tube X3. A notable example is the measurement of lift, drag, and pitching moment at superorbital conditions in X3 using a three-component stress-wave force balance (Abdel-Jawad et al. 2001). The test article was a scale model of the MUSES-C reentry capsule, which was tested at flow velocities from 7.5 to 8.5 km/s. The results showed that the measurement technique is suitable for assessing the stability of reentry capsules at superorbital speeds.

At the HIEST free-piston shock tunnel, a stress-wave force balance was developed in which low-frequency vibrations of the sting and the model are corrected with the aid of acceleration compensation. In one example of its application, the effects of oxygen dissociation on aerodynamics were studied for the Japanese spaceplane HOPE. Chemical nonequilibrium effects during reentry are known to be more pronounced for small vehicles such as HOPE. A 2.5% scale model of HOPE, which has a fuselage length of 0.4 m, was tested at total enthalpies of 8–24 MJ/kg, and the effects of oxygen dissociation on normal forces and pitching moment were observed (Itoh et al. 2002b).

The short-duration nature of the test process can be exploited by the use of a simple and cost-effective method rather than the force balance. A measurement technique that uses accelerometers to measure model responses to motion was originally developed at the CAL in the 1960s for hypersonic shock tunnels (Duryea & Sheeran 1969) and is described in Bernstein (1975). In this technique, accelerometers are used to measure the acceleration of a test model that is "weakly restrained" (i.e., partially in free flight during the test period). This can be accomplished by suspending the model with thin wires so that the restorative forces due to model support are negligible within the short test period. The aerodynamic force can then be obtained as the product of the model mass and the measured acceleration. The technique was further developed by the HIEST team to measure axial-only forces. A signal recovery process based on frequency domain de-convolution is applied to remove oscillations caused by mechanical vibrations associated with insufficient rigidity of the test model. The technique was used to quantify the thrust of a large-size scramjet model 2.9 m in length and a mass of 290 kg (Tanno et al. 2005). The model was originally designed to study combustion characteristics under hypervelocity conditions. It has a side-wall compression inlet and a side-wall expansion nozzle. The combustor is a simple rectangular duct configuration, and the inlet entrance of the cross section is 0.2 m × 0.25 m. The model was suspended using four 2-mm diameter wires. Figure 4.15 shows the scramjet model installed in the HIEST test section. The freestream velocity was 3.5 km/s, equivalent to Mach 12 flight, and the dynamic pressure was 120 kPa. The measurements agreed with numerical simulation to within 5%.

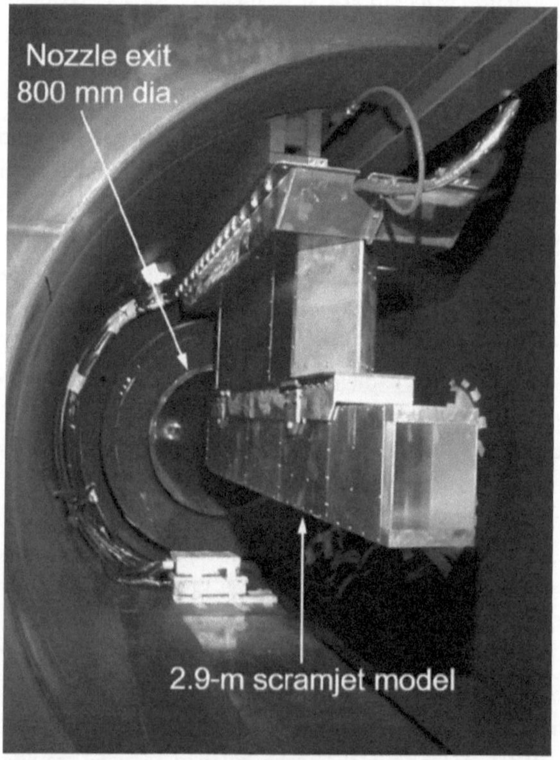

Figure 4.15 Axial force measurement on a scramjet model in HIEST. (Photo courtesy of H. Tanno, JAXA.)

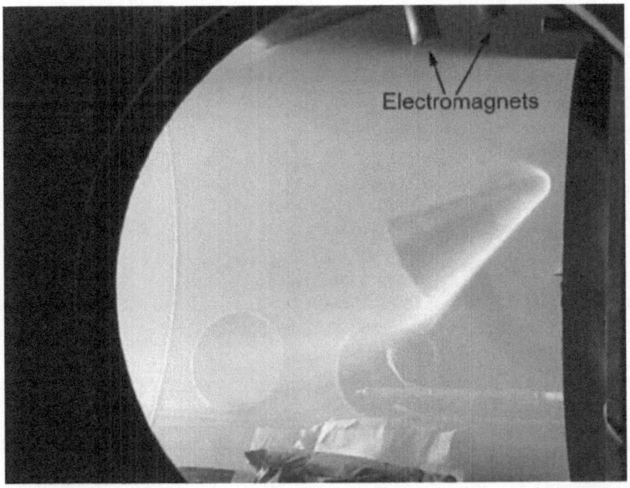

Figure 4.16 Free-flying blunt cone model released from electromagnets during an experiment in HIEST, $V_\infty = 6$ km/s. (Photo courtesy of H. Tanno, JAXA.)

Various alternative free-flight force measurement techniques have been devised by early researchers to take advantage of the short-duration nature of hypersonic test conditions and the minimal aerodynamic interference (Bernstein 1975). A notable recent implementation of those techniques that used electromagnets for suspension and launching of a free-flying model and accelerometers installed on the model for force measurements has been employed in the HIEST shock tunnel (Tanno et al. 2012). The test model is a blunt cone that has a base diameter of 20 cm, a length of 31.6 cm, and a mass of 19.2 kg. The model is initially supported by electromagnets, and its release is timed to synchronize with the arrival of the test flow. Figure 4.16 shows a photograph of the free-flying model during an experiment at a freestream velocity of 6 km/s. The results demonstrated that the pitching moment coefficient on the blunt cone yielded a significantly larger pitch-up at 16 MJ/kg than the low enthalpy condition of 4 MJ/kg (Tanno 2013).

In HEG, there have been efforts to move away from stress-wave-based methods toward nonintrusive free-flight techniques and to develop flow visualization-based force measurement methods. Free-flight force and moment measurements using a simpler suspension technique (with weak suspension threads) have been developed and a free-flying reentry capsule model has been tested in HEG (Laurence et al. 2018). The free-flying model incorporates internally mounted accelerometers that allow the measurements of five components of motion (three translational and two rotational). A weak-thread suspension technique was employed that relies on the arrival of the test flow itself to detach the supports, thus eliminating the need for a sophisticated timed release mechanism. The model was suspended prior to each experiment by nylon threads that are weakened at the junctions with the model body so that they would break on flow arrival. A catcher mechanism is used to terminate

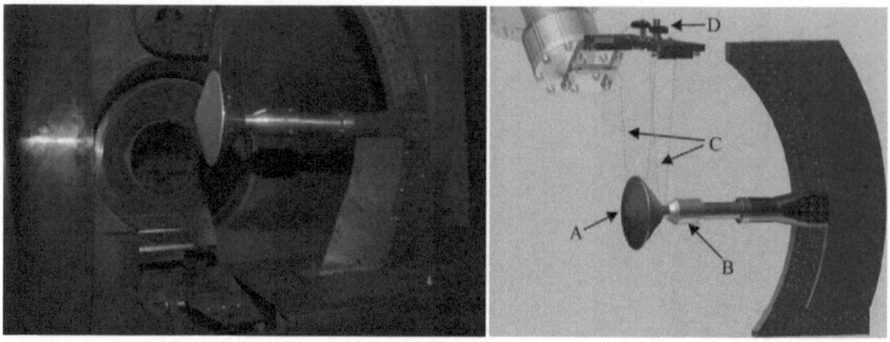

Figure 4.17 Experimental setup for force and moment measurements on a free-flying capsule model in the HEG. Left: photograph; right: schematic, (A) capsule model, (B) catcher mechanism housing, (C) suspension threads, and (D) model incidence/height adjustment mechanism. Reproduced with permission from Laurence et al. (2018).

the model motion. A short protuberance was attached to the rear of the model to form part of the mechanism. The other part of the mechanism consisted of a housing fixed to a conventional sting, inside which the model protuberance was placed before the experiment. The housing allowed for a model displacement of approximately 15 mm in the axial direction and several millimeters in the lateral direction and prevented the model from falling to the floor at the end of the test. Shown in Figure 4.17 is the experimental setup in the HEG test section. The stainless-steel capsule model has a diameter of 18 cm and a mass of 3.1 kg. The freestream velocity is 2.41 km/s, approximating Mach 8 flight at 28 km altitude. Forces and moments were measured using both the internally mounted accelerometers and visualization-based tracking techniques. For flow visualization, focus shadowgraphy was used. Overall, there is very good agreement between the measurement techniques. When applied to pitching measurements, the visualization-based techniques appear capable of higher accuracy than accelerometer-based measurements.

4.7.3 Hypersonic Boundary-Layer Transition

The state of the boundary layer is a crucial design parameter for hypersonic vehicles. Laminar–turbulent boundary-layer transition severely increases the heat flux and skin friction drag that can greatly affect the survivability of hypersonic vehicles. In hypersonic flow, real-gas effects can influence the state of the boundary layer as dissociation reactions can absorb the flow energy and have a stabilizing effect.

It is known that turbulence transition generally occurs at lower Reynolds numbers in pulse facilities than those in flight. It was thus thought to be difficult to study transition in those facilities. Being in an academic environment, T5 has been used in many fundamental studies relating to high-enthalpy aerothermodynamics, and great advancement has been made in using T5 to study turbulence transition (Adam 1997, Rasheed 2001). An interesting outcome from the studies conducted in T5 is that transition to

Figure 4.18 Photograph of the 7° half-angle cone model that was used in HIEST and HEG to study hypersonic boundary-layer transition at Mach 7.5. Reproduced with permission from Wagner et al. (2011).

turbulence at hypervelocities can be replicated in shock tunnels if the Reynolds number characterized by the boundary layer is used instead of the freestream value. This is because shock-tunnel flows often produce a higher freestream static temperature than that occurs in flight, but the characteristic boundary-layer temperature and the associated boundary-layer quantities (such as those evaluated at the reference temperature) can match those in flight well. In addition, the noise spectrum in shock tunnels is relatively quiet in the very high frequency range of the Mack mode instabilities that dominate the transition process at high enthalpies. These imply that pulse facilities can indeed be useful for studying boundary-layer transition at these conditions.

Various studies have also been carried out in other free-piston shock tunnels to investigate hypersonic boundary-layer transition. A 7° half-angle cone model was used to study hypersonic boundary-layer transition in both HEG and HIEST (Wagner et al. 2011). A photograph of the model is shown in Figure 4.18. The cone model has a length of 1,100 mm and was tested with nose radii of 2.25 and 5 mm. The freestream unit Reynolds number was varied over the range of 1.2×10^6/m to 6.2×10^6/m at Mach 7.5 and a total enthalpy of 3.3 MJ/kg. The transition location was determined by means of surface heat-flux measurements. The pressure fluctuations associated with Mack mode instabilities in the boundary layer were measured using pressure transducers and compared with numerical stability calculations. The good agreement between the experimental data and the calculations demonstrates the applicability of free-piston shock tunnels to study the flow instabilities in hypersonic turbulence transition. The high-pressure capability of the free-piston drivers also allows HEG and HIEST to meet the high Reynolds number requirements in the experiments.

4.7.4 Hypersonic Propulsion

HEG and HIEST were designed to reach total enthalpy levels for suborbital earth reentry conditions, but both are also capable of generating lower enthalpy conditions for studying supersonic combustion and testing scramjets. The study of the effectiveness of streamwise vortices on fuel mixing enhancement in a supersonic combustor is an example. A HyShot-IV scramjet model equipped with hypermixer injectors to generate streamwise vortices in the combustor was tested in HIEST. The engine flow

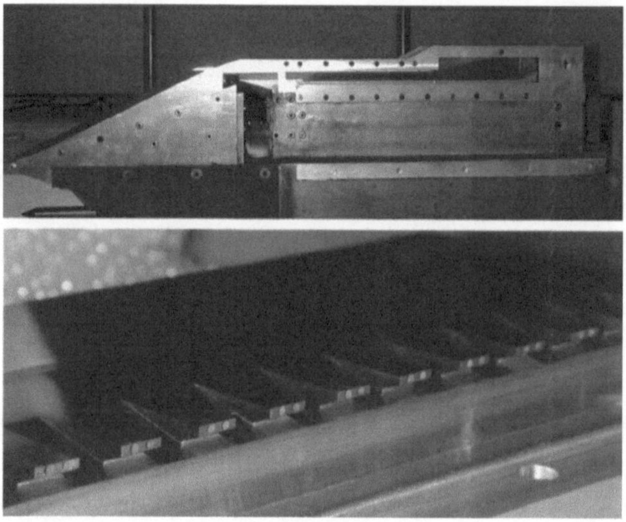

Figure 4.19 Top: The HyShot-IV scramjet model tested in HIEST. Bottom: Hypermixer injectors. (Photos courtesy of H. Tanno, JAXA.)

path, including the combustor, is basically two dimensional and has a total length of approximately 0.9 m, height of 0.13 m, and width of 0.105 m. The inlet ramp is wider than the engine body with a width of 0.196 m. The model was tested at a freestream Mach number of 8, and the total enthalpy was 3 MJ/kg. The Mach number at the combustor entrance was 2.9. The results confirmed that streamwise vortices were effective for the improvement of combustion performance. Figure 4.19 shows the photographs of a HyShot-IV scramjet model and the hypermixer injectors tested in HIEST.

More advanced scramjet engine concepts, such as the M12 scramjets, have also been investigated in HIEST. The M12-04 engine was tested in HIEST over a wide range of Mach numbers from 9 to 14. The external inlet has a multistage compression ramp. The internal flowpath has a two-dimensional configuration with a rectangular cross section. The model has a total length of 2.756 m, a width of 0.2 m, and a total mass of 331 kg. Figure 4.20 is a photograph of the engine model that shows a three-ramp inlet design. Combustion and thrust performance were studied. Positive net thrust over the engine internal drag was obtained up to the Mach 10 condition under a fuel equivalence ratio of unity (Takahashi et al. 2009).

The T4 RST at the University of Queensland is a facility in the academic environment. It has been used in a wide range of applications, although the main emphasis has been on supersonic combustion and scramjets (Stalker et al. 2005). Several axisymmetric scramjet models have been designed and tested in T4, one of which is shown in Figure 4.21 (UQ Scramjet 2013). The model being shown is approximately 0.3 m in length and consists of 6 identical combustor flow paths symmetrically placed in the circumferential direction around an axisymmetric center body. The model was tested over a range of total enthalpy from 2.8 to 8.5 MJ/kg, corresponding to flight velocities

Figure 4.20 The M12-04 scramjet engine model used in HIEST. (Photo courtesy of H. Tanno, JAXA.)

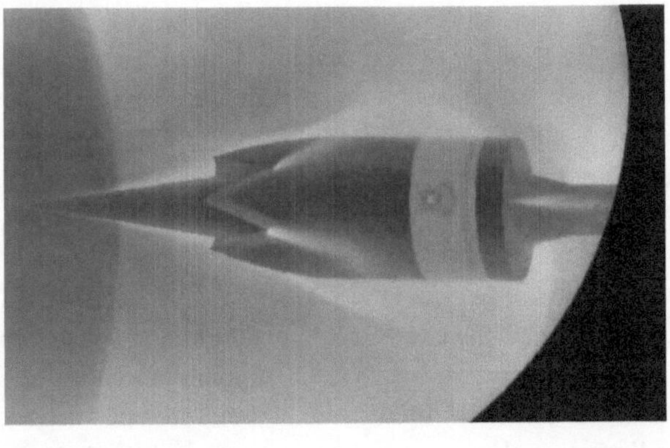

Figure 4.21 Luminosity photograph of an axisymmetric scramjet in T4. (Photo courtesy of UQ Centre of Hypersonics.)

of up to approximately 4 km/s. The model was fueled with a mixture of 13% silane and 87% hydrogen at equivalence ratios of up to approximately 0.8. The measurements involved the axial force on the model and positive net thrust was obtained up to 3.7 MJ/kg (Paull et al. 1995).

Scramjet engines designed with rectangular-to-elliptical shape transition (REST) inlets and elliptical combustors have been the recent focus in T4. One example was a REST engine model designed for Mach 8 flight (Smart et al. 2008). The model with an approximate length of 1 m includes an external inlet and an exhaust nozzle. It was fueled with gaseous hydrogen and tested at conditions equivalent to Mach 8 flight at 30 km altitude and a dynamic pressure of 0.5 atmospheres. A clear pressure rise was seen in the measurements, indicating combustion in the combustor and nozzle.

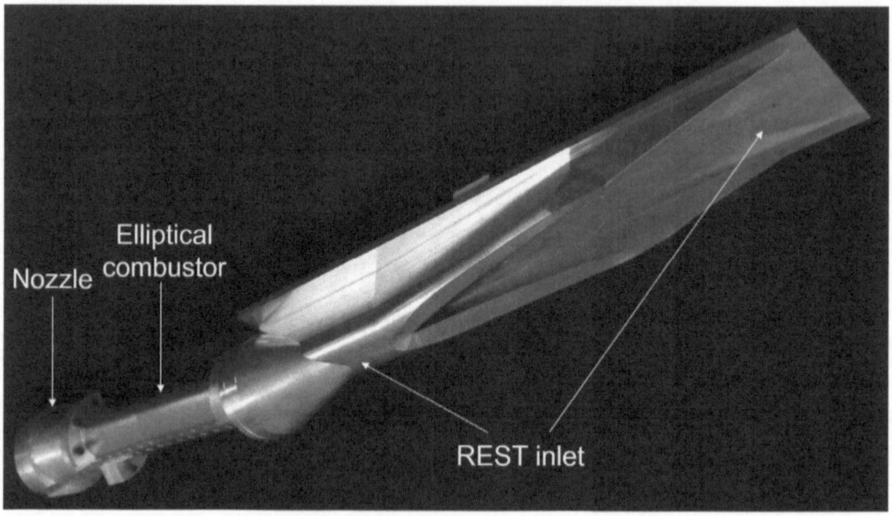

Figure 4.22 The Mach 12 RESTM12 half-scale scramjet model tested in T4. (Photo courtesy of Dylan Wise and UQ Centre of Hypersonics.)

Another REST scramjet, RESTM12, was designed to operate over a range of Mach numbers from 6 to 12, and at a constant dynamic pressure of 50 kPa has also been tested in several research studies. The engine model shown in Figure 4.22, which is half the design scale, has a total length of 0.94 m. In one study, it was tested at the Mach 12 condition in T4 (Wise 2014). However, as T4 is an RST, it was incapable of generating sufficient total pressure to simulate the flight condition. As a compromise, the model was tested in a semi–free-jet mode in which the freestream test flow only simulates the conditions behind the engine's forebody oblique shock wave. Even in this testing mode, the target dynamic pressure of 50 kPa could not be reached. The actual test conditions correspond to flight dynamic pressures of 27.1 and 32.2 kPa. There was also the limitation that the test flow was contaminated by the driver gas during the test period, which is an inherent issue in RSTs at high-enthalpy test conditions. The driver gas contained a mixture of helium and argon, and the presence of these inert gases would inhibit combustion. Nevertheless, robust combustion was observed when fuel was injected in the inlet in addition to injection in the combustor (Wise 2014).

To perform ground testing of scramjets at high Mach numbers, the test facility needs to have a sufficiently high total pressure capability. This is especially so if the model is subscale so that pressure-length scaling is required (i.e., a small-scale model is compensated for by a higher pressure). The free-piston-driven SET is a preferred facility choice to meet the test requirements as it can generate much higher total pressures than those by RSTs. Although the X3 facility at the University of Queensland was originally designed to simulate superorbital flight conditions, its test capability has been extended to include free-jet testing of scramjets at Mach 10–12 conditions (Toniato et al. 2015). The Mach 12 REST engine model (RESTM12) was

Figure 4.23 Mach 12 REST scramjet model in the X3 shock-expansion tunnel. (Photo courtesy of P. Toniato and UQ Centre of Hypersonics.)

tested in X3, which was configured to operate in the SET mode (i.e., the acceleration tube exit was fitted with a nozzle that expands the flow to the test section). The test model, shown in Figure 4.23, has a total length of 1.44 m and includes a forebody plate to simulate the vehicle forebody (Toniato 2019). Tests were carried out at Mach 11.1 and dynamic pressures of 52 and 104 kPa, corresponding to total pressures of 233 and 450 MPa, respectively. The combustion performance of two fueling configurations – combustor-only and inlet-combustor injection – was investigated. Pressure measurements were taken and were compared with numerical simulations. The results showed that combustion performance in the three-dimensional engine was robust and inlet-combustor injection led to a significant increase in combustion than combustor-only injection. The work also serves to demonstrate the applicability of SETs for high Mach number free-jet testing of scramjets, which overcomes the limitations of inadequate total pressure and driver gas contamination in RSTs at those conditions. Currently, SETs are the only gasdynamic facilities that can duplicate the required total pressure for free-jet hypersonic air-breathing engine tests at Mach numbers above 10, and free-piston-driven SETs are especially suitable for such tasks.

4.7.5 Fundamental Flow Physics

Free-piston-driven high-enthalpy shock-heated facilities are invaluable tools to study fundamental flow physics that involves high-temperature thermochemical processes. The short duration nature of the experiment also provides an advantage as there is much less concern of overheating of the test article and instrumentation, making the design of experiments simpler.

Free-piston shock tunnels have been used extensively to study flow problems in the hypervelocity regime. The research carried out in the T5 facility has elucidated a number of fundamental problems. One of the problems of interest is the shock-wave stand-off distance on a blunt body in hypervelocity flow. An analytical theory that can accurately determine the shock stand-off distance on cylinders and spheres in high-enthalpy conditions was developed and verified in T5 (Wen 1994, Hornung & Wen 1995). Under the high Mach number and high temperature conditions, the deceleration of the flow around a sphere or cylinder causes the gas to undergo nonequilibrium chemical reactions. The theory includes the total enthalpy and a reaction rate parameter that can accommodate reactions in complex gas mixtures. A previous correlation, which was investigated in another free-piston shock tunnel T3 (Hornung 1972), did not take into account the effects of recombination reactions and could only deal with the simplest case of dissociation of a single diatomic gas. The redeveloped theory was verified in T5 using nitrogen, air, and carbon dioxide as test gas media. The high facility pressure capability of T5 allows it to produce sufficient freestream pressure to conduct high-quality visualization.

Another fundamental problem is the shock-on-shock interaction, which occurs when an oblique shock impinges on a strong bow shock formed by a blunt body in hypersonic flow. This type of shock-on-shock interaction, also known as an Edney type IV interaction, causes intense enhancement of the local heat transfer rate and surface pressure and therefore has practical consequences for the thermal protection of hypersonic flight vehicles. An experimental study was performed in T5, and a theoretical model was developed to elucidate the finite-rate thermochemical processes occurring in the interaction region (Sanderson & Sturtevant 1995, Sanderson et al. 2004). Nitrogen was used as the test gas in all experiments, and the flow velocity ranges from 2.5 to 5.3 km/s, corresponding to total enthalpies of 3.9–19 MJ/kg. Real-gas effects are generally observed to increase the range of shock impingement points for which enhanced heating occurs. Holographic interferometry and heat flux measurements in T5 showed heat flux amplification in the dissociating flow that characterizes high speed, although the effect at a flow velocity of 5 km/s with nitrogen is less severe than had been predicted by extrapolation from tests in "cold" hypersonic facilities.

A similar experimental investigation that also used nitrogen as the test gas was conducted in the HEG at total enthalpies (~20 MJ/kg) sufficient to cause significant dissociation (Kortz et al. 1995). This work included the first quantitative interferometric visualization of shock impingement flows. It was found that the jet flow behind the bow shock, which is typical for the Edney type IV interaction, was not observed and the peak pressure and heat loads were much less severe than was expected from previous ideal gas investigations. Computations carried out to simulate the experimental results indicated a suppressed reaction rate in the impinging jet fluid at the condition tested (Hannemann et al. 1995).

Besides the interest in hypersonic flow, free-piston shock tunnels have also been used to study gaseous detonation. The problem of detonation initiation and stabilization by hypersonic projectiles was studied in the T5 shock tunnel (Bélanger et al. 1996, Kaneshige & Shepherd 1996, Kaneshige 1999). The ability to generate high-enthalpy stagnation conditions allows the shock tunnel to be used in a novel way other than a wind tunnel. The shock-tube end of T5 was converted into a powerful gas gun that was used to launch 25.4-mm diameter projectiles at speeds of around 3,000 m/s into various detonable mixtures contained in a chamber in the facility's test section. A clear transition between shock-induced combustion and detonation is observed. For the first time, initiation and stabilization have been observed with fuels other than hydrogen.

4.8 Summary

The use of a high-velocity piston to compress and raise the temperature of the driver gas has been discussed in this chapter. The use of piston compression is not only very efficient in generating high driver gas temperature and sound speed, it also has the advantage of pressure amplification as the resulting driver gas pressure far exceeds the initial compressed air pressure in the reservoir. The free-piston driver is thus capable of generating very high test-flow stagnation pressures.

Because the driver gas is compressed to a very small volume, the opening of the diaphragm and the flow of driver gas into the driven tube can cause rapid variation in the driver gas conditions, thus leading to the formation of expansion waves that could reach the test section before the experiment is completed. Therefore, it is much more preferable to operate the free-piston driver by maintaining an approximately constant-pressure level in the driver gas, as opposed to the constant-volume type.

Analytical models of constant-volume and constant-pressure modes of piston operation have been formulated. For the constant-volume mode driver, the equations are greatly simplified by assuming the gas properties behind and ahead of the piston to be spatially uniform and the flow is adiabatic. Simple closed-form solutions are obtained to predict the maximum piston velocity, maximum piston displacement at full stroke, and the final driver gas conditions attained.

For the constant-pressure driver, the piston motion and the processes at and after diaphragm rupture must be carefully controlled so that the driver gas pressure remains relatively constant to delay the rapid variation in the driver gas condition and the formation of expansion waves. The theory of tuned piston operation is presented that would yield optimum driver performance by ensuring the piston to come to rest and avoid a strong impact with the compression tube at the end of the piston stroke. A closed-form analytical solution is obtained for the approximate case when the pressure behind the piston is neglected. A unique relationship between the nondimensional piston velocity β at diaphragm rupture and the characteristic parameter ω that describes the piston mass and initial driver condition has been established in order to achieve tuned piston operation.

The free-piston driver has been shown to be a highly capable technique to drive high-enthalpy shock-heated facilities. In comparison with the heated light-gas and detonative combustion drivers, the free-piston driver has a clear advantage in delivering high shock strength, and therefore, high total enthalpy test conditions. Because of this, it has been widely used in the hypersonic test community. Several examples of major free-piston high-enthalpy test facilities around the world, and also their use in hypersonic testing, have been presented.

It is interesting to note that most of the major free-piston-driven facilities are RSTs designed for testing reentry and hypersonic propulsion. The free-piston driver is highly capable of generating the required total enthalpies for these applications when used in RSTs. However, RSTs do suffer from the effects of driver gas contamination and test gas dissociation when simulating flight conditions above Mach 10. They are also incapable of reaching the high total pressures required for free-jet testing of air-breathing propulsion systems above Mach 10. One solution to avoid these issues is to perform testing in an SET. As the SET is more capable of generating high total enthalpy and total pressure than the RST, it allows the use of other less complex driver techniques – the heated light-gas and detonation drivers – that can achieve the extreme test conditions but with less flow degradation due to contamination and dissociation.

Nomenclature

a	sound speed of driver gas
a_1	sound speed of pre-shocked gas in driven tube
A	cross-sectional area
A^*	cross-sectional area of driven tube (shock-tube) entrance
D	denominator; diameter (Table 4.2)
f	p_A/p_{Ar}
k	$= p_r V_r/(m u_r^2)$
L	initial length of compression tube; tube length (Table 4.2)
L_A	initial length of reservoir tube
m	piston mass
M_{s1}	primary shock Mach number
n	$= (\gamma + 1)/2$
p	pressure in driver gas in front of piston
p_A	pressure in reservoir gas (air) behind piston
R	gas constant
t	time
T	temperature in the driver gas
u	piston velocity
U_r	reference piston velocity to match driver gas outflow at diaphragm rupture
V	driver gas volume
v	$= V/V_r$, nondimensional driver gas volume
X	piston displacement
x	$= \rho V/(\rho_r V_r)$, nondimensional driver gas mass
y	$= u/u_r$, nondimensional piston velocity
β	$= u/U_r$, nondimensional piston velocity at diaphragm rupture
ϵ	$= (p_A/p)_r$, pressure ratio across piston at diaphragm rupture
γ	ratio of specific heats of driver gas
γ_A	ratio of specific heats of reservoir gas (air)
λ	volumetric compression ratio

θ	$= k\beta^2$
ρ	driver gas density
σ	$= 1/(n-1)$
ω	characteristic parameter of piston motion

Subscripts:

0	initial state
1	initial state in driven gas in conventional shock tube
4	initial driver gas state in conventional shock tube
A	reservoir gas (air)
f	final state
i	initial state
r	condition at diaphragm rupture

Superscript:

*	condition at driven tube entrance

References

Abdel-Jawad, M. M., Mee, D. J., Morgan, R. G., Jacobs, P. A., and Philpot, J. A. (2001). *Transient Force Measurements at Superorbital Speeds.* Paper 5358, 23rd International Symposium on Shock Waves, Fort Worth, Texas.

Adam, P. (1997). *Enthalpy Effects on Hypervelocity Boundary Layers.* Ph.D. Thesis. California Institute of Technology. (Unpublished).

Atkins, W. W. and Halperson, S. M. (1960). Hypervelocity Projection Techniques and Impact Studies, January 1959–June 1960. *NRL Memorandum Report No. 1116.*

Bélanger, J., Kaneshige, M., and Shepherd, J. E. (1996). Detonation Initiation by Hypervelocity Projectiles. In *Proceedings of the 20th International Symposium on Shock Waves*, Vol. II, ed. B. Sturtevant, J. E. Shepherd, and H. G. Hornung. Pasadena, CA: World Scientific, pp. 1119–1124.

Bernstein, L. (1975). *Force Measurements in Short-Duration Hypersonic Facilities.* AGARDograph No. 214, ed. R. C. Pankhurst. London: AGARD.

Bi, Z., Zhang, B., Zhu, H., et al. (2018). Experiments and Computations on the Compression Process in the Free Piston Shock Tunnel FD21. Presented at the *5th International Conference on Experimental Fluid Mechanics*, Munich, Germany.

Bird, K. D. (1962). *48-Inch Hypersonic Shock Tunnel, Description and Capabilities.* Cornell Aeronautical Laboratory, May.

Bray, K. N. C., Pennelegion, L., and East, R. A. (1959). A Progress Report on the University of Southampton Hypersonic Gun Tunnel. *Aeronautical Research Council Technical Report. C.P. No. 457.*

Charters, A. C., Denardo, B. P., and Rossow, V. J. (1957). Development of a Piston-Compressor Type Light-Gas Gun for the Launching of Free-Flight Models at High Velocity. *NACA TN 4143.*

Chinitz, W., Erdos, J. I., Rizkalla, O., Anderson, G. Y., and Bushnell, D. M. (1994). Facility Opportunities and Associated Stream Chemistry Considerations for Hypersonic Air-Breathing Propulsion. *Journal of Propulsion and Power*, **10**(1), 6–11.

Dann, A. G., Morgan, R. G., Gildfind, D. E., et al. (2012). Upgrade of the X3 Super-orbital Expansion Tube. Presented at the *18th Australasian Fluid Mechanics Conference*, Launceston, Australia, December 3–7.

Duryea, G. R. and Sheeran, W. J. (1969). Accelerometer Force Balance Techniques. Presented at the *3rd IEEE International Congress on Instrumentation in Aerospace Simulation Facilities*, Brooklyn, USA.

Eggers, A. J. Jr. (1955). A Method for Simulating the Atmospheric Entry of Long-Range Ballistic Missiles. *NACA Report 1378*.

Glass, I. I. and Hall, J. G. (1959). Handbook of Supersonic Aerodynamics, Section 18, Shock Tubes. *NAVORD Report 1488* (Vol. 6), December.

Glössner, C. and Olivier, H. (2005). Force and Moment Measurements on the Hopper-Configuration for High Enthalpy Conditions. In *Proceedings of the 24th International Symposium on Shock Waves*, ed. Z. L. Jiang. Beijing: Springer, pp. 203–208.

Hannemann, K., Brück, S., and Brenner, G. (1995). Numerical Simulation of Reacting Flows Related to the HEG. In *Shock Waves @ Marseille II*, ed. R. Brun and L. Z. Dumitrescu. Berlin: Springer, pp. 251–256.

Hannemann, K., Schramm, J. M., Wagner, A., and Camillo, G. P. (2018). The High Enthalpy Shock Tunnel Göttingen of the German Aerospace Center (DLR). *Journal of Large-Scale Research Facilities*, **4**, A133.

Hertzberg, A., Wittliff, C. E., and Hall, J. G. (1961). Summary of Shock Tunnel Development and Application to Hypersonic Research. Cornell Aeronautical Laboratory. *Report No. AD-1052-A-12, AFOSR TR 60-139*.

Hornung, H. G. (1972). Non-equilibrium Dissociating Nitrogen Flow Over Spheres and Circular Cylinders: Part 1. *Journal of Fluid Mechanics*, **2**(53), 149–176.

Hornung, H. G. (1988). The Piston Motion in a Free-Piston Driver for Shock Tubes and Tunnels. *GALCIT Report FM 88-1*, California Institute of Technology.

Hornung, H. G. (2010). Ground Testing for Hypervelocity Flow, Capabilities and Limitations. In *Aerothermodynamic Design, Review on Ground Testing and CFD*, ed. O. Chazot and K. Bensassi, RTO-EN-AVT-186. RTO/NATO, paper 1.

Hornung, H. G. and Wen, C.-Y. (1995). Nonequilibrium Dissociating Flow Over Spheres. *AIAA Paper 95-0091*.

Itoh, K. (1996). Tuned Operation of a Free-Piston Shock Tunnel. In *Proceedings of the 20th International Symposium on Shock Waves*, ed. B. Sturtevant, J. E. Shepherd, and H. G. Hornung. Vol. I. Pasadena, CA: World Scientific, pp. 43–51.

Itoh, K. (2002). Characteristics of the HIEST and Its Applicability for Hypersonic Aerothermodynamic and Scramjet Research. In *Advanced Hypersonic Test Facilities*, ed. F. K. Lu and D. E. Marren. Vol. 198 of *AIAA Progress in Astronautics and Aeronautics*. Reston, VA: AIAA, pp. 239–253.

Itoh, K., Komuro, T., Sato, K., et al. (2002a). Characteristics of Free-Piston Shock Tunnel HIEST (1st Report, Tuned Operation of Free-Piston Driver). (In Japanese). 日本機械学会論文集 B編, J-STAGE, 68 巻 (2002) 675 号.

Itoh, K., Tanno, H., Komuro, T., Sato, K., and Ueda, S. (1994). Tuned Operation Theory for Free-Piston Driver of High-enthalpy Shock Tunnel. In *Symposium on Shock Waves Japan'94*, pp. 29–32.

Itoh, K., Ueda, S., Komuro, T., et al. (1998). Improvement of a Free Piston Driver for a High-enthalpy Shock Tunnel. *Shock Waves*, **8**, 215–233.

Itoh, K., Ueda, S., Tanno, H., Komuro, T., and Sato, K. (2002b). Hypersonic Aerothermodynamic and Scramjet Research Using High Enthalpy Shock Tunnel. *Shock Waves*, **12**, 93–98.

Jenkins, R. C. (1974). *Shock Tube Operation Using a Compression-Heated Driver Gas*. Grumman Research Department Report RE-481, Grumman Aerospace Corporation, Bethpage, New York, August.

Jessen, C. and Grönig, H. (1989). A New Principle for a Short-duration Six Component Balance. *Experiments in Fluids*, **8**, 231–233.

Kaneshige, M. J. (1999). *Gaseous Detonation Initiation and Stabilization by Hypervelocity Projectiles*. Ph.D. Thesis. California Institute of Technology. (Unpublished).

Kaneshige, M. J. and Shepherd, J. E. (1996). Oblique Detonation Stabilized on a Hypervelocity Projectile. *Symposium (International) on Combustion*, **26**(2), 3015–3022.

Kortz, S., McIntyre, T. J., and Eitelberg G. (1995). Experimental Investigation of Shock-on-Shock Interactions in the High-Enthalpy Shock Tunnel Göttingen (HEG). In *Shock Waves @ Marseille I*, ed. R. Brun and L. Z. Dumitrescu. Berlin: Springer, pp. 75–80.

Laurence, S. J., Butler, C. S., Martinez Schramm, J., and Hannemann, K. (2018). Force and Moment Measurements on a Free-Flying Capsule Model in a Shock Tunnel. *Journal of Spacecraft and Rockets*, **55**(2), 403–414.

Morgan, R. G. (1997). Superorbital Expansion Tubes. *21st International Symposium on Shock Waves*, Australia, Paper 9000.

Morgan, R. G. (2000). Development of X3: A Superorbital Expansion Tube. *AIAA Paper 2000-0558*.

Morrison, W. R. B., Stalker, R. J., and Duffin, J. (1989). New Generation of Free-Piston Shock Tunnels. In *Current Topics in Shock Waves/17th International Symposium on Shock Waves and Shock Tubes, Bethlehem, PA 1989*; ed. Y. W. Kim. New York: American Institute of Physics, pp. 582–587.

Nagamatsu, H. T., Geiger, R. E., and Sheer, R. E. (1959). Hypersonic Shock Tunnel. *Journal of the Aerospace Sciences*, **29**(5), 332–340.

Nagamatsu, H. T., Sheer, R. E. Jr., Osburg, L. A., and Cary, K. H. (1961). Design Features of the General Electric Research Laboratory Hypersonic Shock Tunnel. GE Research Lab. *Report No. 61-RL-2711C*, May.

Paull, A., Stalker, R. J., and Mee, D. J. (1995). Supersonic Combustion Ramjet Propulsion Experiments. In *Shock Tunnel Studies of Scramjet Phenomena 1994*. NASA Contractor Report NASA-CR-199445.

Porat, H., Morgan, R. G., and McIntyre, T. J. (2012). Study of Radiative Heat Transfer in Titan Atmospheric Entry. *28th International Congress of the Aeronautical Sciences*, Brisbane, Australia, September, *Paper ICAS 2012-3.8.3*.

Rasheed, A. (2001). *Passive Hypervelocity Boundary Layer Control Using an Ultrasonically Absorptive Surface*. Ph.D. Thesis. California Institute of Technology. (Unpublished).

Roffe, G. A. (1963). The Free Piston Shock Tube Driver: A Preliminary Theoretical Study. *Technical Report No. 32-560*. Jet Propulsion Laboratory, Pasadena, CA, December 15.

Sanderson, S. R., Hornung, H. G., and Sturtevant, B. (2004). The Influence of Non-equilibrium Dissociation on the Flow Produced by Shock Impingement on a Blunt Body. *Journal of Fluid Mechanics*, **516**, 1–37.

Sanderson, S. R. and Sturtevant, B. (1995). Shock Wave Interactions in Hypervelocity Flow. In *Shock Waves @ Marseille I*, ed. R. Brun and L. Z. Dumitrescu. Berlin: Springer, pp. 69–74.

Scott, M. P. (2006). *Development and Modelling of Expansion Tubes*. Ph.D. thesis. University of Queensland. (Unpublished).

Smart, M., Stalker, R., Morgan, R., and Paull, A. (2008). Hypersonics Research in Australia. In *Advances on Propulsion Technology for High-Speed Aircraft*. Educational Notes RTO-EN-AVT-150. Neuilly-sur-Seine, France: RTO/NATO, Paper 11.

Stalker, R. J. (1961). *An Investigation of Free Piston Compression of Shock Tube Driver Gas*. National Research Council of Canada (NRC), Division of Mechanical Engineering, Mechanical Engineering Report MT-44, Ottawa, Canada.

Stalker, R. J. (1967). A Study of the Free-Piston Shock Tunnel. *AIAA Journal*, **5**(12), 2160–2165.

Stalker, R. J. (2006). Modern Developments in Hypersonic Wind Tunnels. *The Aeronautical Journal*, **110**(1003), 21–39.

Stalker, R. J., Paull, A., Mee, D. J., Morgan, R. G., and Jacobs, P. A. (2005). Scramjets and Shock Tunnels: The Queensland Experience. *Progress in Aerospace Sciences*, **41**(6), 471–513.

Steward, B. (2004). *Predicted Scramjet Testing Capabilities of the Proposed RHYFL-X Expansion Tube*. Ph.D. Thesis. University of Queensland. (Unpublished).

Takahashi, M., Komuro, T., Sato, K., et al. (2009). Scramjet Combustor Characteristics at Hypervelocity Condition over Mach 10 Flight. In *Proceedings of the 6th European Symposium on Aerothermodynamics for Space Vehicles*, ESA SP-659, ed. H. Lacoste and L. Ouwehand. Noordwijk, Netherlands: European Space Agency, ID.45.

Tanno, H. (2013). *Current Research Activities on Hypersonic Aerothermodynamics in JAXA-HIEST*. Invited Seminar at Nanyang Technological University, Singapore, March 19.

Tanno, H., Komuro, T., Sato, K., and Itoh, K. (2005). Improvements of the Accelerometer Force Measurement Technique. Presented at *the 21st International Congress on Instrumentation in Aerospace Simulation Facilities*, Sendai, Japan, August 29–September 1.

Tanno, H., Komuro, T., Sato, K. et al. (2012). Free-flight Force Measurement Technique in Shock Tunnel. *AIAA Paper 2012–1241*.

Tanno H., Komuro T., Sato K., et al. (2014). Aeroheating Measurement of Apollo Shaped Capsule with Boundary Layer Trip in the Free-piston Shock Tunnel HIEST. *AIAA Paper 2014–0434*.

Toniato, P. (2019). *Free-Jet Testing of a Mach 12 Scramjet in an Expansion Tube*. Ph.D. Thesis, University of Queensland. (Unpublished).

Toniato, P., Gildfind, D. E., Jacobs, P. A., and Morgan, R. G. (2015). Development of a New Mach 12 Scramjet Operating Capability in the X3 Expansion Tube. Presented at *the 7th Asia-Pacific International Symposium on Aerospace Technology*, Cairns, Australia, 25–27 November.

UQ Scramjet. (2013). *Axisymmetric Scramjet*. https://bit.ly/3lKZJ3C

UQ X3. (n.d.). *X3 Expansion Tube*. https://mechmining.uq.edu.au/research/hypersonics/x3-expansion-tube (Accessed: May 9, 2023).

Wagner, A., Laurence, S., Martinez Schramm, J., et al. (2011). Experimental Investigation of Hypersonic Boundary-Layer Transition on a Cone Model in the High Enthalpy Shock Tunnel (HEG) at Mach 7.5. *AIAA Paper 2011–2374*.

Wen, C.-Y. (1994). *Hypervelocity Flow over Spheres*. Ph.D. Thesis. Graduate Aeronautical Laboratories, California Institute of Technology. (Unpublished).

Wise, D. (2014). *Experimental Investigation of a Three Dimensional Scramjet Engine at Hypervelocity Conditions*. Ph.D. Thesis. University of Queensland. (Unpublished).

Wittliff, C. E., Wilson, M. R., and Hertzberg, A. (1959). The Tailored Interface Hypersonic Shock Tunnel. *Journal of the Aerospace Sciences*, **26**(4), 219–228.

5 Detonation-Driven High-Enthalpy Shock Tunnels

The detonation-driven shock tunnel is one of three important classes of hypersonic and high-enthalpy ground-testing facilities that are based on the shock-heated principle and widely used over the world (Lu & Marren 2002, Ozer & Friedrich 2016). The theory and methods for developing the detonation-driven shock tunnels aiming at hypervelocity flow generation with high quality are summarized in this chapter. At first, the primary concepts for detonation drivers are presented to demonstrate its unique advantages for aerodynamic ground-based testing. The difficult problems arising from the development of hypervelocity shock tunnels for simulating flight conditions are identified and discussed in detail to address critical issues underlying the high-enthalpy shock-tunnel design (Jiang & Yu 2017, Jiang et al. 2020b). Then, three kinds of detonation-driven shock tunnels are introduced, and their key techniques and performances are reviewed and discussed in detail. These shock tunnels are: (1) the JF-10 detonation-driven high-enthalpy shock tunnel based on a forward-running detonation mode (the JF-10 shock tunnel in LHD laboratory series) (Jiang et al. 2011); (2) the JF-12 hypersonic-flight-duplicated shock tunnel operating with a backward detonation mode (the JF-12 shock tunnel) (Jiang & Yu 2014); (3) the JF-16 detonation-driven orbital-speed shock-expansion tube with the forward detonation cavity (FDC) driver (the JF-16 expansion tube) (Jiang et al. 2015). In this chapter, several important aspects of the detonation-driven shock tunnels will be presented; the information provided here is useful not only for understanding the detonation-driven high-enthalpy shock tunnel but also for supporting its future development. Finally, some experiments are summarized to demonstrate the capability of the detonation-driven hypersonic shock tunnel and the importance of the measurement techniques for hypersonic and high-temperature flow experiments. Both are the frontiers of high-enthalpy flow research for developing hypersonic vehicles.

5.1 Introduction to Detonation-Driven Shock Tunnels

Hypersonic technology in the foreseeable future will play an important role in the domain of space access and hypersonic flight (Hirschel 1991, Bertin & Cummings 2006). The hypersonic vehicle encounters an aerothermodynamic environment characterized by strong shocks and high-temperature flows. The surrounding airflow in the hypersonic regime behaves quite differently from ones at supersonic Mach numbers.

The kinetic energy associated with hypersonic flight is converted into internal energy that significantly increases the air temperature and gives rise to endothermic reactions, such as molecule vibration and dissociation, and atom ionization of the air near the vehicle surface. Thus, the airflow becomes a reacting gas flow with significant energy exchange and heat transfer. The energy exchange takes place between the reacting gases and the main flow, which results in variations of the flow temperature and pressure around hypersonic vehicles. The heat is transferred from the surrounding reacting flow to the vehicle surface. The heat flux depends upon many factors, including the freestream condition, the configuration of the vehicle and its orientation to the flow, the temperature difference between the air and vehicle surface, and the surface catalysis. The aforementioned phenomena are different from supersonic flows and have come to be known as real-gas effects that vary with the flight Mach number and altitude. Hypersonic technology reflects *closely* the coupling of the aerodynamic forces with the heating environment and its research is the frontier of gas dynamics (Anderson 1989, Park 1990, Bertin 1994).

Hypersonic vehicles have been designed and built since the 1950s. In some of these research programs, flight tests had been completed, and the flight data revealed various problems that were predicted neither through analysis/computation nor through ground-based testing. These problems exposed hypersonic flow phenomena that were not previously identified; we shall call them "unknown unknowns." The severity of the hypersonic flight environment strongly depends on the mission profile and the vehicle configuration. Figure 5.1 shows typical trajectories of various spacecraft and three main regions of different physical and chemical processes in the shock layer near such vehicles. It is important to note that the hypersonic flight regime covers a large

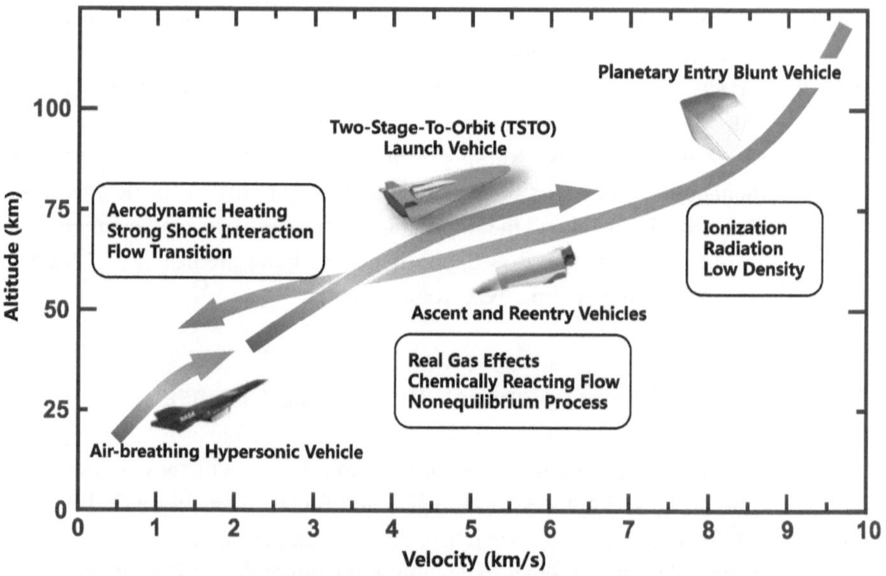

Figure 5.1 Typical trajectories of various spacecraft and major aerothermodynamic effects.

range of speed and altitude. As an example, an air-breathing hypersonic vehicle must operate at relatively low altitudes to maintain the relatively high dynamic pressures required for maximum engine performance; therefore, it often flies at Mach numbers of 5–10 and at altitudes from 25 to 50 km, as indicated in Figure 5.1 where the aerothermal environment is dominated by fluid dynamics effects. Although the real gas effect is weak in such flight conditions, vehicle heating is high at leading edges, in the regions with shock interactions, and in downstream locations where flow transition to turbulence occurs. Since there is considerable uncertainty in boundary-layer transition prediction, this transition uncertainty presents critical challenges to researchers because the convective heating from a turbulent boundary layer is much greater than that from a laminar boundary layer. On the other hand, a high-speed planetary entry blunt vehicle, such as the Galileo probe entering Jupiter or a high-speed Earth return capsule, encounters an aerothermal environment dominated by high-enthalpy effects. The high-enthalpy flow in planetary entry is characterized by an elevated level of ionization and may result in significant radiation and heating. The strong real-gas effects and low-density effect play an important role in this flight regime. As depicted in Figure 5.1, the middle part of the typical trajectories is for two-stage-to-orbit (TSTO), and ascent and reentry vehicles. Strong real-gas effects dominate flows around the vehicles, in which molecular dissociations take place significantly. The hypersonic flow is actually a chemically reacting gas flow where the local flow temperature is a dominating parameter varying from the bow shock to the flow along the vehicle surface and the wake behind the vehicle, and the flow pressure imposes another important variable to reactions. Moreover, hypersonic propulsion adds combustion process to the high-enthalpy flow motion, and this makes hypersonic research a very challenging problem that has never been encountered before (Anderson 1989, Bertin 1994, Jiang et al. 2020a).

In order to design a new hypersonic vehicle and make it fly as required, it is important to develop hypervelocity test facilities for ground-based experimental research. So far, no single ground-test facility can fully simulate all aspects of hypersonic flight, that is, the test duration, flow velocity, gas chemistry, Mach number, Reynolds number, surface temperature, and ablation effects (Hirschel 1991, Lu & Marren 2002, Bertin & Cummings 2006, Ozer & Friedrich 2016). The quality of the hypersonic freestream flow cannot be controlled simultaneously in any single facility, if at all. Nearly all wind tunnels suffer from noise levels much higher than that in flight, and all the tunnels simulating gas chemistry effects also have freestream chemistry contaminants. Ground test simulations are an exercise in the art of combining partial simulations, with each having unique advantages and disadvantages.

After more than 60 years' developments, high-enthalpy and hypervelocity test facilities suitable for studying aerothermochemistry are still based on shock-heated wind tunnels. Three kinds of high-enthalpy shock tunnels have been developed over the world during the past several decades (Lu & Marren 2002, Ozer & Friedrich 2016, Jiang et al. 2020a). The first is the heated light-gas shock tunnel, the second is the free-piston-driven high-enthalpy shock tunnel, and the third is the detonation-driven high-enthalpy shock tunnel. With those facilities, a lot of valuable experimental data have been produced, and the understanding of hypersonic flow benefits greatly from such precious data.

For high-enthalpy shock tunnels, one of the challenges is how to create a true hypersonic environment at a reasonable scale on the ground due to the high-energy output requirement for hypersonic shock tunnels. This key fact is important because the chemically reacting flow is not scalable. And unfortunately, the chemistry is the core of the hypersonic flow, and it makes hypersonic gas dynamics standout from conventional gas dynamics (Anderson 1989, Park 1990, Bertin 1994). To provide the high energy for shock tunnels, the detonation-driven shock tunnel appears to be quite a promising power supply.

The detonation driver was first proposed by Bird in a computational analysis (Bird 1957). A 13.3-m-long detonation driver was built by Yu in 1998 for performance tests (Yu 1998). Yu et al. developed the TH2-D in Aachen, Germany, in 1992 (Yu et al. 1992), and the high-enthalpy shock tunnel is operated successfully with the backward-running detonation mode. Actually, the work was performed as a joint research project between the Shock Wave Laboratory, RWTH Aachen University; and the Institute of Mechanics, Chinese Academic of Sciences (CAS) (Olivier et al. 2002). Then in 1994, a large detonation-driven shock tunnel, the so-called HyPulse, was set up in GASL, USA. It could also be operated as an expansion tube (Chue et al. 2003). These detonation-driven shock tunnels can be operated either in the backward-running detonation mode to achieve stable test flows with long test time at a relatively low enthalpy level or in the forward-running detonation mode to obtain high-enthalpy test flows but with a short test duration. These progresses demonstrated that the detonation-driven hypersonic test facilities could become a powerful facility for hypersonic vehicle ground tests.

In this chapter, the primary concept and equations for detonation drivers are presented first to demonstrate their unique advantages for high-enthalpy flow testing. Difficult problems arising from the development of hypervelocity shock tunnels to simulate flight conditions are identified and discussed to address critical issues underlying these problems. Then, the research progress on the detonation-driven shock tunnels developed in the State Key Laboratory of High Temperature Gas Dynamics (LHD), Institute of Mechanics, CAS is summarized and discussed in detail. These facilities include the JF-10 shock tunnel, the JF-16 expansion tunnel, and the JF-12 shock tunnel. The JF-10 shock tunnel is operated with a forward detonation mode for flow Mach numbers from 10 to 20. The JF-12 shock tunnel with an endearment name of Hyper-dragon I, equipped with a 2.5-m diameter nozzle, covers flow Mach numbers from 5 to 9. The JF-16 expansion tunnel is capable of producing the orbital-speed flow at Mach numbers from 20 to 25 or even higher. Considering unscalable problems at high Mach numbers, the JF-22 Hypervelocity Shock Tunnel (Hyper-dragon II) designed based on the FDC driver will be briefly introduced since its calibration is still ongoing. The Hyper-dragon II equipped with three 2.5-m diameter nozzles is capable of covering flow March numbers from 10 to 25 at altitudes ranging from 30 to 90 km (Jiang et al. 2020b). With these facilities, the typical trajectories of various spacecraft could be covered so that the research on aerothermodynamics and hypersonic propulsion could be carried out in the laboratory to gain reliable experimental data for the fundamental understanding on the physical and chemical phenomena of high-enthalpy hypersonic flows.

5.2 Fundamentals of Detonation-Driven Shock Tunnels

The theoretical analysis of shock tunnels indicates that it is feasible to duplicate flight conditions in a laboratory if the strong incident shock wave can be generated. However, there are still two important issues that have to be considered for ground-based hypersonic tests. The first one is how to minimize the nonequilibrium process effect, and the second is how to obtain the interface-matching condition in the shock-tunnel operation. A great deal of work has been done during the development of both light-gas-heated shock tunnels and free-piston-driven shock tunnels for decades, but there is still a gap between engineering requirements and wind-tunnel performances that could be achieved. These two issues will be discussed from several aspects of detonation-driven shock tunnels.

5.2.1 Requirements for Powerful Shock-Tunnel Drivers

We first discuss the nonequilibrium process simulation with high-enthalpy shock tunnels. Regarding the reentry problem of a spacecraft at a speed of 6,000 m/s, the dissociation of a nitrogen molecule is induced by the bond breakdown of two nitrogen atoms due to strong molecular vibration. The relaxation time at the atmospheric pressure is about 1/1,000 s at the temperature around 2,000 K and about 10^{-7} second at the temperature around 7,000 K. The equilibrium length is 6 m according to the former relaxation time but 0.06 cm for the latter one. The physical fact indicates that if the fuselage is 6 m in its total length, the hypersonic flow outside the fuselage cannot reach its equilibrium states. This is because the post-shock temperature is very high behind the normal part of the bow shock wave, but the temperature is getting lower and lower behind the oblique shock as the airflow is getting further and further away from the normal shock wave. Therefore, there will be flow regions downstream of the normal shock where a temperature of around 2,000 K exists. So, it is necessary for a reliable ground-test facility to produce a sufficiently large test flow region to accommodate a test model that is large enough to minimize the nonequilibrium process effect on aerodynamic features of hypersonic vehicles (Anderson 1989, Jiang et al. 2020b). As a result, it is a critical requirement for shock tunnels to generate not only the pure air flow with the right temperature to simulate correct chemistry but also the large-scale test flow field for the chemical reaction process to complete. With regard to the size of test flow fields, a nozzle diameter from 2.5 to 3.5 m is necessary for most of the hypersonic vehicles in the current flight tests over the world, so that the full-scale test model could be accommodated in the nozzle flow. Actually, the predominant flow process in hypersonic flows is high-temperature-induced thermochemical reactions, and its characteristic scale does not change when a hypersonic vehicle is scaled to a small test model. It means that the model-scaling theory widely applied in traditional supersonic and subsonic wind-tunnel experiments is no longer applicable for hypersonic experiments (Jiang et al. 2020a).

In the hypersonic research area, there is a binary scaling for nonequilibrium processes involving two-body molecular collisions (Anderson 1989). This binary scaling parameter was demonstrated for the blunt-body stagnation shock layer where the nonequilibrium process is dominated by two-body dissociation reactions rather than three-body recombination reactions (Anderson 1989, Park 1990). However, the binary scaling is not applicable for air-breathing hypersonic vehicles like the X-43a and trans-atmospheric vehicles like X-37b. The X-43a has an integrated configuration involving a scramjet engine where the supersonic combustion is governed by much more complex chemical reactions than two-body dissociations. For X-37b, the flow field right behind its bow shock is dominated by two-body dissociations, but in its boundary layer, recombination reactions take place since the flow temperature decreases downstream from the stagnation region. Furthermore, for the hypersonic boundary-layer transition, the surface roughness and catalytic reactions are also not scalable. Hypersonic flight tests in recent years showed obvious discrepancies from wind-tunnel experiments but agree well with full-scale model tests performed in the JF-12 shock tunnel. This is a challenge not only for hypersonic ground test facilities but also for scaling law research on hypersonic flows (Jiang et al. 2020b). However, if we could construct hypervelocity wind tunnels to accommodate the full-scale test models of the present flight-tested hypersonic vehicles, and then draw the correct correlation between flight test data and wind-tunnel experimental results for scaled test models, the work will be of significant importance for developing future hypersonic vehicles that may have much larger dimensions than before.

In summary, for reliable hypersonic ground tests, there are four requirements that must be considered carefully in wind-tunnel development: (1) the test gas, instead of any substitute, must be the pure air to accurately simulate chemical reaction mechanisms; (2) the stagnation temperature and total pressure have to be achieved to excite correct chemical reactions; (3) the scale of the test model must be large enough to ensure that chemical reactions occurring at the correct reaction rates at the right locations on the test model are the same as those in flight tests; and (4) sufficiently long test time is necessary for aerodynamic forces and supersonic combustion tests. The fourth requirement is important not only for the test flow to reach a stable combustion but also for improving the experimental data accuracy of aerodynamic forces and moments. Meeting the four requirements at the same time results in the flight condition duplication in high-enthalpy shock tunnels, and this has been a challenge for developing hypersonic test facilities for decades.

The first difficulty arising from the four requirements discussed above is the power-supply requirement for shock-tunnel drivers. If one wants to produce the flight condition at the 40-km altitude for Mach number 15, the output power of shock tunnels will be about 1,056 MW to generate a nozzle flow being 2.5 m in diameter. This output power is about 2,054 MW at the altitude of 35 km. The input power for the shock tunnel is even much higher than its output power. Such a power demand is necessary not only for achieving high-enthalpy gas states but also for compressing a large amount of air mass to a high thermal state and maintaining a long test duration.

We discuss the free-piston driver first since the driver is widely applied to high-enthalpy shock tunnels over the world. This class of the shock tunnels was used to generate most of valuable data for the high-enthalpy flow research (Stalker 1967, Hornung 1988, Itoh et al. 1998, 2002). However, the free-piston-driven shock-tunnel technique has three limitations to meet the four requirements as mentioned above. The first limitation is the power requirement. The total energy of a 1,000-kg mass piston moving at about 300 m/s is about 45 MJ. Although the output power is about 450 MW, but for a 100-ms test duration, this value is still much lower than the required value because the efficiency of the energy transfer from the piston to the test gas is not considered. For flights at low altitudes, the requirement of the total energy carried by a free piston could be even bigger. Actually, the piston mass and speed in the given example are already the maximum that were reached in the HIEST operation. Therefore, the free-piston driver works well for small shock tunnels to produce high-enthalpy flows with a short test duration, but it has its limitation to meet the need for large-scale hypersonic test facilities that are to operate for a long test duration. The second one is the test duration that is limited not only by the output power requirement but also by the piston motion. For the free-piston-driven shock tunnel, the pressure in front of the moving piston will fall rapidly once the diaphragm ruptures and the phenomenon results in a rapid decay of the incident shock wave. The tuned operation mode can be applied to improve the incident shock decay and to extend the test duration of the free-piston-driven shock tunnels, but its test time is still measured in a few milliseconds. For example, the test duration of the HIEST is between 2 and 5 ms. The last is the moving part of the free-piston driver. Heavy pistons moving at high speeds lead to technical difficulties in operation of piston launching and stopping. The heavier the piston, the more difficult the shock-tunnel operation.

5.2.2 On Detonation Drivers

A detonation wave is a supersonic combustion wave across which the thermodynamic states increase sharply. It can also be considered as a reacting shock wave where reactants transform into products at its shock propagating speed. For a freely propagating detonation, the expansion waves behind the detonation front will reduce the flow pressure and particle velocity to match the real boundary condition. The detonation with a sonic condition behind it can be modeled as the Chapman–Jouguet (C-J) detonation if it is assumed to be steady, one-dimensional, and with an instantaneous heat release. This is the theoretical base when we consider the detonation driver as a power supply for shock tunnels (Lee 2008, Jiang and Teng 2022).

In order to explain the detonation driver and its working principle, we take a circular steel tube with one end closed as a detonation tube and fill it with a detonable gas mixture. A detonation wave is initiated at its closed end and propagates toward the other end. According to the Taylor rarefaction wave theory, the pressure and velocity distributions along the detonation tube are shown schematically in Figure 5.2. Due to the stationary boundary condition at its closed end, an expansion wave

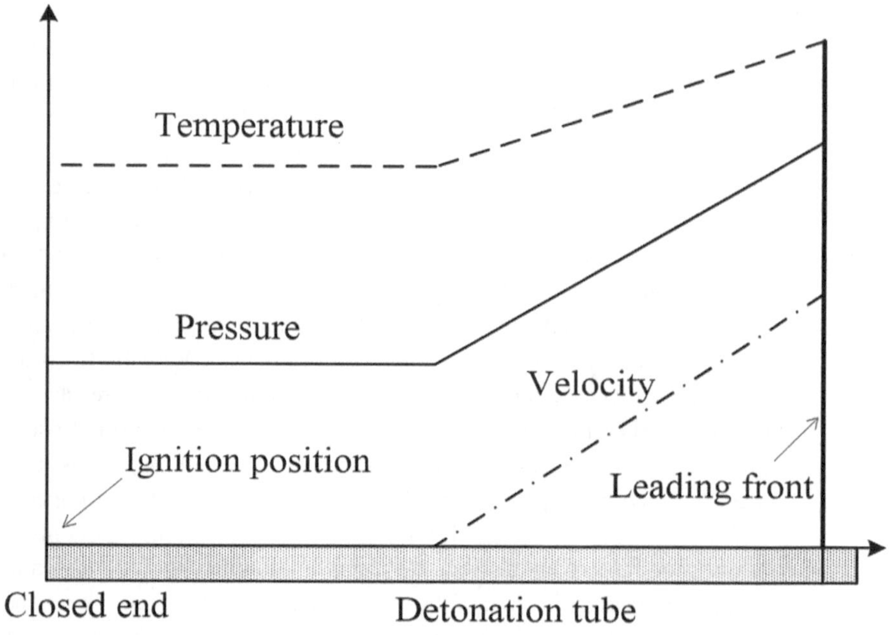

Figure 5.2 Schematic diagram of the distribution of detonation wave properties along the length of a detonation tube and calculated with the C-J theory.

system is developed behind the detonation front through which the gas flow velocity gradually decreases to a stationary state. There exists a column length of the detonated gas at rest, which is about half of the propagation distance of the detonation front, and the products in the column have very high temperature and pressure, as shown in Figure 5.2.

Hertzberg and Smith found the detonation occasionally when they were trying to explore a method to generate a strong shock wave with combustion, and a constant volume combustion was supposed to take place in their experiment (Hertzberg & Smith 1954). The finding was demonstrated later in experiments. There are possibly two working modes of detonation drivers, depending on where the ignition takes place. It is called the backward-running detonation driver if the igniter is installed between the detonation tube and the driven section. Otherwise, the forward-running detonation driver is established when the igniter is placed at the closed end of the detonation tube, as shown in Figure 5.2. These two drivers are named simply as the forward and backward detonation drivers, respectively.

Bird made the first try to numerically predict the performance of the two modes of detonation drivers (Bird 1957). It is found that the forward detonation driver may not be suitable for shock tunnels since the incident shock wave in the driven section is damped continuously. The incident shock wave becomes stable when the backward detonation driving mode is used, but an extremely high pressure will be generated when the detonation front reflects at the closed end of the detonation tube, which will pose a problem for the safe operation of the detonation-driven shock tunnel. These are two primary physical issues when the detonation drivers are developed.

5.2.3 Backward Detonation Drivers

The backward detonation driver is good at producing a high-quality incident shock wave since a column of the compressed detonation products with uniform pressure is generated regardless of the fact that it is less capable of generating a powerful incident shock than the forward detonation driver. However, it is fully capable of generating hypersonic flows with Mach numbers 5–10 because the required incident shock is not very strong. Furthermore, since the total temperature of the test gas is about 1,500–2,000 K for flows at the Mach numbers of 5 and 6, the detonation driver with a hydrogen–oxygen mixture is usually too powerful to apply directly, and the detonable gas mixture has to be diluted with inert gases, such as nitrogen, to reduce the detonation pressure and temperature, which are the measures of the drivers' capability. The driving pressure and temperature can be calculated with the Taylor rarefaction wave theory.

In order to avoid the extremely high pressure from the detonation-front reflection, Yu proposed a new concept for the backward detonation driver by appending a damping section to the open end of a detonation tube for accommodating the huge energy carried by the detonation front (Yu 1989). This damping section is separated from the detonation tube with a diaphragm and is evacuated to quite a low pressure before shock-tunnel operation. The volume of the damping section needs to be large enough to receive all the detonated products that follow the detonation front at an acceptable pressure-increasing level so that shock-tunnel operation safety could be ensured. Then, a driven section is appended to the closed end of the detonation tube, as shown in Figure 5.2, to generate a stable incident shock wave after the main diaphragm ruptures. This concept was experimentally verified, and the result demonstrated that the backward-running detonation is applicable to high-enthalpy shock tunnels with superior performance than the conventional driver technique. The backward detonation-driven shock tunnel with a damping section was applied successfully first in the TH2-D high-enthalpy shock tunnel in the Aachen University of Technology (Yu et al. 1992). Shortly after, the JF-10 shock tunnel was constructed based on the concept developed in the Institute of Mechanics, CAS. The backward detonation driver has become a promising driver technique being comparable with both the free-piston driver and the light-gas-heated driver since then.

According to detonation physics, different detonable gas mixtures have different detonation temperature and pressure. Numerous detonable gas mixtures were tested to examine the performance of a backward detonation driver by using the JF-10 shock tunnel. The experimental data of two test cases are presented in Figure 5.3, showing pressure variations during shock-tunnel operation (Olivier et al. 2002). Figure 5.3(a) shows the stagnation pressure P_5 at the end of the driven section in the front of the diaphragm when a hydrogen–oxygen gas mixture at an initial pressure of 3.5 MPa is used in the driver in the first test case, and the thus-generated incident shock has a Mach number of 2.5 at the condition P_1 = 1.85 kPa. Figure 5.3(b)

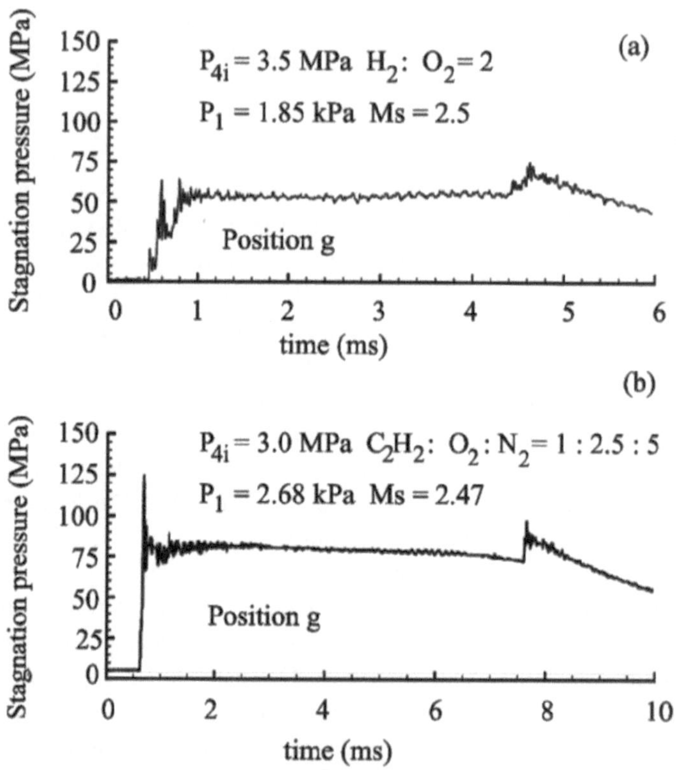

Figure 5.3 P_5 variations obtained with a backward detonation driver at the same measurement position in two test cases: (a) the driver gas: hydrogen–oxygen gas mixture; (b) the driver gas: acetylene–oxygen gas mixture (Anderson 1989).

presents the experimental data obtained with a acetylene–oxygen gas mixture in the driver at an initial pressure of 3.0 MPa, and the obtained incident shock Mach number is about 2.47. It is observable that the pressure history shown in Figure 5.3(b) reaches about 75 MPa, even though the P_1 is higher than that in the first test case. The P_5 value in the second case is 50% higher than that in Figure 5.3(a). Therefore, the backward detonation driver with the acetylene–oxygen gas mixture is suitable for generating hypersonic flows at the low Mach number region, that is, the Mach numbers from 5 to 7, where the high total pressure is required, but the stagnation temperature is not very high.

5.2.4 Forward Detonation Drivers

For a backward detonation driver, the driving pressure, P_2 is about eight times higher than P_{4i} according to the Taylor rarefaction wave theory. By referring to Figure 5.2, it can be seen that the detonation front has much higher energy than the stationary gases behind it. This is because not only the detonation pressure is much higher than

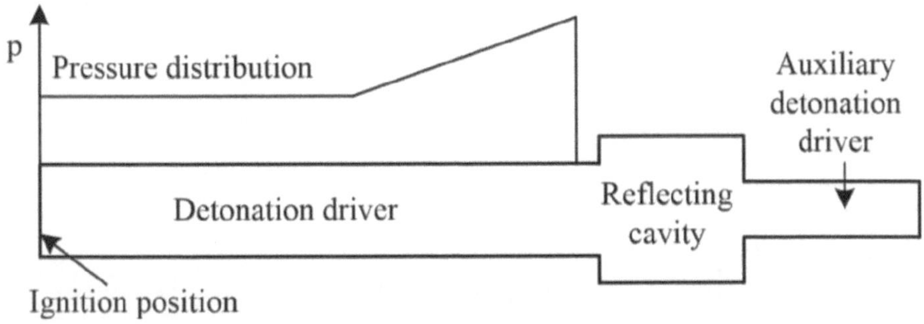

Figure 5.4 Schematic diagram of the FDC driver with pressure distribution.

the pressure in the stationary part, but also the kinetic energy is large. Attaching a driven section to the open end of a detonation tube makes a forward detonation driver.

Theoretically speaking, the driving pressure of the forward detonation driver is about five times higher than the backward detonation driver at the same initial conditions as measured by the C-J detonation parameters. However, the Taylor rarefaction wave leads to the gradual attenuation of the incident shock wave and results in the fact that the forward detonation driver is not acceptable for practical applications to shock tunnels.

To solve the above-mentioned attenuation problem, Jiang et al. had proposed the FDC driver, as schematically shown in Figure 5.4, based on the shock-reflection concept (Jiang et al. 2002). The FDC driver consists of three parts, that is, the detonation section, the reflecting cavity, and the auxiliary detonation section. The diameter of the auxiliary detonation section is bigger than that of the detonation section. When the detonation wave passes through the reflecting cavity, as shown in Figure 5.4, its middle part propagates directly into the auxiliary detonation section, while the part that has entered the cavity is reflected back from the end wall of the cavity. As a result, a reflected shock wave develops and travels upstream. As the detonated gas flow passes through the reflected shock wave, its pressure and temperature will be elevated because of the flow converging effect that can support the reflected shock wave. Considering the shock-wave system developed in the FDC driver, it is understood that the stronger upstream-traveling shock wave and the weaker downstream-traveling shock wave from shock interaction are favorable for enhancing the FDC driver performance. This is also the principle of configuration optimization of the FDC driver.

The FDC driver is successfully applied to the JF-10 shock tunnel in the Institute of Mechanics, CAS, and the shock tunnel can be used to generate high-enthalpy flows with a total temperature up to 7,000 K. One of the P_5 curves from the JF-10 shock-tunnel performance test is presented in Figure 5.5. The uniformity of the pressure distribution shows that the FDC driver does eliminate the flow attenuation caused by the Taylor rarefaction wave successfully, and fluctuations around the average pressure

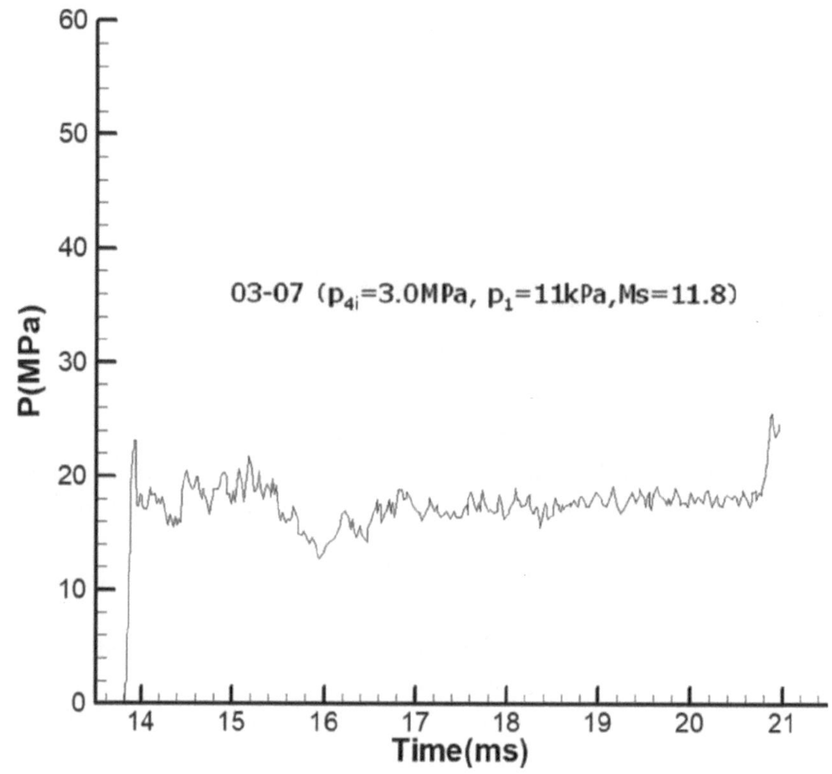

Figure 5.5 P_5 of the JF-10 high-enthalpy shock tunnel with the FDC driver (Jiang et al. 2002).

could damp rapidly in the driven section. The 6-ms test time demonstrated the excellent performance of the JF-10 shock tunnel that is only about 25 m long (Zhao et al. 2005).

5.2.5 Special Features of Detonation Drivers

The strength of the incident shock wave in the driven sections depends on two key parameters of the driver gases. One is the specific heat ratio and the other is the driver-gas temperature. The higher the sound speed of the driver gas, the more powerful the shock-tunnel driver. This is the reason why light gases are usually selected and then heated as driver gases. Therefore, the incident shock Mach number is determined from the driver-gas composition and its thermodynamic state in region 4 once the driven gases are given.

Figure 5.6 shows the relation between the pressure ratio P_4/P_1 and the incident shock Mach number for different driver gases. If air at room temperature is used as the driver gas at initial pressure P_1, the pressure ratio P_4/P_1 will increase dramatically when the required incident shock wave is higher than 2. This means that increasing the pressure ratio is an effective method to increase the incident shock Mach number, but the Mach number above 2.7 is not practical to obtain in engineering applications

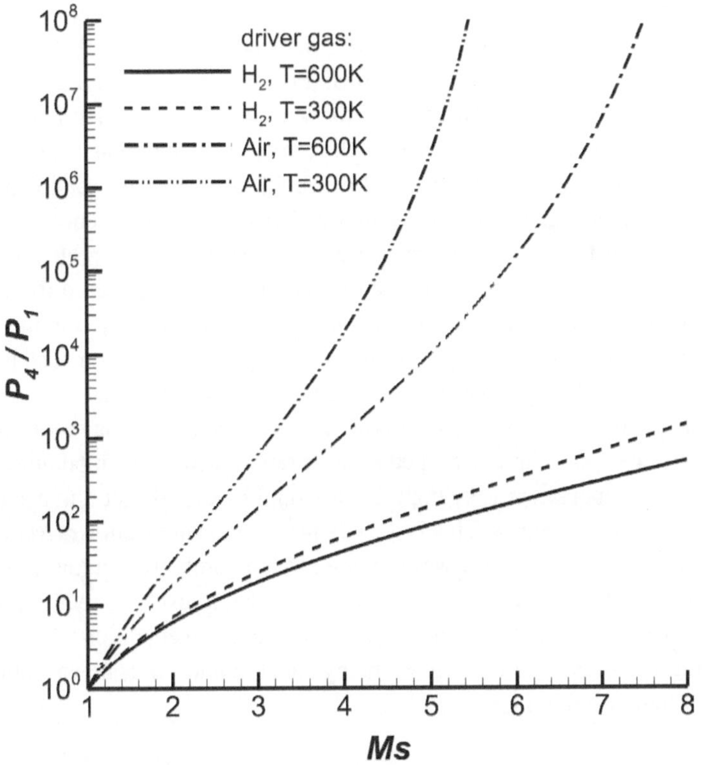

Figure 5.6 Driving pressure ratio versus incident shock Mach number for different driver gases at two air-heated temperature levels.

because the required pressure ratio will approach infinity rapidly, as indicated in Figure 5.6. Heating air to 600 K can help increase the sound speed of the driver gas, resulting in an increase of the incident Mach number, but the practical Mach number of the heated air is not higher than 3.5.

The light gas has a higher sound speed than air, and its sound speed can also be increased if it is heated. Figure 5.6 shows that the incident shock Mach number can easily be obtained to more than 5 by using hydrogen as a driver gas. Heating hydrogen to 600 K could help obtain the incident shock Mach number as high as 6. However, heating and discharging a large amount of hydrogen bring a serious problem for safely operating shock tunnels. The heated helium is widely accepted to replace hydrogen as the driver gas to solve the potential risks arising from heating hydrogen. The LENS facilities are shock tunnels that have been operated successfully with the light-gas-heated driver with the help of powerful compressor techniques (Holden 1994, Holden et al. 1995, 2004). This conclusion is drawn under the condition that the initial pressure in driven sections cannot be made as low as possible because a certain amount of test gases is needed to generate test flows with a certain test duration. The more the test gas is needed, the larger the amount of the driver gas is needed.

The helium-heated driver was considered once as one of the JF-12 shock-tunnel operation modes, but simple calculations showed that the shock tunnel operation would be quite expensive and not practical. For example, to generate a 2.5-m diameter nozzle flow of 5-MPa total pressure for Mach 9 at the 40-km attitude, the helium mass required for each shot is about 187 kg mass for the light-gas-heated driver. This imposes a limit for the helium-heated driver to be an affordable technology for large-scale high-enthalpy shock tunnels. Actually, 7 kg of hydrogen is enough for a similar run if a backward detonation driver is used with the hydrogen–oxygen mixture as its driver gas. It is easy to understand that the chemical energy released from detonation drivers is more powerful than the mechanical energy provided by compressors. When developing large-scale and long-test-duration hypersonic shock tunnels, it becomes a huge difficulty, but is not a problem at all for small shock tunnels.

For detonation drivers, for example, the product of hydrogen–oxygen detonations is water vapor with a low specific heat ratio but with a temperature as high as 3,000 K. The temperature is so high that it would be very difficult to achieve by current gas-heating techniques. This is the key fact that the detonation driver can be used to generate the incident shock wave of high Mach numbers. In addition, the product composition and the detonation temperature can be adjusted as required by diluting with inert gases to meet the interface-matching operation condition. This advantage is of significant importance for high-enthalpy shock tunnels to achieve stable reservoir states for long test durations.

5.3 Development of Detonation-Driven Shock Tunnels

In order to introduce the techniques and performances of detonation-driven shock tunnels, three shock tunnels are described and discussed in the following subsections. The first one is the JF-10 shock tunnel that is operated with a forward detonation driver and capable of generating high-enthalpy flows at Mach numbers from 10 to 20, The second one is the JF-12 hypersonic flight-duplicated shock tunnel, and it is built based on the backward detonation driver. For the time being, the JF-12 shock tunnel is the largest shock tunnel with the best performance in the world. The last one is the JF-16 expansion tube that was designed based on the FDC driver with an acceleration tube. With the JF-16 expansion tube, the orbital-speed flow at velocity from 7 to 10 km/s can be generated. These shock tunnels are typical ones with the same working principles and have been operated successfully for years.

5.3.1 The JF-10 Shock Tunnel: The Detonation-Driven High-Enthalpy Shock Tunnel with the FDC Driver

The JF-10 shock tunnel was developed by Yu for exploring reentry physics and could be operated with both the forward and backward detonation drivers (Yu 1998). In order to achieve high-quality test flows at high-enthalpy levels, the JF-10 shock tunnel

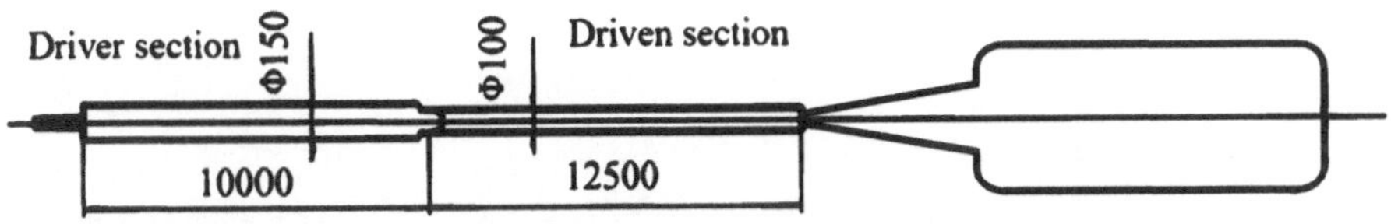

(a) The original JF-10 shock tunnel driven by a converging forward detonation driver

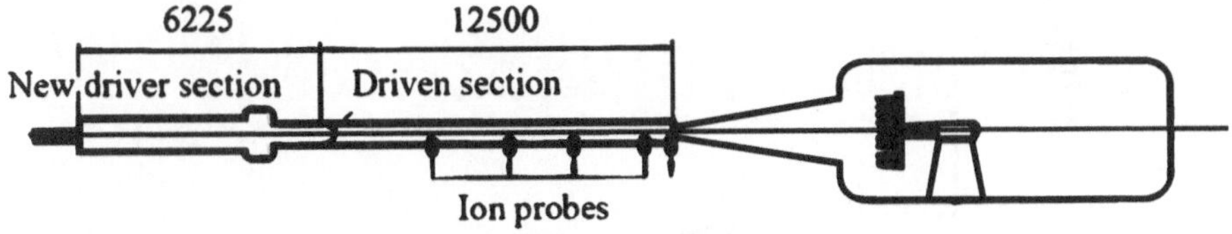

(b) The improved JF-10 shock tunnel driven by the FDC driver

Figure 5.7 Schematic of the original and the improved JF-10 shock tunnels (length unit: mm).

was improved by replacing its forward detonation driver with the FDC driver (Jiang et al. 2002, Zhao et al. 2005) in 2003. Its performance tests were completed during the following two years, and some calibration results are presented in this chapter to demonstrate the capability of the FDC driver, including P_5 pressure variations, the experimental data of nozzle flow uniformity, and the incident shock decay along the driven section. The experimental data show that the improvement on the JF-10 shock tunnel is successful, and the research work on high-enthalpy flow physics will benefit from the thus obtained high-quality test flows.

5.3.1.1 Layout of the JF-10 Shock Tunnels

The original and the improved JF-10 shock tunnels are schematically shown in Figure 5.7. The improved JF-10 shock tunnel, as shown in Figure 5.7(b), consists of three main parts. The first part is the FDC driver being about 6.225 m in length and 150 mm in diameter, and its cavity is 300 mm in diameter and 2 m in length. The second part is a driven section that is 12.5 m in length and 100 mm in diameter. The third part is a conical nozzle with a 500 mm diameter exit. The original JF-10 shock tunnel is shown in Figure 5.7(a), and its driver section is 150 mm in diameter and 10 m in length. It is about 4 m longer than the FDC driver. Other parts are the same as the improved JF-10 shock tunnel. Using a large-diameter driver to power a small-diameter driven section is one way to approach forward detonation applications. The FDC driver could behave better from its operation principle than the converging driver. The research work on the FDC driver was reported in detail by Jiang et al., and the improvement seems very promising (Jiang et al. 2002, Zhao et al. 2005). As a high-enthalpy shock tunnel operated on a forward-running detonation mode, its test duration is as long as 6 ms. Considering the improved JF-10 shock tunnel is less than 20 m in length, the key performance is quite remarkable.

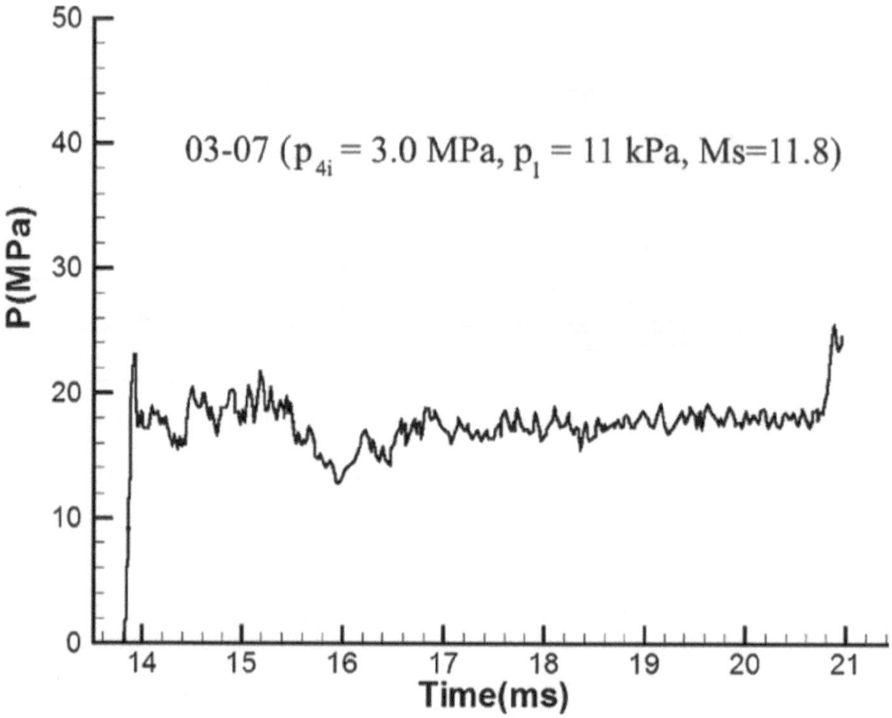

Figure 5.8 Pressure history measured at the end of the improved JF-10 shock tunnel.

5.3.1.2 Pressure Variations at the End of the Driven Section

For shock tunnels, P_5 variation at the end of the driven section is a key parameter that indicates the test flow quality and stands for one of the most important performance parameters of shock tunnels. To obtain P_5, calibration experiments were carried out in the condition of $P_{4i} = 3.0$ MPa and $P_1 = 11$ kPa, where P_{4i} and P_1 are the initial pressures of the driver and driven sections, respectively. The incident shock Mach number is found to be 11.8, and the P_5 pressure history is plotted in Figure 5.8.

Figure 5.8 shows the pressure history measured at the end of the driven section. This pressure profile acts as a measure of the reservoir pressure variation during nozzle working duration. The more stable the reservoir pressure is, the more uniform the nozzle flow will be. From this figure, it is observable that the pressure level maintains approximately constant for more than 6 ms. By comparing it with the 2-ms test time achieved with the free-piston-driven shock tunnel (HIEST), it is obvious that this pressure variation is quite remarkable since the FDC driver extends effectively the test duration and demonstrates well the performance of the improved JF-10 shock tunnel.

Figure 5.9 shows the pitot pressure variation measured in the nozzle flow during the same experiment as mentioned above. The pressure profile looks quite similar to the reservoir pressure variation as shown in Figure 5.8. The good agreement between the two pressure profiles indicates not only experimental data verification

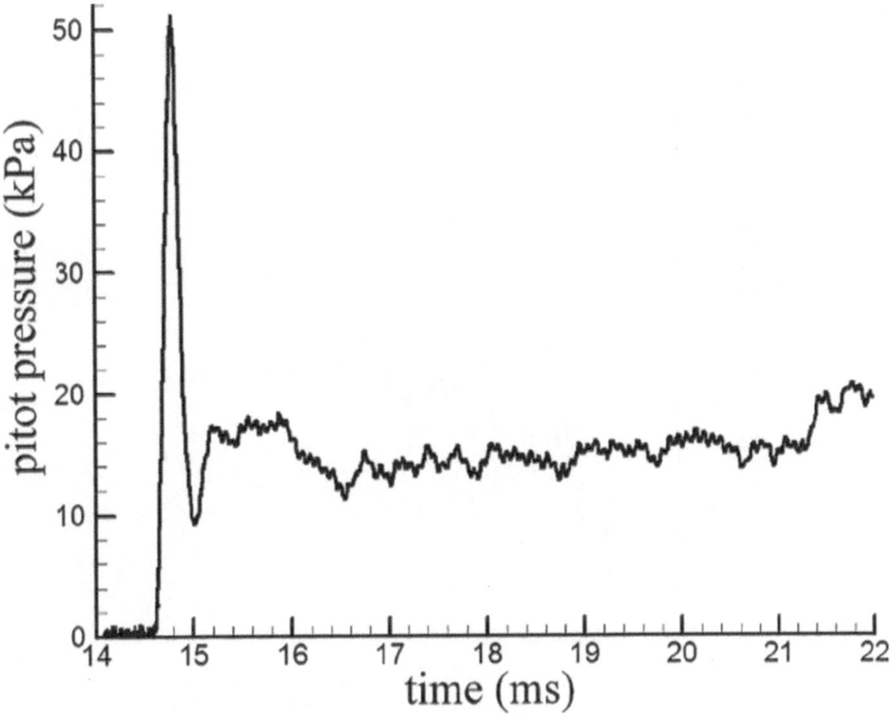

Figure 5.9 Pitot pressure variation measured in the nozzle flow.

but also the effective test time possibly available from the improved JF-10 shock tunnel. This time duration is not only long enough for both flow pressure and heat transfer measurements but also provides a sufficient test period for aerodynamic force and moment tests with existing measurement instrumentations. As is well known, stress force balances are very difficult to apply for shock-tunnel force measurements because of the short test duration. The results of aerodynamic force measurement will be more reliable if shock tunnels could provide with long test duration. Therefore, a 6-ms long test duration is very useful for conducting aerodynamic force experiments.

Similar performance tests were also conducted with the original JF-10 shock tunnel, and the P_5 variation measured at the end of the driven section is presented in Figure 5.10. From the figure, it is observable that the test duration is about 3 ms before the pressure level decreases significantly. Considering that the 6-ms test duration was achieved in the improved JF-10 shock tunnel equipped with the FDC driver, one can appreciate the progress achieved in the detonation-driven high-enthalpy shock tunnel by using the FDC driver technique.

5.3.1.3 Nozzle Flow Uniformity

Another parameter to measure test flow quality is the nozzle flow uniformity. To examine the nozzle flow distribution, five measurement positions in the JF-10 shock-tunnel

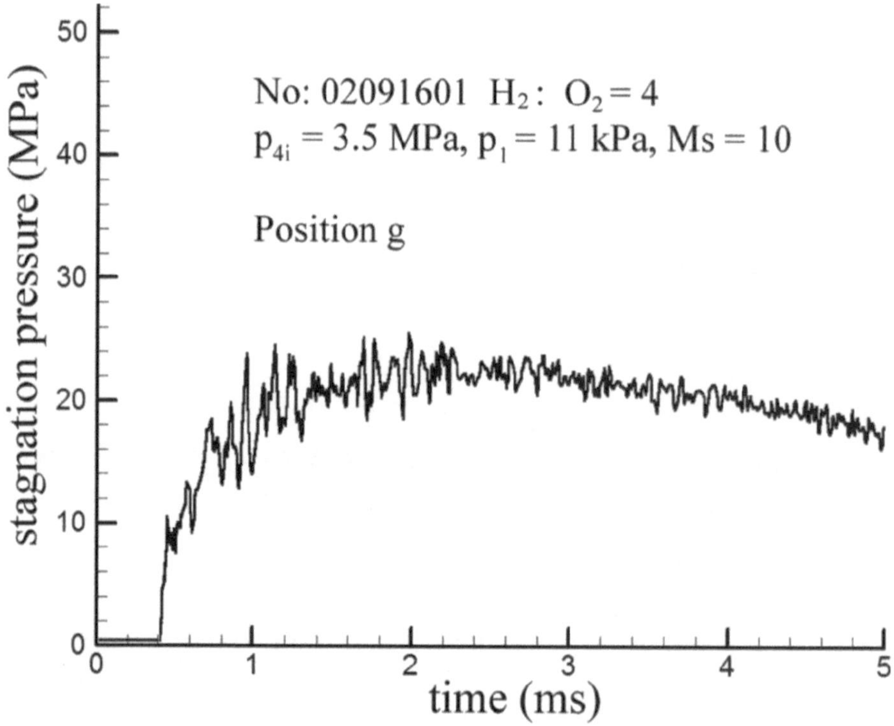

Figure 5.10 Pressure variation at the end of the driven section of the original JF-10 shock tunnel.

nozzle flow are selected along the nozzle centerline, and the starting position is set at the nozzle exit. These positions are (a) $L = -50$ mm; (b) $L = 100$ mm; (c) $L = 265$ mm; (d) $L = 400$ mm; and (e) $L = 600$ mm, respectively. The experimental setup of the pitot pressure measurement is shown in Figure 5.11. Experiments were carried out at each measurement position in both the vertical and the horizontal directions. The measurement data were plotted together to demonstrate the nozzle flow uniformity.

Figure 5.12 shows the pitot pressure distributions in nozzle flows at the position $L =$ 100 mm, and Figure 5.13 does that at the position $L = 400$ mm. It is observable that the hypersonic flow within the nozzle central area appears quite uniform if evaluated from a viewpoint of the pitot pressure distribution. This uniform flow area is found to be approximately 700 mm in length and 400 mm in diameter. Although the pressure uniformity does not fully confirm the hypersonic flow uniformity, these results are quite promising and very encouraging for showing improvement on the nozzle flow quality.

5.3.1.4 Incident Shock-Wave Decay

The incident shock decay in the driven section is also an important parameter for evaluating the performance of hypersonic shock tunnels since the test flow quality is affected considerably. It is associated not only with driver capability but also with

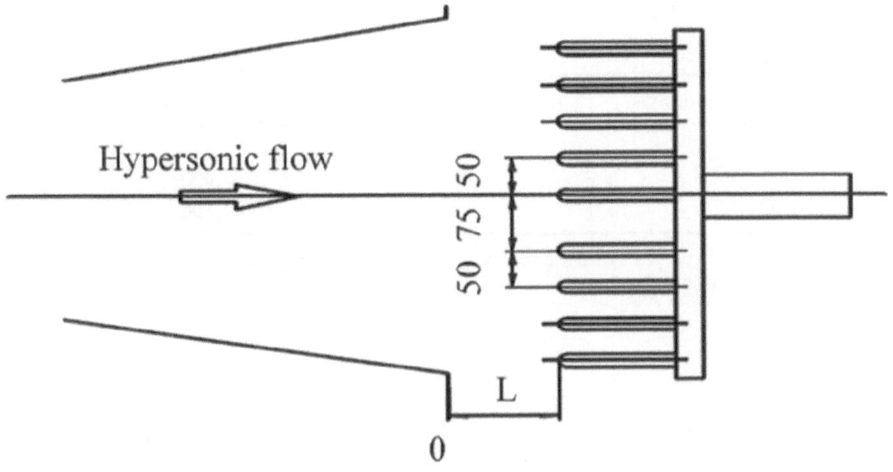

Figure 5.11 Experimental setup of pitot pressure measurements in the nozzle flow of the improved JF-10 shock tunnel.

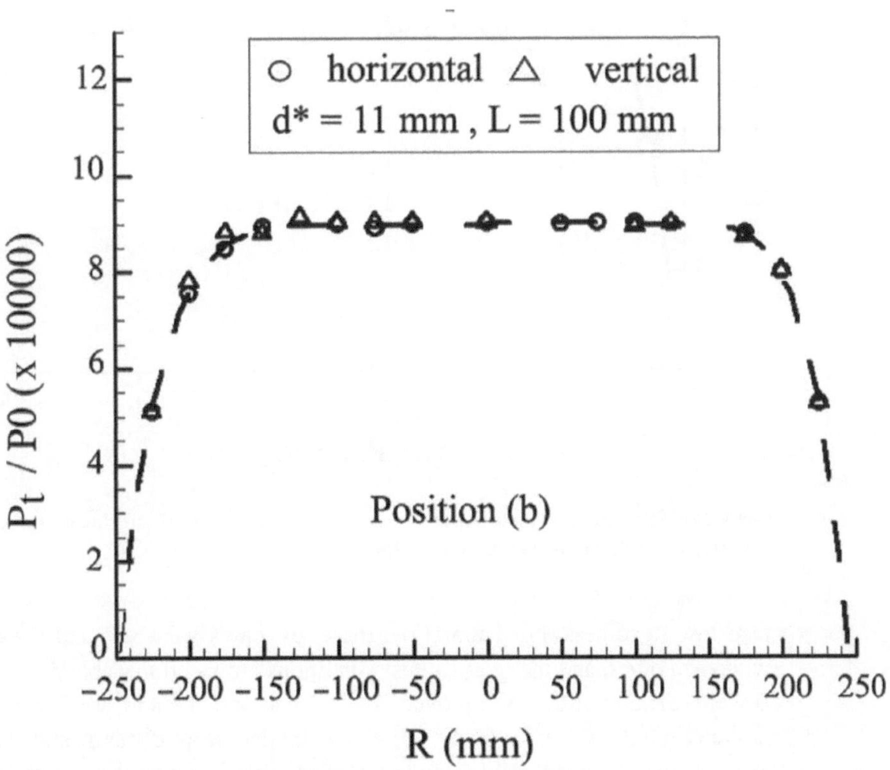

Figure 5.12 Pitot pressure distributions in the nozzle flow of the JF-10 shock tunnel in two directions, $L = 100$ mm and the nozzle throat diameter $d^* = 11$ mm.

Table 5.1 Test conditions for investigating the incident shock decay in the driven section

Test conditions	P_{4i} (MPa)	P_1 (kPa)	P_{50} (MPa)	T_{50} (K)
A	1.5	4.5	8.0	7,480
B	3.0	11	19.4	7,920
C	1.5	4.7	8.3	7,200

Note: P_{50} and T_{50} are the reservoir pressure and temperature, respectively.

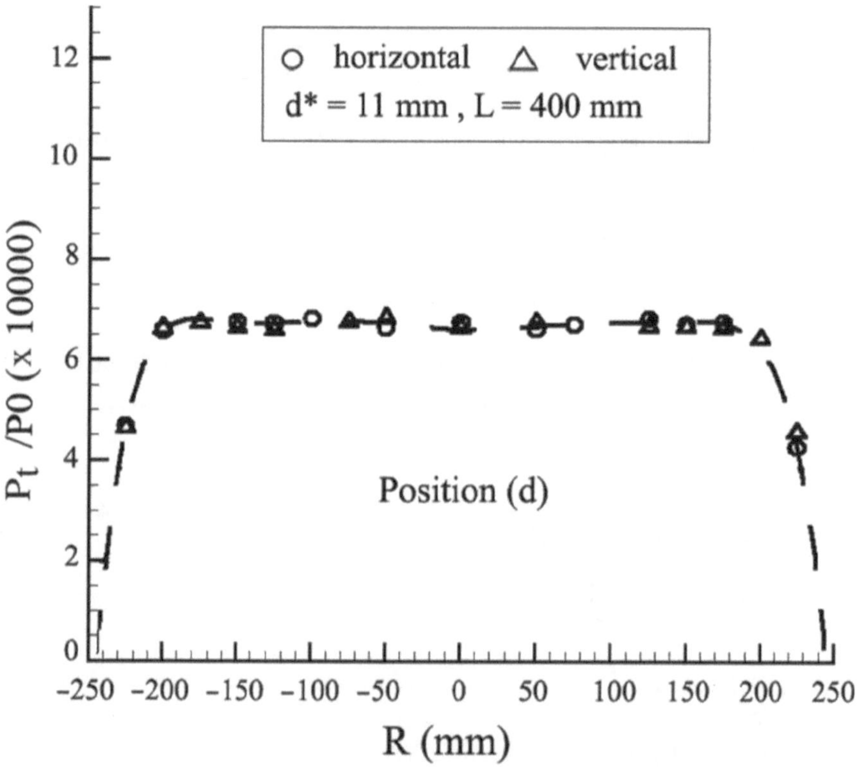

Figure 5.13 Pitot pressure distributions in the nozzle flow of the JF-10 shock tunnel in two directions, $L = 400$ mm and the nozzle throat diameter $d^* = 11$ mm.

test flow steadiness. For calibration, three test cases were selected for comparison with each other, and the relevant test conditions are listed in Table 5.1. Cases A and B were carried out in the improved JF-10 shock tunnel, and Case C was carried out in the original one, respectively. Experimental data from Cases A and B are used to examine the incident shock decay at different initial conditions when the FDC driver is applied, and Case C is used for the comparison between two detonation drivers. The total flow enthalpy is kept to be almost the same in the three cases.

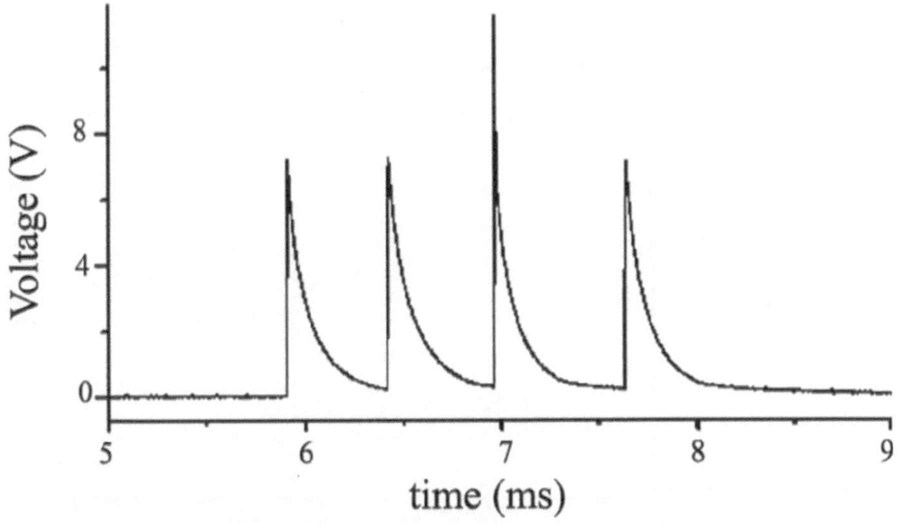

Figure 5.14 Ion probe signals for incident shock speed measurement.

The incident shock speed is measured with several ion probes distributed along the driven section with an equal interval. Results of the qualified tests for these ion probes are given in Figure 5.14. The signal jump occurs sharply, which indicates the arrival of shock waves. Therefore, the incident shock speed could be calculated from experimental data with high accuracy.

Figure 5.15 shows the incident shock decay rate along the driven section, and the decay rate is determined by calculating the Mach number drop per tube diameter, that is, the diameter of the driven section. By examining Cases A and C, it can be seen that stronger incident shock waves can be generated with the FDC driver than the original driver, and the incident Mach number in its early stage could be 10% higher than that in the original shock tunnel in similar operating conditions. The incident shock decay in the improved JF-10 shock tunnel is about 30% less; actually, the decay rate is 3.04% in Case A but 3.94% in Case C. The experimental data of Cases A and B are almost the same with each other, and this indicates that the initial pressure ratio of the driver and driven section has no significant effect on the incident shock decay if the total flow enthalpy is the same. It is more encouraging that the shock-wave decay is weaker in Case A than the one in Case C, and the result demonstrate the good performance of the FDC driver.

5.3.1.5 Concluding Remarks

From the above discussions based on experimental data from the JF-10 shock tunnel, the research progress on the forward detonation driver for high-enthalpy shock tunnels is summarized as follows:

(a) The improvement on the JF-10 shock tunnel with the FDC driver was demonstrated to be very successful. The FDC driver can generate a much stronger

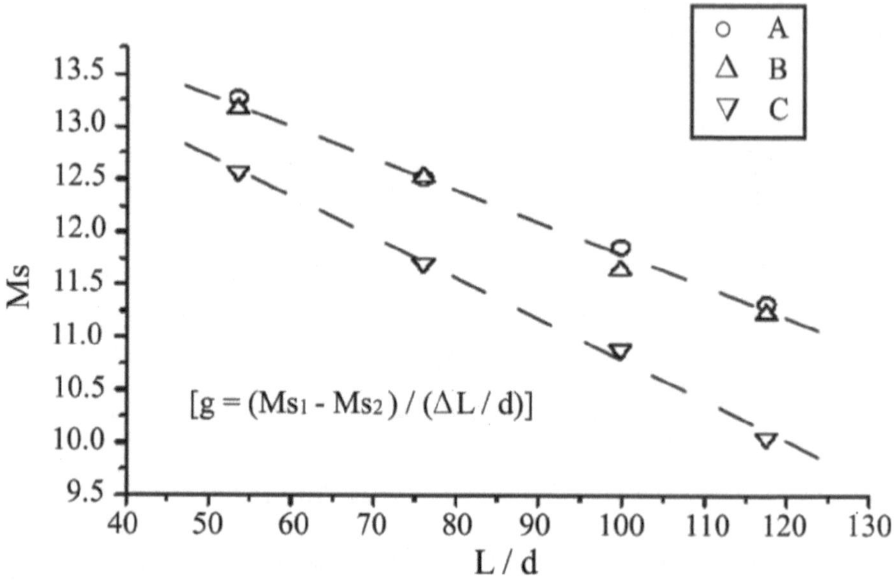

Figure 5.15 Incident shock-wave decay measured in two shock tunnels for three test cases at a similar enthalpy condition.

incident shock with a lower decay rate at similar initial conditions than the original JF-10 shock tunnel with a converging forward detonation driver based on the concept that the large diameter detonation tube is directly connected to a small-diameter driven section with a diameter ratio of 1.5.

(b) The test duration of the improved JF-10 shock tunnel was extended to 6 ms, which is two times longer than the original JF-10 shock tunnel. This extension is remarkable if considering the FDC driver is 40% shorter than the original JF-10 forward detonation driver. The 6-ms test duration is of significant importance because it will make aerodynamic force measurements less difficult than before. So far, in the high-enthalpy flow research area, the aerodynamic force measurement accuracy is in doubt because of the short test time, and this has been a challenge to the present aerodynamic balances for decades.

(c) Calibration of nozzle flow uniformity showed that the uniform core flow of the improved JF-10 shock tunnel is about 700 mm in length and 400 mm in diameter. It is a quite a useful characteristic so that large test models could be used in the wind-tunnel test, and more heat flux sensors can be installed in the models to gain high spatial accuracy. For the time being, the only flow visualization that can be carried out for such high-enthalpy flow tests is achieved by using ballistic guns.

5.3.2 The JF-12 Shock Tunnel: Hypersonic Flight-Duplicated Shock Tunnel with a Backward Detonation Driver

Theoretically speaking, the output power of the backward detonation driver is only about one-fifth of the FDC driver, but the driver gas generated from the backward detonation driver is quite uniform in pressure and temperature. Moreover, the column of the reservoir gas is as long as the half-distance that the detonation front has propagated. This special feature is very useful to produce a long-test-duration shock tunnel if the required flow Mach numbers is not very high, for example, about Mach numbers 5–10. This flight region is of special importance for air-breathing hypersonic vehicles, and it is still very difficult to carry out Mach 8 propulsion tests. The following subsections will present the introduction to the JF-12 shock tunnel, including why it is made, how it works, and what performance it has.

5.3.2.1 Requirements from Hypersonic Propulsion Testing

To maintain the relatively high dynamic pressure required for engine performance, air-breathing hypersonic vehicles usually operate at high Mach numbers at relatively low altitudes. This requirement leads to the need for high total pressure test flows. Moreover, the scaling law is not valid due to chemical reactions in the engine, and a large-diameter nozzle flow is also necessary to accommodate the entire vehicle that is integrated with the engine. Advanced hypersonic test facilities have been developed for more than 60 years, but there is still a lack of facilities for generating high-enthalpy flows with Mach numbers higher than 7 for hypersonic propulsion tests due to technological barriers in wind-tunnel techniques, especially the facility damage due to severe heat transfer problems during the long test duration, and the large energy needed to supply to shock tunnels.

In order to meet the ground test requirement of air-breathing hypersonic vehicles, several types of facilities have been developed so far, including combustion-heated wind tunnels, arc-heated hypersonic wind tunnels, conventional air-heated wind tunnels, and high-enthalpy shock tunnels. To clarify real-gas effects on hypersonic vehicle performance and test-gas contamination effects on supersonic combustion, the high-enthalpy shock tunnel appears to be a unique test facility to provide hypersonic propulsion tests with pure airflow at hypervelocity, the required total pressure and temperature. However, the test duration of high-enthalpy shock tunnels is usually short and often measured in a few milliseconds for most such tunnels. The short test-duration imposes severe technical barriers on supersonic combustion diagnosis and aerodynamic force and moment measurements. Moreover, it becomes even much more difficult for hydrocarbon fuel combustion since the fuel injection, evaporation, mixing, and ignition take time. Therefore, the development of the long-test-duration large-scale shock tunnel is of significance for testing air-breathing hypersonic vehicles.

5.3.2.2 Introduction to the JF-12 Shock Tunnel

Aiming at air-breathing engine testing, a research project has been launched in the LHD, Institute of Mechanics, CAS, since 2002 to create key techniques for the long test duration of hypervelocity shock tunnels. Some progresses are summarized in this chapter as a brief introduction to the JF-12 shock tunnel (Jiang & Yu 2014). From the viewpoint of shock-wave dynamics, Li et al. investigated the wave–interface interaction during shock-tube operation, and a new tailored condition was proposed for detonation-driven shock tunnels to achieve long test duration (Li et al. 2008b). The

Figure 5.16 Photo of a hypersonic flight-duplicated high-enthalpy shock tunnel with a backward detonation driver: the JF-12 shock tunnel.

shock/boundary-layer interaction near the end of the driven section was examined numerically in detail, and the motion of the driver gas to the nozzle throat via the separated boundary layer is clarified (Li et al. 2008a). And then, a steel ring of the required diameter is proposed to install around the nozzle throat so that the driver-gas mixing with the test gas could be delayed effectively. Considering the precursor shock reflection in the damping section, a new concept was proposed for wave-reflection damping in the vacuum tank, and an E-shaped aerodynamic configuration of the vacuum system is adopted in developing the JF-12 shock tunnel (Jiang et al. 2012). With these new techniques, a large-scale hypervelocity detonation-driven shock tunnel has been designed based on the backward-running detonation mode and installed in 2012, as shown in Figure 5.16, and its calibration was completed successfully during the following three years.

The detonation-driven hypersonic flight-duplicated high-enthalpy shock tunnel is the name from the National Major Scientific Research Instrument Project. The JF-12 shock tunnel is named after LHD shock tunnels under the serial number. After visiting the LHD Laboratory, Professor Sasoh from Nagoya University suggested another name for the JF-12 shock tunnel – Hypervelocity Detonation-Driven Real-Gas Shock Tunnel. The name could be abbreviated as Hyper-Dragon. The JF-12 shock tunnel is about 268 m in length, and its nozzle is 2.5 m in diameter. The American Institute of Aeronautics and Astronautics (AIAA) reported it as the largest shock tunnel in the world in the annual review of Aerospace America in 2015. Its performance tests demonstrated that the JF-12 shock tunnel is capable of reproducing the pure airflow with Mach numbers from 5 to 9 at altitudes of 25–50 km with 100-ms test duration.

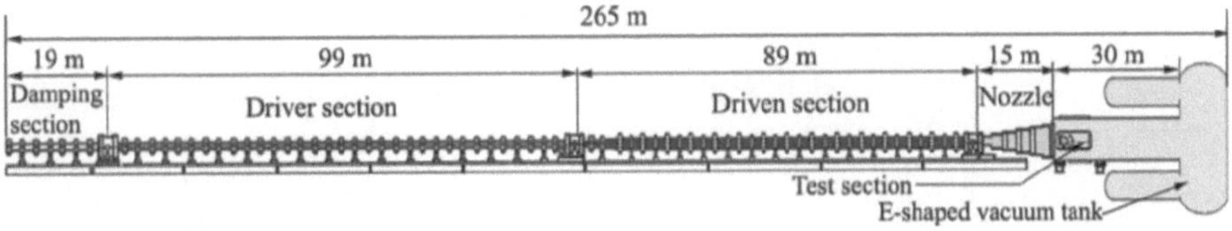

Figure 5.17 Schematic of the JF-12 shock tunnel and its dimensions of the main parts.

The JF-12 shock tunnel will be a useful tool for testing engine/frame-integrated hypersonic vehicles and investigating the fundamental physical issues in hypersonic and high-temperature gas dynamics.

Figure 5.17 shows schematically the entire system of the JF-12 shock tunnel. From its right to left, the first part is the vacuum system for damping wave reflection from the nozzle starting process, and its vacuum tank of volume of 600 m^3 is 43 m in length. The second part is the test section that is 15 m in length and 3.5 m in diameter. The contoured nozzle is 15 m in length and 2.5 m in diameter. Another nozzle is relatively smaller, having an exit of 1.5 m in diameter, is especially designed for Mach numbers from 5 to 7. Next to the nozzle, there is the driven section being 89 m in length and 720 mm in diameter. The detonation driver is 99 m in length and 400 mm in diameter. The JF-12 shock tunnel operates in the backward-running detonation mode, that is, the detonation is ignited at the right end of the driver section and propagates toward the left. The detonation driver and the driven section are connected with a transient part in which the tube diameter is gradually reduced from 720 to 400 mm. Between the detonation driver and the transient part, there is a diaphragm rig that is used to produce the incident shock wave into the driven section after the direct detonation initiation. The damping section is located at the left end of the facility and is 19 m in length and 400 mm in diameter.

For experimental measurements, the JF-12 shock tunnel is now equipped with a 640-channel digital-data-acquisition system that is able to acquire, amplify, digitize, and store the experimental data in real time. Immediately after each test run, the data are transferred to the computer in the central control room for processing. Several six-component force and moment balances are designed based on the 100-ms test duration, and calibration data showed that the balances work pretty well, and the experimental data are reliable.

5.3.2.3 Tailored-Interface-Matching Condition for the JF-12 Shock Tunnel

In order to describe the key issues for developing a long-test-duration shock tunnel, a typical wave diagram for the detonation-driven shock tunnel operated in the backward-running detonation mode is shown in Figure 5.18. Detonation is initiated between the detonation driver and the driven section, and its wave system propagates both forward and backward. The nozzle starts once the incident shock wave is reflected from the end of the driven section after the reservoir state is developed. However, the reservoir state will be interfered with if the reflected shock wave returns from the left of the shock tunnel, and the test duration may be shortened if wave-reflection effects are significantly strong.

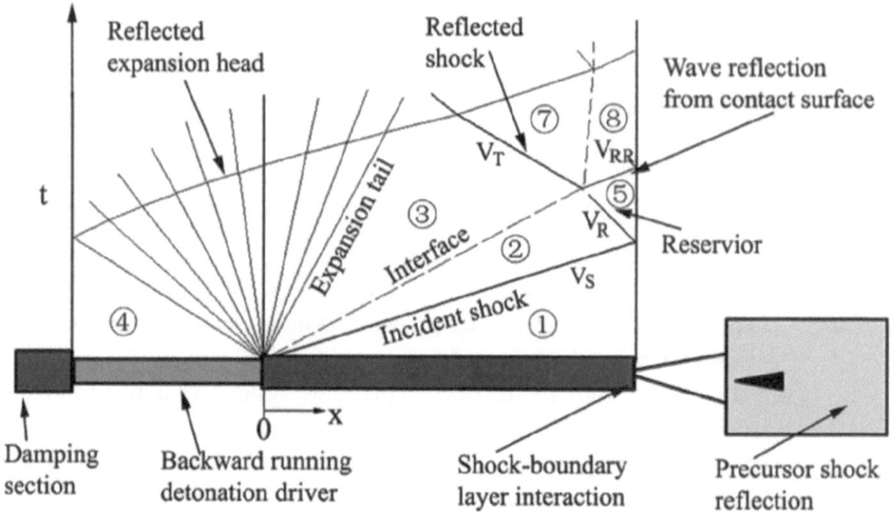

Figure 5.18 Physical key issues for developing a long-test-duration shock tunnel.

When building a long-test-duration and hypervelocity shock tunnel for Mach numbers ranging from 5 to 9, there are several physical issues that affect the test time in different ways. These physical issues are briefly described below, and a detailed discussion will be presented in Chapter 6. The first issue is the powerful detonation driver; the shock-tunnel performance will benefit from the detonation driver with high pressure and temperature. The second issue is the tailored-interface-matching condition because wave reflections take place from the interface separating the driver gas from the test gas. The reflections will ruin the test flow quality if a shock tunnel is operated in an untailored operation condition. The third issue is the precursor wave reflection in the vacuum tank during nozzle operation. The test duration will be terminated once the reflected waves propagate into the test flow region. A large amount of laboratory space may be occupied since the vacuum tank must be long enough to accommodate the wave propagation for the long test duration. The shock/boundary-layer interaction is another key issue for extending the shock-tunnel test time. Immediately after the incident shock reflection, the driver gas moves within the boundary layer near the end of the driven section, which will lead to driver–driven gas mixing and results in test gas contamination. The last issue arises from the damping section. The extremely high pressure from the detonation front reflection must be carefully dealt with because the resulting huge mechanical load will affect shock-tunnel operation safety. These physical issues will be discussed briefly in the following sections, and the corresponding techniques for constructing shock tunnels having a long test duration will also be introduced.

5.3.2.4 Calibration Data of the JF-12 Shock Tunnel

Figure 5.19 presents the reservoir pressure (P_5) variation at the end of the driven section for Mach number 7 during shock-tunnel operation. The incident shock Mach

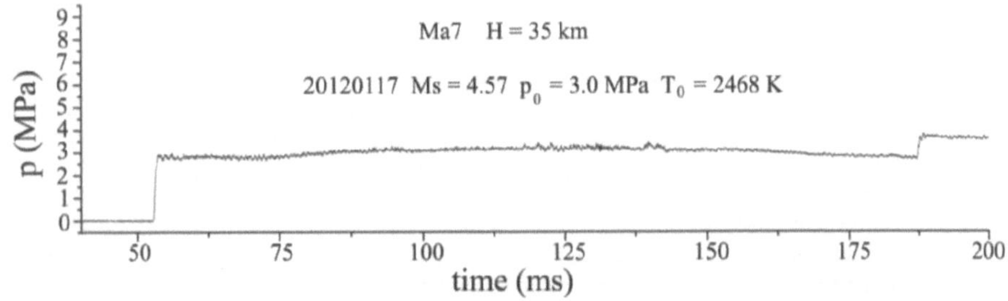

Figure 5.19 Reservoir pressure (P_5) variation at the end of the driven section for Mach 7.

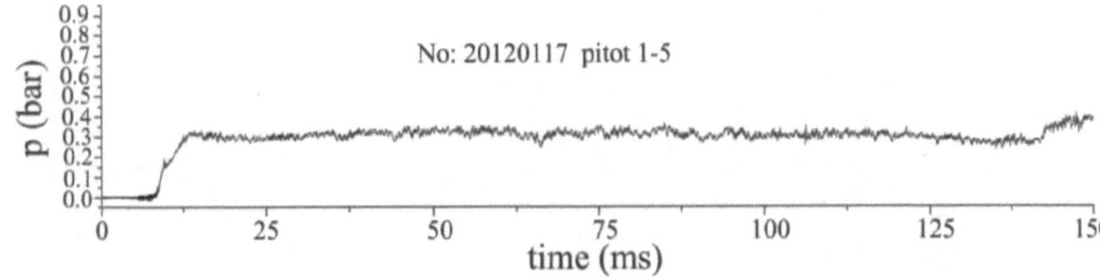

Figure 5.20 Pitot pressure variation at the nozzle exit in the test section for Mach 7.

number is measured to be 4.57, and the total temperature of the test gas is about 2,468 K. It is observable that the uniform pressure is maintained as long as 130 ms. Figure 5.20 shows the pitot pressure variation at the nozzle exit in the test section from the same shot. The pitot pressure profile looks quite similar to the P_5 curve, and the JF-12 test duration is confirmed because the test gas that remains in the driven section is much more than the mass flow into the nozzle during the test period.

To evaluate the test flow quality, the flowfield uniformity is a meaningful parameter. A survey rake is made for the pitot pressure measurement, and it is 2-m long in both vertical and horizontal directions. Forty pressure transducers are distributed along the rake in equal intervals. The transducers along the horizontal branch are numbered from the nozzle center to the nozzle wall. The pressure profile recorded by transducer No. 5 is plotted in Figure 5.20.

The experimental data from the pitot pressure survey are presented in Figure 5.21. It is observable from the figure that the pitot pressure near the boundary layer of the test flow is slightly higher than the ones recorded near the center axis, and the phenomenon is common in high-enthalpy shock tunnels. From the figure, it is concluded that the test flow region at the nozzle exit plane is about 2 m in diameter.

The JF-12 shock tunnel had been considered to operate with the light-gas-heated driver, and two driver techniques were evaluated in its design stage. Some information is provided here for reference. If the flight condition, for instance, for Mach number 8 at an altitude of 40 km, is required to be duplicated in the JF-12 shock tunnel, 7 kg of hydrogen gas must be filled into the detonation driver section to reach the initial pressure of 2.5 MPa for the hydrogen–oxygen gas mixture. The hydrogen

Table 5.2 Driver-gas mass estimated for the detonation and light-gas drivers at the condition of $H = 40$ km and $M_\infty = 8$

		Hydrogen	
	Initial pressure	Weight	Bottles
Detonation driver	2.5 MPa	7 kg	18
Light-gas driver	14.2 MPa	152 kg	380

Hydrogen bottle standard in China: $V = 0.04$ m^3; $p = 12$ MPa; $m = 0.4$ kg.

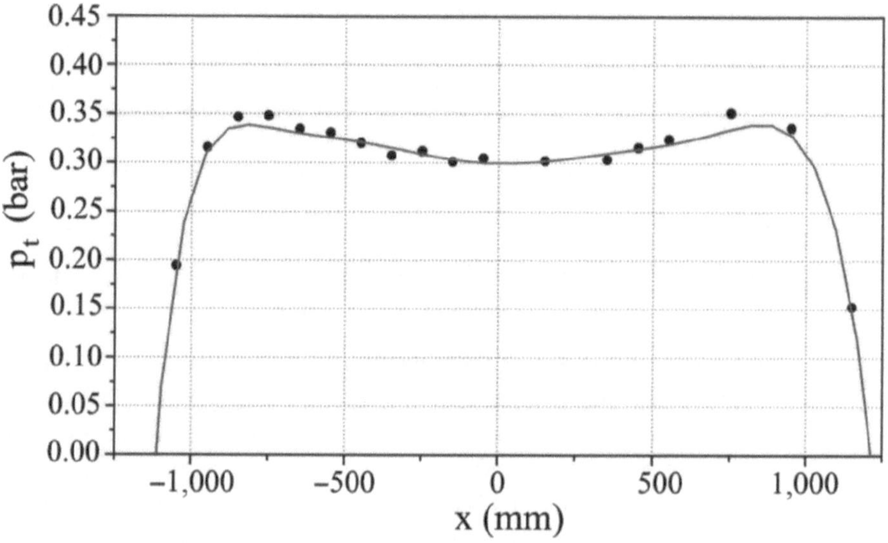

Figure 5.21 Pitot pressure profile at the nozzle exit plane for Mach 7.

consumed for such a shot is about 18 bottles of compressed hydrogen gas. If the light-gas-heated driver is adopted in the JF-12 shock tunnel, the same shot would need about 380 bottles of hydrogen as the driver gas to reach an initial pressure of 14.2 MPa. The consumed driver gases estimated for the two drivers are given in Table 5.2. The light-gas-heated driver is an important technique for shock tunnels; however, its extension to large-scale shock tunnels may cause some problems; not only the operation expenses are high, but also heating, storing, and transporting a huge volume of light gases are very tough problems to contend with.

5.3.2.5 Concluding Remarks

The success of the JF-12 shock-tunnel development was demonstrated with its calibration tests. The remarkable performance had been achieved by using several critical new techniques, such as the backward detonation driver, the acetylene–oxygen mixture, the tailored-interface-matching condition, the shock/boundary-layer interaction control at the end of the driven section, and the E-shaped aerodynamic configuration of the vacuum tank. These shock-tunnel techniques make the JF-12 shock tunnel capable of duplicating the flight conditions for Mach number 5–9 at altitudes of 25–50 km. Its large-scale test flow with about 130-ms test duration at reasonable operation expenses is invaluable for investigating fundamental physics in hypersonic propulsion. Moreover, the hypersonic boundary-layer development, the scaled-test-model effect, and vehicle-engine-integrated models are possible to be tested reliably. Most hypersonic tests are not scalable because of the reacting gas flow. The JF-12 shock tunnel provides with a chance for us to reveal how and why the high-enthalpy flow behaves differently from conventional air dynamics.

5.3.3 The JF-16 Expansion Tube: Detonation-Driven Orbital-Speed Expansion Tube with the FDC Driver

Among the developed hypersonic test facilities, the shock-expansion tube appears to be a unique facility for generating orbital-speed flows for the reentry problem study. In the shock-expansion tube, the unsteady flow expansion is applied to further speed up the test gas in the acceleration tube to achieve a higher flow velocity, lower static flow temperature, and less gas dissociation than RSTs. The concept of the shock-expansion tube was first proposed by Resler and Bloxsom in 1952 to generate high-enthalpy flows (Resler & Bloxsom 1952). Theoretical research was carried out later by Trimpi, but his experimental results showed that steady test flows were not achieved as predicted (Trimpi 1962). Morgan, from the Center for Hypersonics at the University of Queensland, applied the free-piston driver to shock-expansion tubes, and three expansion tubes, referred to as X1, X2, and X3, were set up, and a series of orbital- and superorbital-speed experiments were carried out (Neely & Morgan 1995, Morgan 2001). The JX-1 expansion tube based on the free-piston driver was built later by Sasoh at Tohoku University, Japan (Sasoh et al. 2000). Applying the FDC driver, Jiang and Gao set up the detonation-driven orbital-speed shock-expansion tube, referred to as the JF-16 expansion tube, in 2006. The experimental data show that the velocity of the test gas reaches 8,300 m/s, the test duration is 50–100 μs, and the total flow enthalpy is about 40 MJ/kg (Gao et al. 2008, Jiang et al. 2015, Zhou et al. 2016a, 2016b). In this chapter, the JF-16 expansion tube is taken as an example to explain the third type of high-enthalpy shock tunnels.

5.3.3.1 Descriptions of the JF-16 Expansion Tube

The JF-16 expansion tube consists of five main parts, as shown schematically in Figure 5.22. These parts are the FDC driver referenced as the detonation-driven section, a shock tube, an expansion acceleration section, a test section, and a vacuum section. This FDC driver is 6 m in length and 150 mm in its inner diameter. A detonation igniter is installed at its left end to initiate a forward-running detonation wave. The shock tube is 2.75 m in length and 68 mm in inner diameter, and it can be replaced with a 4.75-m-length tube with the same diameter for different experimental requirements. The primary diaphragm is located between the FDC driver and the shock tube,

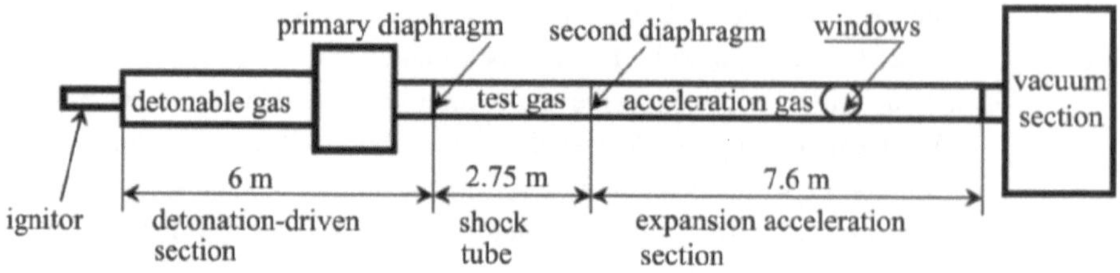

Figure 5.22 Schematic of the JF-16 expansion tube.

separating the detonable gas mixture from the test gas. The expansion acceleration section is 7.6 m in length with a square section of 60 mm × 60 mm. The second diaphragm is placed between the shock tube and the expansion acceleration section, separating the test gas from the acceleration gas. For performance study, a set of flow visualization windows opposing each other is installed in the expansion acceleration section. A vacuum section is added to the right end of the facility to weaken wave reflections for minimizing their effects on the generated hypervelocity flow.

Surface pressure measurement was carried out along the shock tube and the expansion acceleration section. By setting the position of the second diaphragm as a starting point, the distance to each pressure transducer is set to be 1612, 2447, 3247, and 3847 mm. These measurement stations are named as B, C, D, and E from left to right in order. Ten surface ion probes were also distributed along the shock tube and the expansion acceleration section with an equal interval to measure the incident shock speed. The detonable gas used is a hydrogen–oxygen mixture at a ratio of 4:1. The igniter is a small diameter tube filled with a detonable gas mixture to provide a certain amount of hot gases to the FDC driver to ensure a direct initiation. The gas mixture in the igniter is initiated with explosion wires. The facility is set up first as an expansion tube and improved later to be a shock-expansion tunnel (Gao et al. 2008, Zhou et al. 2016a, 2016b).

5.3.3.2 Test Flow Quality Evaluation

In performance tests of the JF-16 expansion tube, the shock tube and the expansion acceleration section were filled with air at 20 mmHg and 0.15 mmHg, respectively, as the base test case. For other cases, the changes will be explained explicitly if there is any difference in initial conditions. The FDC driver was filled with a H_2–O_2 mixture at an initial pressure of 1.0 MPa and 1.5 MPa, respectively, for two test cases. The wall pressure variations in the shock tube, related to the test flow quality, are presented in Figures 5.23 and 5.24, respectively, for discussion.

Figure 5.23 shows the pressure histories measured at four measurement points in the first test case where the initial pressure in the FDC driver is set to be 1.0 MPa. It is observable that a pressure plateau appears after the sharp pressure jump that indicates the arrival of the incident shock wave at each measurement point. There is a pressure overshoot at point B after expansion acceleration, and the overshoot disappears at point C. The uniform test flow is achieved from the position marked with C and D, and the test duration is measured to be about 50 μs.

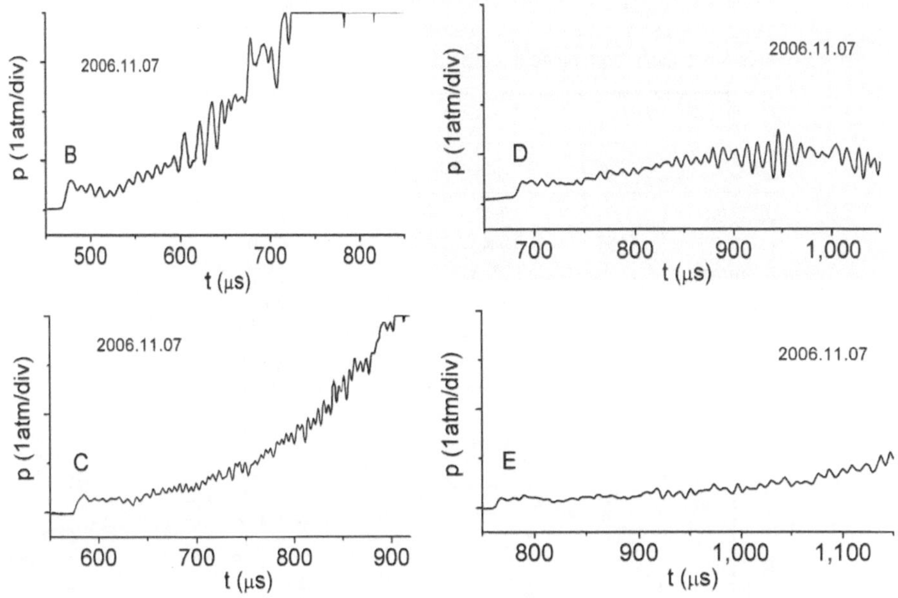

Figure 5.23 Wall pressure histories at four measurement points under the initial conditions of 1.0 MPa in the FDC driver, the initial pressures of 20 mmHg in the shock tube, and 0.15 mmHg in the acceleration section.

Figure 5.24 shows pressure histories measured at four measurement points in the second case where the initial pressure in the FDC driver is set to be 1.5 MPa, higher than the first case, but other initial conditions are kept to be the same. It is observed that the pressure plateau appears at measurement point C. This result implies that the uniform test flow is achieved from this position, and the test duration is about 70 μs. Actually, it is not sufficient to determine the test time based only on the pressure plateau because there is no pressure change across the contact surface that separates the test gas from the driver gas. In the following chapter, the flow visualization will be presented in a series of conical shock photographs. The test time can be determined there because the temperature will vary across the contact surface.

By comparing Figure 5.23 with Figure 5.24, it is found that the pressure plateau appears later in the second case. This means that the stronger flow expansion leads to a longer development time of the test flow in the acceleration section. Therefore, the high-quality test flow can be obtained at different positions for different initial test conditions. According to the observed phenomena, the test section for the expansion tube should be movable along the expansion section so that experiments could be carried out with more uniform test flows. Actually, the test flow needs space to accelerate: the higher the incident shock Mach number, the longer the flow expansion distance needed to develop.

By using the measured incident shock speeds, some key flow parameters behind the incident shock wave can be calculated with Mirels' theory. The results at 2- and

Table 5.3 Key parameters during flow acceleration evaluated along the expansion section at the distance of 2 and 4 m from the second diaphragm along the expansion tube

	2 m		4 m	
	Case1	Case2	Case1	Case2
u_s (m/s)	7,512	8,472	7,168	8,163
l / l_m	0.51	0.83	0.81	0.99
l_m (mm)	89	25.3	90	25.3
l (mm)	45	21	73	25
u_c/u_s	0.957	0.964	0.973	0.973
u_c (m/s)	7,190	8,336	6,971	8,108

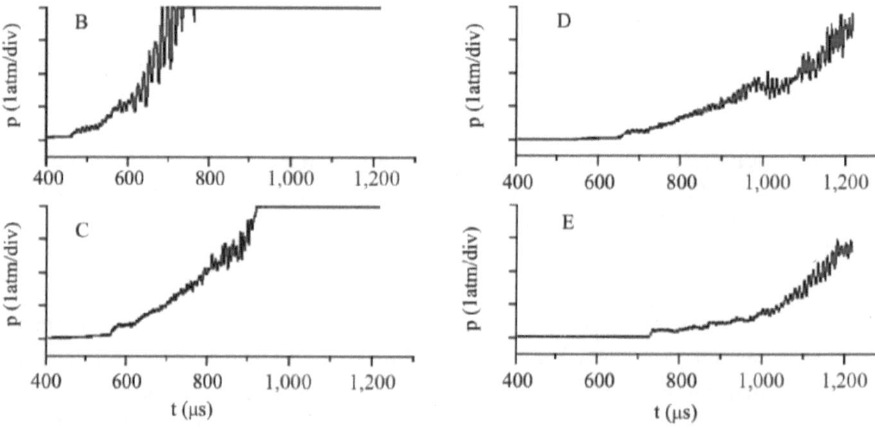

Figure 5.24 Wall pressure histories at four measurement points under the initial conditions of 1.5 MPa in the FDC driver, the initial pressures of 20 mmHg in the shock tube, and 0.15 mmHg in the acceleration section.

4-m positions starting from the second diaphragm along the expansion tube are presented in Table 5.3. In this table, u_s is the shock speed, u_c is the flow velocity, and l_m and l are the maximum and measured distances between the incident shock wave and the gas interface behind the incident shock, respectively. From this table, it can be seen that the flow velocity at 2 m reaches 7,190 m/s in Case 1 and 8,336 m/s in Case 2. At the 4-m position, the flow velocity is a little lower, but it still reached 6,971 m/s in Case 1 and 8,108 m/s in Case 2. This result implies that the incident shock wave will decay along the expansion tube due to viscous attenuation.

5.3.3.3 Other Parameters for the JF-16 Expansion Tube

There are several parameters being important in developing expansion tubes, and it is found that the performance of the JF-16 expansion tube is affected more or less by these parameters in some ways. These parameters were investigated in detail, and the relative results are presented in the chapter for reference.

The first parameter is the initial pressure in the shock tube, and the initial pressure was found to play an important role in test flow stability. Setting the initial pressure in the FDC driver to be 1.0 MPa, several experiments were carried out by varying the initial shock-tube pressure from 10 to 50 mmHg. The pressure variations measured at the same position in the shock tube are presented in Figure. 5.25. It can be seen that the higher the initial shock-tube pressure, the stronger the flow pressure fluctuation. The pressure ratio across the incident shock wave changes as the initial shock-tube pressure varies because of the constant pressure in the FDC driver. This means that the sound speed ratio (the driver gas to the test gas) will change across the incident shock wave, for example, the sound speed ratio is 0.826 in the case shown in Figure 5.25(a), 0.877 in the case shown in Figure 5.25(b), and 1.086 in the case shown in Figure 5.25(c). It is believed that the flow perturbation is generated mainly in the driver section, and the higher sound speed ratio will lead to stronger perturbations in the driver gas. The perturbation could propagate rapidly into the test gas and results in pressure fluctuation. Therefore, having the sound speed ratio being less than unity is a good choice for achieving high-quality test flows.

The second parameter is the incident shock-wave decay ratio. Results of the incident shock decay along the expansion acceleration section at different initial shock-tube pressures are presented in Figure 5.26. It is observed that the decay ratio is almost the same for the three cases, approximately 3% per meter, but the incident shock speed reaches its maximum value when the shock-tube pressure attains 20 mmHg. The maximum incident shock speed actually results from the pressure ratio across the primary diaphragm, the flow expansion after the second diaphragm, and their interaction. For the time being, the mechanism governing this phenomenon is not clear, and more attention should be paid to such problems when tuning an expansion tube.

The last parameter is the thickness of the second diaphragm. It is found that the shock velocity is closely related to the second diaphragm when tuning the JF-16 expansion tube. By using three kinds of diaphragms with different thicknesses, that is, 90-μm terylene, 60-μm terylene, and 25-μm cellophane, respectively, the results of the incident shock speed in the expansion tube are measured and presented in Figure 5.27. It is observed that the expansion tube performance benefits from the 25-μm cellophane, and the velocity decay ratio is small. Thick diaphragms can lead to the shock speed decay upon diaphragm rupture, and then the flow re-acceleration takes place later, as shown in Figure 5.27.

5.3.3.4 Concluding Remarks

The JF-16 expansion tube was developed successfully by combining up-to-date techniques of the detonation driver with an expansion acceleration section. The hypervelocity test flow of over 8.3 km/s and a total enthalpy up to 40 MJ/kg were generated with this facility for the experiment requiring orbital speeds. From its performance tests, the hypervelocity flow was demonstrated to be relatively uniform, and this indicates that high-quality test flows can be achieved by using the FDC driver and the

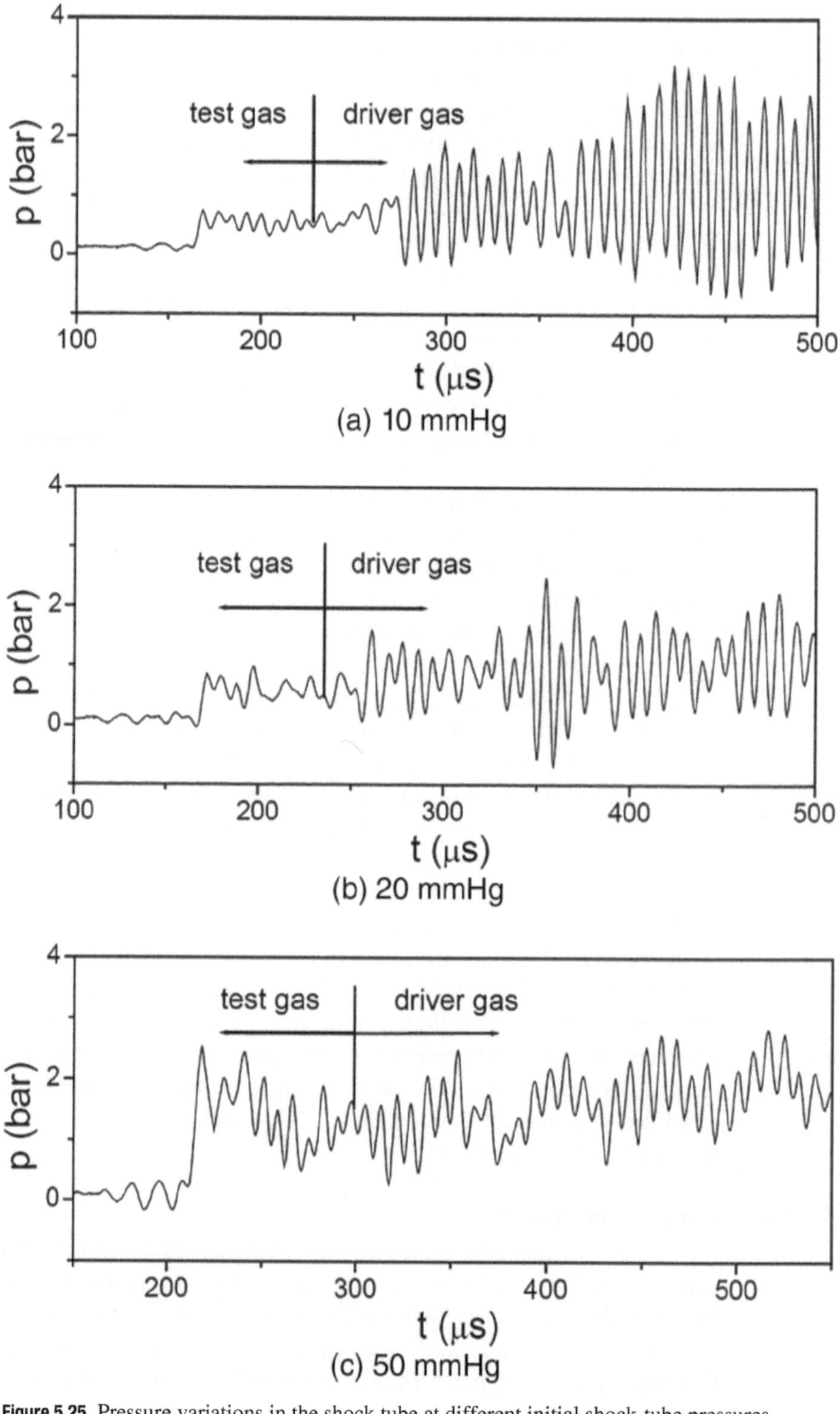

Figure 5.25 Pressure variations in the shock tube at different initial shock-tube pressures.

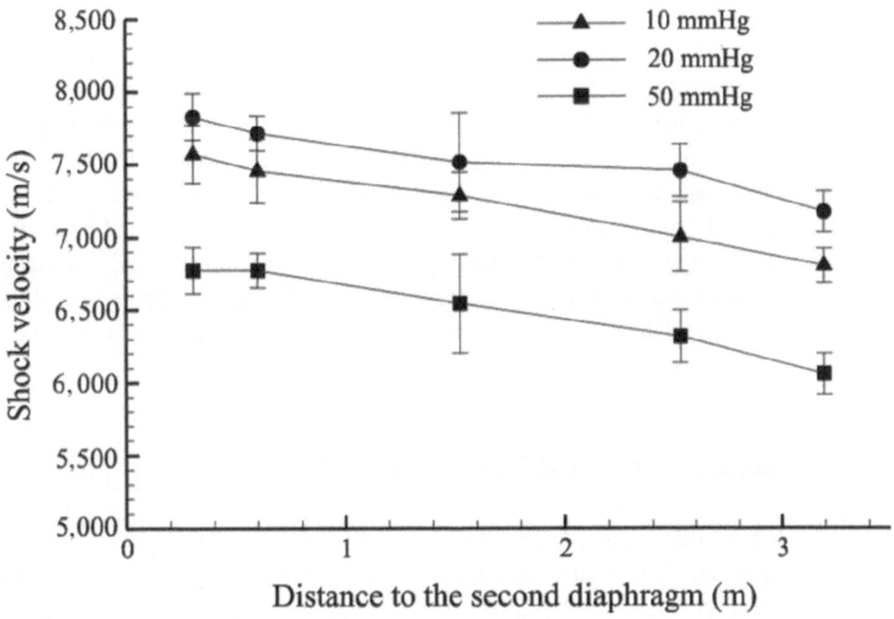

Figure 5.26 Incident shock decay along the expansion acceleration tube at different initial pressures of the shock tube.

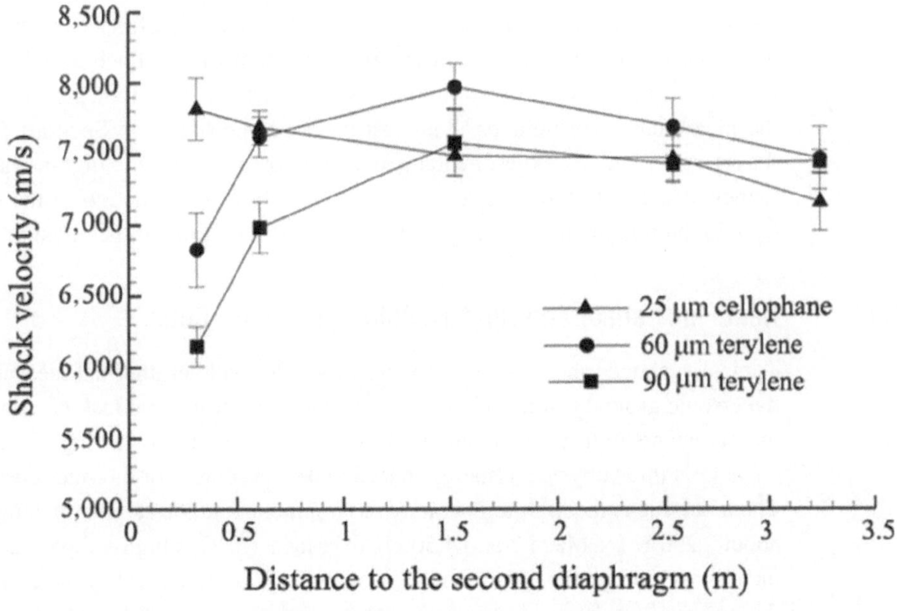

Figure 5.27 Incident shock velocity decay along the expansion tube with different diaphragms.

acceleration tube together. The sound speed ratio across the driver–test gas interface has a critical impact on the magnitude of test flow perturbations, but the perturbations can be suppressed to a certain extent if this ratio does not exceed unity. The maximum velocity of the test flow can be obtained by adjusting the filling pressure of the shock tube while the filling pressures are maintained constant in both the detonation driver and the expansion acceleration section. This implies that, with the optimization of operation conditions, it is possible to achieve the good performance of the JF-16 expansion tube with facility configuration optimization. Experimental data also showed that the second diaphragm affects both the gas dynamics and the chemical kinetics of the test gas, and the JF-16 expansion tube performance benefits from its light and thin diaphragms.

5.4 Some Experimental Results from High-Enthalpy Flow Tests

Generating high-enthalpy flows for reliable ground tests has been a challenging research topic for decades. So far, various types of hypersonic ground test facilities have been developed over the world, including conventional hypersonic wind tunnels, shock-heated high-enthalpy wind tunnels, arc-heated test facilities, and ballistic guns. However, different facilities are used to address different aspects of design problems associated with hypersonic flight because there is no single ground-based facility capable of duplicating hypersonic flight environments. Moreover, experiments in the aerothermodynamic environment of high-enthalpy flows are also challenging tasks to carry out, that is, the measurements of aerodynamic forces and moments, the heat-transfer distribution, and flow structure visualization. Each aerodynamic flow parameter has its unique feature that leads to different requirements for developing the measurement techniques. Four test cases are presented in Sections 5.4.1–5.4.4 for reference, and its experimental results are obtained with the high-enthalpy shock tunnels discussed above. Here, we only give a brief introduction to the challenges from high-enthalpy flow testing. More detailed discussion will be given in Chapter 8.

5.4.1 Heat Flux Sensor and Heat Transfer on a Sharp Cone

Typically, shock tunnels are mainly used for cold-wall aerothermal measurements in hypersonic ground testing. However, it becomes a formidable task to carry out such measurements in true hypersonic flight simulations for a long testing duration. There are at least three physical issues that need to be considered for thermal sensor designs. The first issue is the high-temperature environment, and the stagnation temperature is about 2,250 K for Mach 7 and 4,500 K for Mach 10. This high-temperature environment is very tough for the thermal sensors to survive because they have to withstand the hot gas flow for a certain time. The second one is the strong shear flow at wall surfaces due to the hypervelocity boundary layer, which results in the extremely high heat flux that is proportional to the velocity gradient, and the difference between the stagnation and wall temperatures. The large amount of aerothermal heat resulting

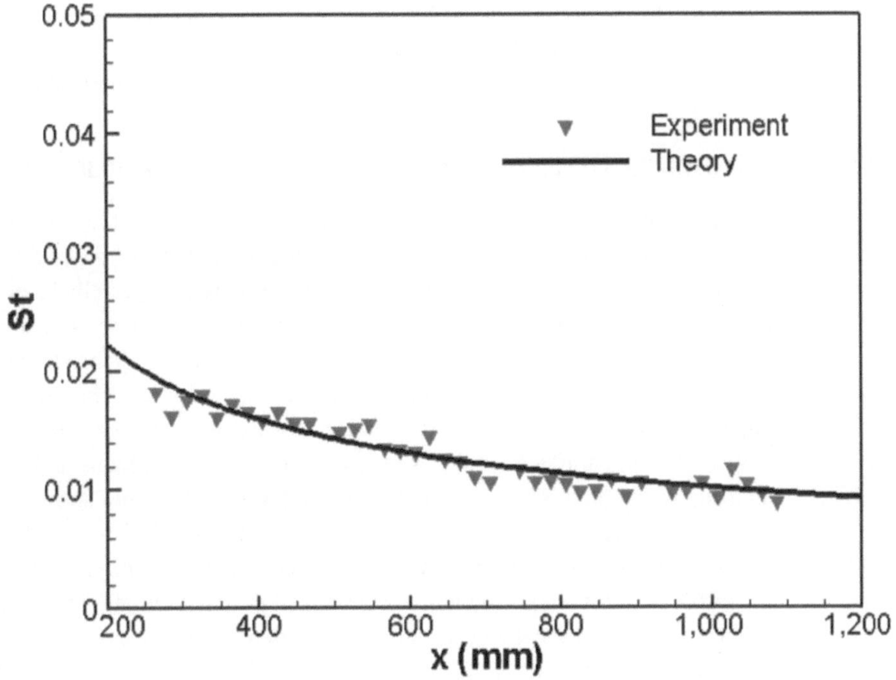

Figure 5.28 Comparison between theoretical and experimental results along the generatrix of the sharp cone model for Mach 7 with a total temperature of 2,250 K.

from the heat flux multiplying by the test time can melt the sensor surface in a few milliseconds. The last is the surface oxidization of the heat flux sensors, and large errors in the experimental data may be induced with repetitive usage.

According to the total temperature and the test duration of the JF-12 shock tunnel, a new type of the thermal couple is proposed based on three physical concepts. The first concept is to make the sensor diameter as small as possible to reduce the heat transfer to the thermal sensors. The second is to make the isolation layer between the two electrodes as thin as possible to increase sensor sensitivity. The last is to choose carefully the two electrode metals with similar thermal properties because this aspect is found to be very helpful to reduce the systematic errors of the thermocouple sensor. Calibration tests were carried out for the new type of thermocouples with a 7° cone test model. The new thermocouple is 1.4 mm in diameter, and the test model is 1,000 mm in length and 528 mm in diameter. For the flow total temperature of 2,250 K at Mach number 7, the experimental data are presented in Figure 5.28. From the figure, it is observable that all the measurement data fall within an error band of ±7%. This is the best result from shock-tunnel tests in the world because the errors are usually about 10%–20% for conventional shock-tunnel measurements (Wang et al. 2016b, 2017). Regarding this significant progress in the year of 2015, Kremeyer of the Hypersonic Technologies and Aerospace Planes Program Committee commented in Aerospace America's Year in Review (Kremeyer 2015) that the CAS's Institute of Mechanics had finished calibrating its 3.5-m diameter JF-12 hypervelocity wind tunnel up to Mach 9, and that heat transfer measurements and simulations yielding accurate results had been obtained for external aerodynamics and supersonic combustion.

5.4.2 Aerodynamic Balance and Force Measurements on a Sharp Cone

The shock tunnel can be used to generate hypersonic flow with Mach numbers as high as required, but its application in aerodynamics experiments is limited by its short test duration. The aerodynamic force and moment testing have especially been a challenge for shock-tunnel experiments for decades. Several kinds of aerodynamic balances have been proposed so far, but the data accuracy is difficult to improve. There are two physical issues that interfere with experimental data accuracy. The first is the pulsed impact on the test model due to the sudden starting of the nozzle flow. The unsteady flow establishment process induces irregular mechanical vibrations in the force measurement system and is quickly damped as the test flow becomes steady. The other is the inertia vibration that depends on the natural frequencies of the aerodynamic force measurement system consisting of the test model, aerodynamic balance, and model support. For typical shock tunnels, the periods of these vibrations at their low frequencies are as long as the test time, resulting in a difficult problem for accurately extracting the aerodynamic signals from the outputs of the measurement system. One way to solve the problem is to increase the rigidity of the measurement system, but its signal output can be too weak to detect, and the data accuracy is also low.

Based on the 100-ms test duration of the JF-12 shock tunnel, a new concept is proposed for developing the aerodynamic force measurement system. The idea goes as follows. At first, the lowest frequency is chosen to be the critical one for the force measurement system so that two or three perfect periods can be detected during the test duration. Then, a test model is designed as large as possible to make sure that the physical processes are truly captured, and also the test model must be made as light as possible to reduce the mechanical loads acting on the balances. And then, the model support rigidity can be increased so that its lowest natural frequencies could be higher than the critical frequency. Finally, the balance is carefully designed so that its lowest frequency can be equal to the critical one that matched with the shock-tunnel test time.

Experimental results are presented in Figure 5.29 for reference, and the detailed introduction could be found in references (Wang et al. 2016c, Liu et al. 2017). From the voltage signal of the pitching moment, the period of the starting vibrations lasts for more than 40 ms, and then the inertia vibrations follow. Three entire cycles are obtained within 100 ms. Therefore, it becomes easy to obtain accurate data from the inertia vibration periods with simple data processing, but it is very difficult to obtain precise data from the flow establishment vibration period. Unfortunately, the test time for all the conventional shock tunnels falls in the 2–40 ms category.

The aerodynamic force measurement in the JF-12 shock tunnel demonstrated that the new concept works perfectly, and the data accuracy reaches that of the blow-down-type conventional hypersonic wind tunnels. It is true that the data accuracy

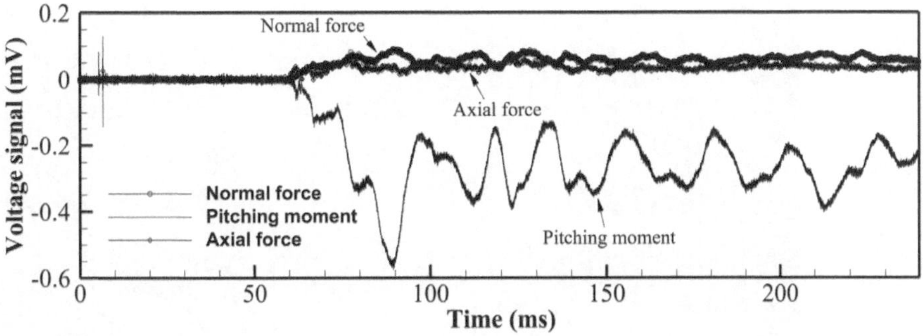

Figure 5.29 Stress balance signals from aerodynamic force calibration tests for Mach number 7 with the JF-12 shock tunnel.

depends not only on the aerodynamic balance but also on the test model, model support, and wind-tunnel test time. These devises must be taken as a whole system when performing force measurement tests.

5.4.3 Shock Visualization of Extreme Hypervelocity Flows

The JF-16 expansion tube can be used to generate orbital-speed test flows. The flow velocity can reach 8.3 km/s, and the total enthalpy is up to 40 MJ/kg, but the test time is less than 100 μs. Under such test conditions, flow measurement is a real challenging problem. In order to demonstrate test flow stability and the effective test time, flow visualization was carried out by using a Schlieren system with a high-speed camera. Photos of the conical shock in front of a sharp cone are presented in time order in Figure 5.30. In order to obtain conical shock photographs, a 15° sharp cone made of steel is mounted in the test section. The test flow is generated at the same condition as Case 2 in Section 5.3.3, and the flow velocity is measured to be about 8,100 m/s. A FASTCAM SA4 camera is used for photographing shock structures from the test section, while the exposure time of 1 μs is applied, and the time interval between successive frames is 25 μs (Gao et al. 2008, Jiang et al. 2015, Zhou et al. 2016a, 2016b).

Figure 5.30(a) shows the photograph of the conical shock at 25 μs after the incident shock passes, and the shock-compressed flow in the photograph appears in a gray shade. By examining the conical shock angle from Figure 5.30(b) to 5.30(d), it can be seen that the shock configuration is the same and the shocked flow turns into a lighter gray shade. The darker gray post-shock flow field in Figure 5.30(a) shows the characteristics of the low-temperature gas filled initially into the expansion acceleration section. The flow field from Figure 5.30(f) to 5.30(h) appears to have the lightest shade, indicating the characteristics of the high-temperature driver gas. Therefore, it is concluded that the grayish post-shock flow from Figure 5.30(b) to 5.30(e) is the test gas filled initially into the shock tube. This fact indicates that the test flow thus

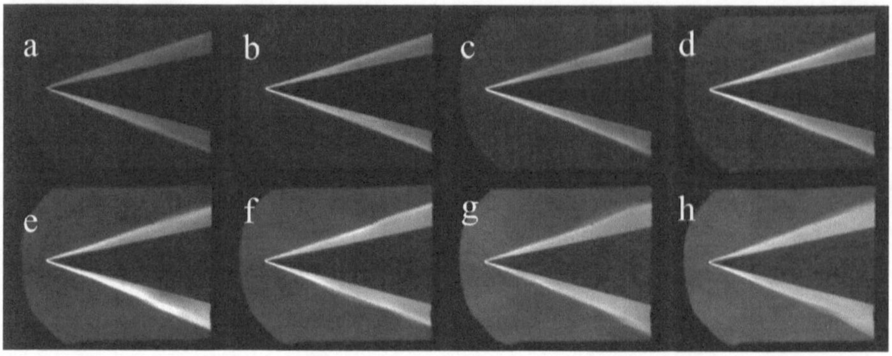

Figure 5.30 Sequential photos of the conical shock wave in front of a 15° sharp steel cone placed in the 8,100 m/s hypersonic flow; the time interval is 25 μs between successive frames, and the exposure time is 1 μs.

generated is uniform since the shock angle has kept constant, and the test duration is not shorter than 50 μs but not longer than 100 μs. Figure 5.30(e) shows that the post-shock flow turns into lighter shades, and the phenomenon indicates the mixing of the test gas with the driver gases.

Test flow quality is well demonstrated by the flow visualization results, and the extremely high-enthalpy flow can be investigated with optical technology. Furthermore, the conical shock configuration obtained is at a flow velocity of 8.1 km/s with a total flow enthalpy of 39 MJ/kg according to the equilibrium theory, and there are no reports on similar data at such orbital speeds. Therefore, these visualization photographs are useful not only for high-enthalpy flow diagnosis but also for CFD validation.

5.4.4 Electron Density Measurements of High-Enthalpy Flows

When vehicles fly at hypersonic speeds, the bow shock wave will be driven out in front of the vehicles. The high-temperature flow is generated due to shock-wave compression and viscous deceleration, resulting in vibrational excitation, dissociation, and even ionization of air molecules in the flow layer around the vehicles. This phenomenon is the so-called high-temperature plasma sheath that affects significantly the radar and radio wave transmission from or to aircraft. The electron density distribution in the high-enthalpy boundary layer flow outside the aircraft is very important as it can interfere with radio communication. It has been one of the interesting topics of high-enthalpy flow research for decades.

The JF-10 shock tunnel can be used to generate test flows of total temperatures up to 8,000 K at velocity of 5,000 km/s for the 6-ms test duration. The high-enthalpy flow is good for measuring the electron density distribution because the boundary layer is similar to that in flight. Wang et al. had developed their electrostatic probe technology and performed an experimental work on the flow field of the 10° wedge model (Wang et al. 2016a). Their needle-shaped probe can be used to obtain sufficient spatial

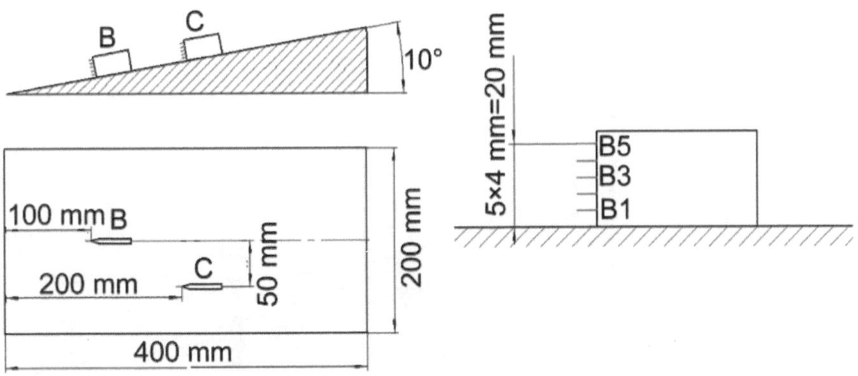

Figure 5.31 Schematic diagram of the 10° wedge test model and sensor arrangement.

resolution without disturbing the test flow field significantly, as shown in Figure 5.31. In the figure, there are two probe holders marked with B and C with a 100-mm interval. Each holder consists of five probes with a 4-mm interval between them. The basic principle of the electrostatic probe is to apply a certain bias voltage to the probe and place the probe in the plasma. When the probe potential is not equal to the plasma potential, the probe will attract electrons or ions. In this way, a current is generated in the needle loop, and the electron concentration can be calculated from the obtained current. The performance test results showed that the developed probe can obtain the spatial electron density distribution and has good repeatability.

The experiment work was carried out at the condition of total enthalpy of 16 MJ/kg and a stagnation temperature of 7,900 K. Figure 5.32 shows the result of the electron density measurement at the probe holder B for five runs. It can be found that the electron density is successively reduced from the wall surface to the outside of the boundary layer, and both good regularity and repeatability exist among the five probes. The experiment demonstrated that electrostatic probe technology is possible to be used to determine the tendency of electron density distributions, but it is unable to determine the quantitative electron density value. When conducting research on communication interruption during hypersonic flight, it is necessary to know the quantitative electron density value to judge whether or not the electron density exceeds the critical one. The present method is applicable but has certain limitations.

As a high-enthalpy shock tunnel, the JF-10 shock tunnel can be used to study the reentry problem. The electron density distribution over the 10° wedge model was measured successfully by using the electrostatic probe that can provide with sufficient spatial resolution without much disturbance on the flow field. However, developing measurement techniques for high-enthalpy flows is full of challenges, and there is still a long way to go for gaining reliable experimental data after high-enthalpy shock tunnels had been developed.

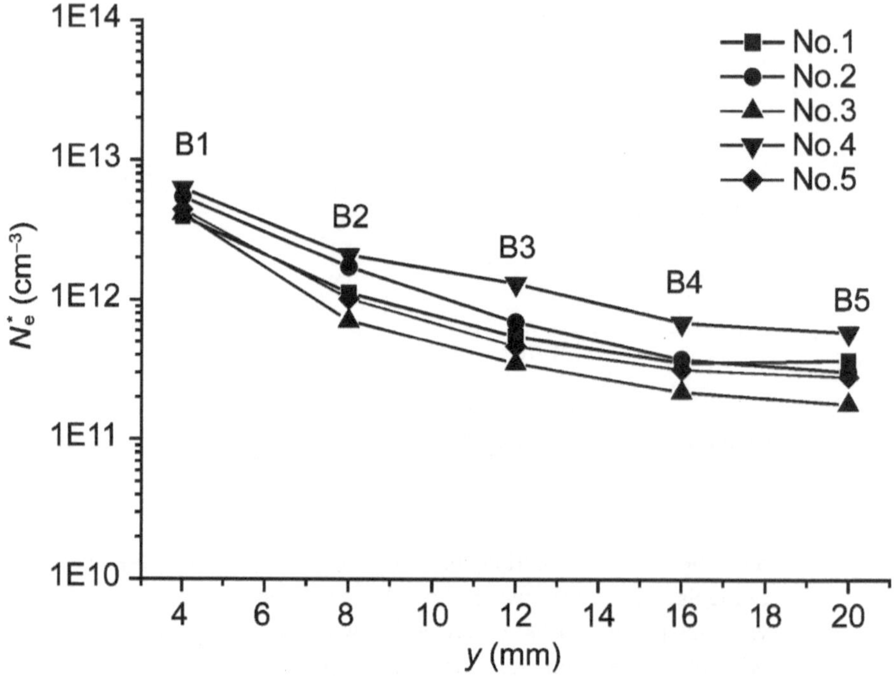

Figure 5.32 The electron density distributions of five runs at the probe holder B.

5.5 Conclusions

The first enclosed wind tunnel was invented in 1871 to measure the drag over a solid body by Wenham, a Council Member of the Aeronautical Society of Great Britain. Many wind tunnels for various purposes have been built in the world since then. Nowadays, to develop hypersonic vehicles, the wind tunnel has to be made larger and larger in size to accommodate test models large enough to minimize the unscalable effects due to chemically reacting flows. Therefore, an increasingly higher level of driving power is required for wind-tunnel operation. For large-scale hypersonic wind tunnels, the power supply needed is so huge that the blow-down wind tunnels could not be employed to replicate the correct velocity or flow enthalpy that are necessary for hypersonic experiments.

Because of the power supply limitation on continuous wind tunnels, shock tunnels were developed for ground-based hypersonic experiments. The detonation-driven high-enthalpy test facilities have been studied and developed in the LHD, Institute of Mechanics, CAS, for years. The facilities include the JF-10 high-enthalpy shock tunnel, the JF-16 detonation-driven orbital-speed shock-expansion tube, and the JF-12 hypersonic flight-duplicated shock tunnel. Since the beginning of 2018, the design and manufacture of the JF-22 Hypervelocity Shock Tunnel have been initialized based on the FDC driver, and now, its installation is completed and calibration is ongoing.

The JF-22 hypervelocity shock tunnel equipped with three 2.5-m diameter nozzles is capable of covering March numbers from 10 to 25 at altitudes ranging from 30 to 90 km. The typical trajectories of various spacecraft can be covered with these facilities, and the research on aerothermodynamics and hypersonic propulsion can be carried out to gain reliable data for the fundamental understanding of hypersonic physics.

Why do we use detonation in shock tunnels? The detonation is the fastest combustion in nature, and its rapid energy conversion is a special property that is very useful in modern society. Generally speaking, a 20-m^2 detonation front operates at a power level equal to the total power the earth receives from the sun. The FDC driver and the backward detonation driver combining with a damping section make use of the detonation successfully in hypersonic ground-test facilities and provide shock tunnels with the required power level at affordable expenses. The techniques proposed for extending shock-tunnel test duration are of significant importance because the long test duration cannot be achieved only by increasing shock-tunnel length. The detonation-driven high-enthalpy shock tunnels developed so far have demonstrated their superior performance, such as the stable reservoir pressure, the high-enthalpy level, the long test duration, and the low operating cost. However, the heat transfer from detonated gases to the facility and the facility damage due to the harsh thermo-environment occurring in the operation of detonation-driven shock tunnels are very difficult to overcome. The problems appear much more severe than others. For advanced hypersonic test facilities, the detonation-driven shock tunnels are very promising, but their improvement both in its design and operation is still necessary in the future.

Nomenclature

d	diameter
H	altitude
L	length
M	Mach number
M_s	shock Mach number
m	mass
P	pressure
St	Stanton number
T	temperature
V	volume
x	axial coordinate

Subscripts:

0	total or stagnation condition
1	initial state in the driven gas
2	state behind incident or primary shock
3	state between the expansion wave and the contact surface

4 initial state in the driver gas
5 state behind reflected shock in the test gas
i initial
t pilot

References

Anderson, J. D. (1989). *Hypersonic and High Temperature Gas Dynamics*. New York: McGraw-Hill Book Company.

Bertin, J. J. (1994). *Hypersonic Aerothermodynamics*, Washington, DC: AIAA Education Series.

Bertin, J. J. and Cummings, R. M. (2006). Critical Hypersonic Aerothermodynamic Phenomena. *Annual Review of Fluid Mechanics*, **38**, 129–157.

Bird, G. A. (1957). A Note on Combustion Driven Shock Tubes. In *Hypersonic Facilities in the Aerodynamics Department, Royal Aircraft Establishment*, ed. P.A. Hufton, AGARD Report 146, May.

Chue, R. S. M., Tsai, C.-Y., Bakos, R. J., Erdos, J. I., and Rogers, R. C. (2003). NASA's HYPULSE Facility at GASL – A Dual Mode, Dual Driver Reflected-shock/Expansion Tunnel. In *Advanced Hypersonic Test Facilities*, Vol. 198, eds. F. Lu and D. Marren. Reston, VA: AIAA, pp. 29–71.

Gao, Y., Zhao, W., and Jiang, Z. (2008). Experimental Study on the Performance of the Detonation-driven High-enthalpy Shock Expansion Tube, *Chinese Journal of Theoretical and Applied Mechanics*, **40**(4), 473–477 (In Chinese).

Hertzberg, A. and Smith, W. E. (1954). A Method for Generation Strong Shock Wave. *Journal of Applied Physics*, **25**(1), 130–131.

Hirschel, E. H. (1991). Viscous Effects. *Space Course*, **12**, 1–35.

Holden, M. S. (1994). Design, Development and Calibration of the LENS Facility. *AFOSR-TR 94-0161*.

Holden, M. S., Chadwick, K. M., and Kolly, J. M. (1995). Hypersonic Studies in the LENS Facilities. In *6th International Aerospace Planes and Hypersonic Technologies Conference*, Chattanooga, Tennessee, April 3–7, *AIAA 95–6040*.

Holden, M. S., Wadhams, T. P., and Candler, G. V. (2004). A Review of Experimental Studies in the LENS Shock Tunnel and Expansion Tunnel to Examine Real-Gas Effects in Hypervelocity Flows. *42nd AIAA Aerospace Sciences Meeting & Exhibit Reno*, NV, January 5–8. *AIAA Paper2004 – 0916*.

Hornung, H. G. (1988). The Piston Motion in a Free-Piston Driver for Shock Tubes and Tunnels. *GALCIT Report FM88-1*.

Itoh, K., Ueda, S., Komuro, T., et al. (1998). Improvement of a Free Piston Driver for a High-enthalpy Shock Tunnel. *Shock Waves*, **8**, 215–233.

Itoh, K., Ueda, S., Tanno, H., Komuro, T., and Sato, K. (2002). Hypersonic Aerothermodynamic and Scramjet Research Using High Enthalpy Shock Tunnel. *Shock Waves*, **12**, 93–98.

Jiang, Z., Hu, Z., Wang, Y., and Han, G. (2020a). Advances in Critical Technologies for Hypersonic and High-enthalpy Wind Tunnel. *Chinese Journal of Aeronautics*, **33**(12), 3027–3038.

Jiang, Z., Li, J., Hu, Z., Liu, Y., and Yu, H. (2020b). On Theory and Methods for Advanced Detonation-driven Hypervelocity Shock Tunnels. *National Science Review*, **7**(7), 1198–1207.

Jiang, Z., Li, J., Zhao, W., Liu, Y., and Yu, H. (2012). Investigation into Techniques for Extending the Test-duration of Detonation-driven Shock Tunnel. *Chinese Journal of Theoretical and Applied Mechanics*, **44**(5), 824–831.

Jiang, Z., Lin, J., and Zhao, W. (2011). Performance Tests of JF-10 High-enthalpy Shock Tunnel with a FDC Driver. *International Journal of Hypersonics*, **1&2**, 29–35.

Jiang, Z. and Teng H. (2022). *Gaseous Detonation Physics and Its Universal Framework Theory*. Singapore: Springer.

Jiang, Z., Wu, B., Gao, Y., Zhao, W., and Hu, Z. (2015). Development of the Detonation-driven Expansion Tube for Orbital Speed Experiments. *Science China*, **58**(4), 695–700.

Jiang, Z. and Yu, H. (2014). Experiments and Development of Long-test-duration Hypervelocity Detonation-driven Shock Tunnel (LHDst), *AIAA Paper 2014–1012* (invited lecture).

Jiang, Z. and Yu, H. (2017). Theories and Technologies for Duplicating Hypersonic Flight Conditions for Ground Testing. *National Science Review*, **4**(3), 290–296.

Jiang, Z., Zhao, W., Wang, C., and Takayama, K. (2002). Forward-running Detonation Drivers for High-enthalpy Shock Tunnels. *AIAA Journal*, **40**, 2009–2016.

Kremeyer, K. (2015). Researchers Aim for Hypersonic Missiles and Transport Vehicles. *Aerospace America*, AIAA, December, **53**(11), 45.

Lee, J. H. S. (2008). *The Detonation Phenomenon*. New York: Cambridge University Press.

Li, J., Feng, H., and Jiang, Z. (2008a). Gas Contamination Induced by the Interaction of Shock/Boundary Layer in Shock Tunnel. *Chinese Journal of Theoretical and Applied Mechanics*, **40**(3), 289–296 (In Chinese).

Li, J., Feng, H., Jiang, Z., and Yu, Y. (2008b). Numerical Computation on the Tailored Shock Mach Numbers for a Hydrogen-Oxygen Detonation Shock Tube. *Acta Aerodynamica Sinica*, **26**(3), 291–296 (In Chinese).

Liu, Y., Wang, Y., Yuan, C., Luo, C., and Jiang, Z. (2017). Aerodynamic Force and Moment Measurement of 10° Half-angle Cone in the JF-12 Shock Tunnel. *Chinese Journal of Aeronautics*, **30**(3), 983–987.

Lu, F. K. and Marren, D. E. (2002). *Advanced Hypersonic Test Facility, Progress in Astronautics and Aeronautics*, Vol. 198. Reston, VA: AIAA.

Morgan, R. (2001). *Free-Piston-driven Expansion Tubes. Handbook of Shock Waves.* New York: Academic Press, pp. 603–622.

Neely, A. J. and Morgan, R. G. (1995). *Hypervelocity Aerodynamics in a Super-orbital Expansion Tube*. Berlin: Springer.

Olivier, H., Jiang, Z., Yu, H., and Lu, F. K. (2002). *Detonation-driven Shock Tubes and Tunnels, Advanced Hypersonic Test Facilities*. Reston, VA: AIAA, pp. 135–203.

Ozer, I. and Friedrich, S. (2016). *Experimental Methods of Shock Wave Research*. New York: Springer.

Park, C. (1990). *Nonequilibrium Hypersonic Aerothermodynamics*. New York: Wiley.

Resler, E. L. and Bloxsom, D. E. (1952). *Very High Mach Number Flows by Unsteady Flows Principles*. Ithaca, NY: Cornell University Press.

Sasoh, A., Morgan, R. G., Littleton, B. N., McIntyre, T. J., and Bishop, A. I. (2000). High-enthalpy Expansion Tube Experiments with Gas Injection. *AIAA Journal*, **38** (12), 2253–2259.

Stalker, R. J. (1967). A Study of the Free-piston Shock Tunnel. *AIAA Journal*, **5** (12), 2160–2165.

Trimpi, R. L. (1962). A Preliminary Theoretical Study of the Expansion Tube: A New Device for Producing High-Enthalpy Short Duration Hypersonic Gas Flows, *NASA TR-133*.

Wang, Q., Huang, J. D, Nie, C. S., et al. (2016a). Research on Electron Density Measurements for High Enthalpy Flows. *Science China Technological Sciences*, **46**, 500–508 (In Chinese).

Wang, Q., Li, J., Zhao, W., and Jiang, Z. (2016b). Comparative Study on Aerodynamic Heating Under Perfect and Nonequilibrium Hypersonic Flows. *Science China Physics, Mechanics & Astronomy*, **59**, 624701.

Wang, Y., Liu, Y., and Jiang, Z. (2016c). Design of a Pulse-type Strain Gauge Balance for a Long-Test-Duration Hypersonic Shock Tunnel. *Shock Waves*, **26**, 835–844.

Wang, Q., Li, J., Zhao, W., and Jiang, Z. (2017). Influence of Thermal Sensor Installation on Measuring Accuracy at Stagnation Points. *Journal of Thermo-physics and Heat Transfer*, **31**(2), 318–323.

Yu, H. (1989). Recent Developments in Shock Tube Application. In *Proceedings of the 1989 Japanese National Symposium in Shock Wave Phenomena*, ed. K. Takayama. Sendai, Japan: Tohoku University, 1–10.

Yu, H., Esser, B., Lenartz, M., and Grönig, H. (1992). Gaseous Detonation Driver for a Shock Tunnel. *Shock Waves*, **2**, 245–254.

Zhao, W., Jiang, Z., Saito, T., Lin, J. M., Yu, H. R., and Takayama, K. (2005). Performance of a Detonation Driven Shock Tunnel. *Shock Waves*, **14**, 53–59.

Zhou, K., Wang, Q., Hu, Z., and Jiang, Z. (2016a). Performance Study of a Detonation-driven Expansion Tube. *Acta Aeronautica et Astronautica Sinica*, **37**(3), 822–828. (In Chinese).

Zhou, K., Yuan, C., Hu, Z., and Jiang, Z. (2016b). Flow Field Analysis of JF-16 Expansion Tube and Its Upgrade. *Acta Aeronautica et Astronautica Sinica*, **37**(11), 3296–3303. (In Chinese).

6 Theory and Methods for Long-Test-Duration Shock Tunnels

Aerodynamic hypervelocity ground testing is an important and indispensable means for the development of hypersonic vehicles. Several types of hypersonic test facilities have been developed so far, such as shock tunnels, air-directly-heated wind tunnels, arc-heated wind tunnels, and combustion-based propulsion test facilities. After more than 60 years of development, it is understood that these facilities are still inadequate for the required hypersonic technology development, such as for hypersonic propulsion tests at Mach numbers higher than 7 as well as for thermal environment identification for large-scale test models of trans-atmospheric vehicles (Lu & Marren 2002, Bertin & Cummings 2003, 2006, Ozer & Friedrich 2016).

The high-enthalpy shock tunnel is capable of producing test flows with both high total temperature and pressure, but its short test duration, usually measured in a few milliseconds, limits its application to cold-wall heat flux measurements. With increased test time, it is possible to test air-breathing propulsion systems, rocket and plume interactions, aerodynamic forces and moments, and even dynamic processes such as stage separation. Based on detonation-driven shock tunnels, several key issues that play important roles in extending the test time are introduced in this chapter, and the corresponding solutions are proposed, evaluated, and discussed in detail, including the tailored-interface-matching condition, the shock-wave/boundary-layer interaction at the end of the driven section, and the precursor shock damping in the vacuum tank. The research work on these issues was carried out to find out the flow physics and application methods to improve the high-enthalpy shock tunnel for meeting the test time requirement for supersonic combustion and scramjet engine experiments (Jiang & Yu 2017, Jiang et al. 2020b).

With application of the theories and methods mentioned above to the high-enthalpy shock tunnel, a large-scale detonation-driven hypersonic flight-duplicated shock tunnel (the JF-12 shock tunnel) was successfully developed, which can provide test times of more than 100 ms and is capable of duplicating hypersonic flight conditions for Mach numbers of 5–9 at altitudes of 25–50 km. With the help of the JF-12 shock tunnel, scramjet engine testing and aerodynamic force and moment experiments were carried out to demonstrate the advantage of the long test duration of the high-enthalpy shock tunnel (Jiang & Yu 2014, Jiang et al. 2020a, 2021).

6.1 Fundamentals of Shock-Tunnel Test Duration

For more than a century, the progress of aeronautics and astronautics has been fueled by the primary desire to fly faster, higher, and more cost-effectively since the Wright Brothers' first flight on December 17, 1903. There is the exponential growth in both the aircraft speed and the flying altitude over the past 100 years, and now the era of hypersonic flight is approaching (Anderson 1989, Bertin 1994, Park 1990). With practical hypersonic airplanes, a two-hour flight to anywhere in the world will be possible, and the expense for space access will be cut by 90% with reusable Two-Stage-To-Orbit (TSTO) techniques. The impact of hypersonic technology on modern society could be extremely huge and results in a revolutionary change in both aviation and aerospace industries. Hypersonic flight is, and in the foreseeable future will be, the domain of national security, and civilian transportation and space access. After 60 years of research in hypersonic, both manned and unmanned hypersonic flights have been successfully achieved so far, however practical hypersonic flight with air-breathing propulsion is still far ahead of us (Anderson 1989, Park 1990, Bertin 1994).

Hypersonic gas dynamics is fundamentally different from subsonic and supersonic ones. Hypersonic flows are usually characterized by the presence of strong shock waves and the equilibrium or nonequilibrium gas chemically-reacting processes where gas molecular vibrational excitations, O_2 dissociations, N_2 dissociations, and atom ionizations take place in succession as the flight speed increases from 1.5 km/s to 10 km/s. The accurate prediction of such chemically-reacting gas flow is a critical issue for developing hypersonic vehicles. Ground testing has been the main means for hypersonic vehicle development for decades because of the limitation on modeling the chemically-reacting gas flow, and especially on the supersonic combustion in air-breathing engine/frame-integrated vehicles. One of the most difficult problems arising from hypersonic vehicle testing is that the test model scaling laws widely applied in subsonic and supersonic aerodynamics are not valid any more for hypersonic flow tests in which thermochemical reactions dominate, that is, the chemical reaction scale remains unchanged when the test model is reduced in its size in wind tunnel tests. Therefore, this similarity problem leads to obvious discrepancies between flight tests and wind tunnel experiments.

6.1.1 Hypersonic Test Facilities

There are several types of hypersonic ground test facilities developed so far, such as shock tunnels, air-heated wind tunnels, arc-heated wind tunnels, and combustion-based propulsion test facilities. These facilities being based on different operating principles are used to address various aspects of the vehicle-designed problems associated with hypersonic flights, but there is no single ground-based wind tunnel facility capable of fully duplicating the hypersonic flight environment. Nowadays, critical hypersonic techniques are usually validated with flight tests. Flight testing is a more reliable means than the others we have, but it is quite expensive, has low data accuracy and is time-consuming for new vehicle development.

If one looks back at the last century, supersonic flights had been achieved within 60 years from 1903 to the time when Concorde began its commercial service across the Atlantic Ocean in 1967. However, practical hypersonic flight is still under exploration after 60 years' work since September 22, 1963, when Robert M. White flew the North American X-15 at a Mach number of 6.7. The situation is due to the many research challenges, such as the high lift-to-drag aerodynamic configuration, hypersonic propulsion, the integrated aero-thermal structure, advanced thermal-protection systems, and advanced ground test facilities. Advanced hypersonic ground test facilities that can simulate true hypersonic flight conditions are considered to be a fundamental one among these challenges. Generally speaking, for the time being, existing wind tunnel facilities are inadequate for certain required technology verifications, such as hypersonic propulsion systems, thermal environment identification, and large-scale integrated thermal-structural testing (Lu & Marren 2002, Bertin & Cummings 2003, 2006, Ozer & Friedrich 2016).

6.1.2 Critical Issues of Hypersonic Tests

To carry out a reliable ground test at hypersonic Mach numbers, four critical parameters are required. The first parameter is the test gas that is a key issue to ensure the correct thermochemistry in high-enthalpy test flows. The second is the flow total pressure, temperature, and the proper nozzle expansion needed so that the static pressure, temperature, and flight speed can be replicated at a given altitude. Consequently, the chemical reaction process can be simulated correctly, as well as the aerodynamic forces and moments. The third is the nozzle size. In other words, the nozzle exit must be large enough to accommodate a large-scale test model so that the chemical reaction time scale can be simulated correctly. The last is the test duration. For air-breathing hypersonic vehicles, the airframe is also a part of the scramjet engine, and the vehicle frame/engine-integrated test model is quite large in size. Therefore, the test duration has to be long enough to sustain stable supersonic combustion in the engine. It is estimated that a 100-ms test time is essential for supersonic combustion and engine tests.

Actually, the duration of the test must also be balanced with the consideration that the facility is not seriously damaged because of the extremely high total temperature that can cause severe thermal loads in hypersonic wind tunnels. It is a more challenging problem if one has to simulate these four parameters simultaneously in a single test facility. This is the reason why there is still no single ground-based facility being capable of duplicating hypersonic flight environments with the above-mentioned parameters after 60 years of development of advanced hypersonic ground test facilities. Different test facilities are developed nowadays to address different aspects of the various design problems associated with hypersonic flight. This idea has worked for many years and a lot of experimental data is generated from high-enthalpy flows, but hypersonic wind tunnel test data suffer from uncertainty and low accuracy, thus resulting in slow development of hypersonic vehicles.

6.1.3 Critical Techniques for Hypersonic Test Facilities

In order to duplicate the true hypersonic flight environment for ground testing, three critical issues have to be solved both theoretically and technically to advance the state-of-art in large-scale hypersonic test facilities. The first issue is the high-power driver. To produce a hypersonic flow of 2.5-m diameter for a Mach number of 9 at a 40-km altitude, the required power for the wind tunnel can be as high as 3000 MW! This power level is almost the same as the total installed power capacity of the Gezhouba hydropower station located in the Yangtze River in western China. It is not practicable to meet such an input power for conventional blow-down wind tunnels operated in a continuous mode. This is the reason why shock tunnels are widely applied to generate high-enthalpy flows. The second issue is the test duration. To meet the requirement for achieving stable supersonic combustion in air-breathing hypersonic vehicles, a 100-ms test period is needed, but several times longer than is available in all existing shock tunnels. The last issue is the flow diagnostic technology. It was demonstrated that the heat flux under true hypersonic flight conditions are so high that it would be a huge challenge to obtain experimental data with reasonable accuracy (Jiang et al. 2020a). Therefore, for chemically-reacting flows, it is very important to develop flow diagnostic technologies with high precision for high-enthalpy shock tunnel tests.

There are several kinds of hypersonic test facilities widely used in the world and these facilities are listed as follows with brief comments. The conventional hypersonic wind tunnels are operated on a blow-down mode where a high-pressure supply gas is separated from a low-pressure vacuum chamber by a valve. The high-pressure gas is expanded through a nozzle into the test section ending in a dump tank. The supplied test gas must be heated to prevent liquefaction in the test section after its expansion, and the static temperature of the thus-generated test flows is about 50 K. This type of wind tunnels operates cold at high Mach numbers with low sound speeds, but does not replicate the correct flow velocity or enthalpy, and of course, chemistry.

The electric arc and combustion-heated facilities operate similar to conventional hypersonic wind tunnels except the way that the test gas is heated. Duplicating the aerodynamic heating for a certain total temperature range is possible with the arc-heated wind tunnel, but this facility is limited in the total pressure and the test flow mass because of the huge power needed to heat the large volume of the test gas as it moves through the arc. Its test gas contamination and non-uniform flow-field result in its application only to material ablation tests. In combustion-heated facilities, the total temperature is limited to a flow Mach number of 7 or even lower if hydrocarbon-based fuel is used in the heater. These facilities are capable of producing a large volume of hot gases with minutes of test times, but the combustion products remain in its test flow, therefore, the chemistry is not correct in the high enthalpy test flows.

Shock tunnels utilize a moving shock to generate high-pressure test gases. The stronger the moving shock is, the higher the test gas enthalpy will be. The benefits of shock tunnels are that flight speed up to Mach 25 can be simulated, but their test duration is measured in a few milliseconds. Duplicating the flight conditions for hypersonic propulsion tests has remained a top challenge for 60 years since Stalker published his study on the free-piston shock tunnel that has become a widely-used high-enthalpy test facilities over the world (Lu & Marren 2002, Ozer & Friedrich 2016).

However, if true hypersonic flight simulation could be realized with a single ground test facility it would be an invaluable tool not only for evaluating hypersonic techniques and propulsion, but also for exploring the aero-thermal physics in hypersonic and high-temperature gas flows. Shock-heated high-enthalpy tunnels can be divided into three main classes according to their driving modes, i.e., the light-gas-heated driver, the free-piston driver, and the detonation driver (Jiang et al. 2020a). These three kinds of the shock tunnels are widely used for high-enthalpy flow tests and have provided us with massive experimental data. One of the critical problems remaining for developing the shock tunnel is the extremely-short test duration which is a big limitation on their applications.

Aiming at the critical techniques for the large-scale advanced hypersonic test facility, the long-test-duration hypervelocity detonation-driven shock tunnel was developed successfully, and named as the JF-12 hypersonic flight-duplicated shock tunnel under the serial number in the laboratory (Jiang & Yu 2014, 2017, Jiang et al. 2020a, 2020b). The JF-12 hypersonic flight-duplicated shock tunnel is 265 m in length and equipped with a 2.5-m hypersonic nozzle. The calibrations demonstrated that the facility is capable of reproducing pure airflow for Mach numbers from 5 to 9 at altitudes of 25–50 km with a test duration longer than 100 ms. The AIAA Awards Committee comments that the JF-12 hypersonic flight-duplicated shock tunnel is the world's largest shock tunnel capable of true hypersonic flight simulation and Jiang's work has advanced the state-of-the-art in large-scale hypersonic test facilities. The outstanding feature of the facility is its long test duration. To achieve this feature, several new techniques are integrated into the JF-12 hypersonic flight-duplicated shock tunnel, including the tailored-interface-matching condition for operating the detonation-driven shock tunnel, the shock-induced boundary-layer separation control at the end of the driven section to avoid driver/test gas mixing, and the precursor shock wave damping in the vacuum tank. The JF-12 shock tunnel calibration tests showed that these techniques work well and a 100-ms test time is achieved being free of test gas contamination and wave reflection distributions. In the following sub-chapters, three key techniques will be introduced and their working principles will be discussed in detail.

6.2 Tailored-Interface-Matching Conditions

There are at least three critical issues being responsible for resulting in the short test time in shock tunnels, and these can be explained with Figure 6.1. The first issue is the wave reflection from the interface that separates the driver gas from the driven gas. If the reflected incident shock reflects again from the interface, the total pressure and temperature of the test gas in region 5 at the end of the driven section will be changed during the nozzle operating process. This is the reservoir pressure for the shock tunnel and its change will result in low quality test flow and early termination of the test duration. The wave reflection from the gas interface depends on the change in sound speed

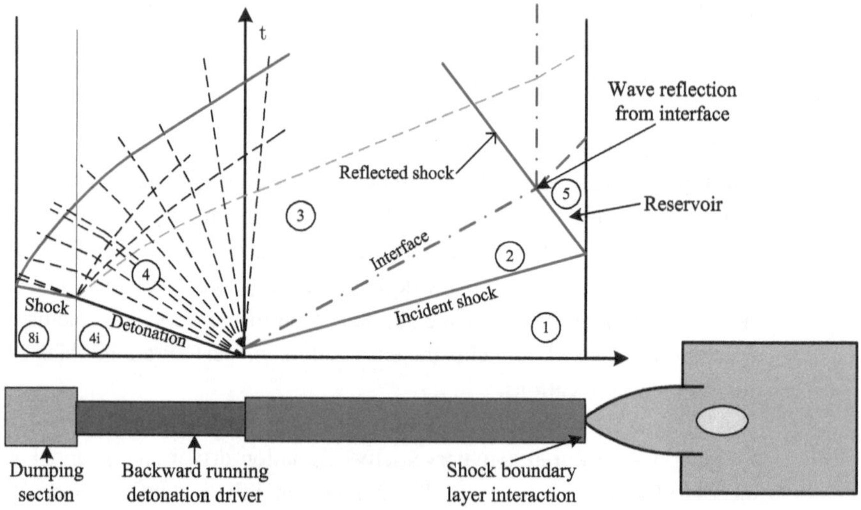

Figure 6.1 Physical issues for developing a long-test-duration shock tunnel.

across the interface. In order to avoid the wave reflection from the interface, the sound speed of the detonated gas products must match with the sound speed of the twice-compressed test gases. This is the so-called tailored-interface-matching condition for operating the reflected shock tunnels to achieve a long test duration (Li et al. 2008a, Jiang & Yu 2014).

6.2.1 Interface-Matching Equation

For the interface-matching condition, it is required that the acoustic resistance for the gases at the two sides of the interface must be identical to each other so that the shock wave could pass freely without any reflection. Under such a requirement, it holds, $p_5 = p_7, u_5 = u_7 \approx 0$. Since there is no pressure gradient across the interface, we can get $p_2 = p_3$, and $u_2 = u_3$. The relation for the reflected shock wave that is a left-propagating shock in region 2 is given by

$$\frac{u_2}{a_2} = \left(\frac{p_5}{p_2} - 1\right) \Big/ \sqrt{\frac{\gamma_2(\gamma_2 - 1)}{2}\left[\frac{(\gamma_2 + 1)p_5}{(\gamma_2 - 1)p_2} + 1\right]} \tag{6.1}$$

The transmitted shock wave is a left-propagating wave in region 3, and

$$\begin{aligned}\frac{u_3}{a_3} &= \left(\frac{p_7}{p_3} - 1\right) \Big/ \sqrt{\frac{\gamma_3(\gamma_3 - 1)}{2}\left[\frac{(\gamma_3 + 1)p_7}{(\gamma_3 - 1)p_3} + 1\right]} \\ &= \left(\frac{p_5}{p_2} - 1\right) \Big/ \sqrt{\frac{\gamma_3(\gamma_3 - 1)}{2}\left[\frac{(\gamma_3 + 1)p_5}{(\gamma_3 - 1)p_2} + 1\right]}\end{aligned} \tag{6.2}$$

Dividing Equation (6.1) by Equation (6.2), and setting $\gamma_2 = \gamma_1$ and $\gamma_3 = \gamma_4$, we can obtain

$$\frac{a_2}{a_3}=\frac{\gamma_1}{\gamma_4}\left[\frac{1+\frac{(\gamma_1+1)}{2\gamma_1}\left(\frac{p_5}{p_2}-1\right)}{1+\frac{(\gamma_4+1)}{2\gamma_4}\left(\frac{p_5}{p_2}-1\right)}\right] \tag{6.3}$$

Substituting the expansion relation between regions 4 and 3, we achieve the interface-matching condition for the driver/driven gas interface and it reads as:

$$\frac{a_4}{a_1}=\frac{2}{\gamma_1+1}\left(M_s-\frac{1}{M_s}\right)$$
$$\left\{\left[\frac{(\gamma_4-1)M_s^2+2}{2\gamma_1\left(M_s^2-1\right)}\right]\left[\gamma_4^2+\frac{\gamma_1\gamma_4(\gamma_4+1)\left(M_s^2-1\right)}{(\gamma_1-1)M_s^2+2}\right]^{0.5}+\frac{\gamma_4-1}{2}\right\} \tag{6.4}$$

where the subscripts follow those in the shock tube problem without explanation. From the above equation, it can be seen that the tailored Mach number *Ms* is a function of the sound speed ratio a_4/a_1, and the specific heat ratios γ_1 and γ_4 of the driver and the driven gases, respectively. As long as the composition and the initial temperature of the driver and the driven gases are known, a_4/a_1 can be determined. And then, the tailored Mach number can be obtained by solving Equation (6.4).

This equation works for the backward detonation driver since the incident shock wave is stable, and it does also for the FDC driver if it operates in its optimized configuration design condition (Jiang et al. 2002, Olivier et al. 2002). It is important to emphasize that the tailored interface is effective only for a given Mach number of the stable incident shock waves, so-called the tailored Mach number. The test gas is air usually at room temperature, therefore, only the driver gas and its aerodynamic state can be adjusted to meet Equation (6.4). The tailored Mach number can be determined when the driver gas and its thermal state are given, and the total enthalpy of the test gas flow can be also determined with the tailored Mach number.

6.2.2 Tailored Mach Numbers for Detonation Drivers

For detonation drivers, the tailored Mach number can be obtained by adjusting the driver gas composition so that the test flow with the required total enthalpy could be generated. It is important to point out that the detonated gases must be used for the calculation instead of the initial detonable gas mixture. Results calculated with Equation (6.4) are presented in Figure 6.2 for detonation-driven shock tunnels. The figure shows that the tailored Mach number increases with increase of the helium concentration of the initial driver gas, and decreases with the increase of the argon concentration fraction.

For the tailored Mach number, similar effects can be achieved by adding hydrogen or nitrogen into the driver gas. Decreasing the argon concentration will increase not only the sound speed of the detonated gases, but also its temperature since the chemical energy release is also increased. The dual-parameter effect on the sound speed for the detonation driver is very helpful to tailor the interface-matching condition. For a stoichiometric hydrogen/oxygen mixture, it is also effective to increase the tailored

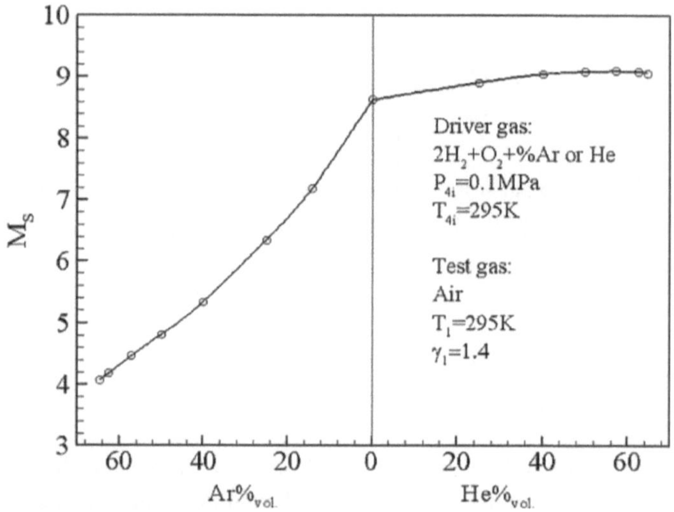

Figure 6.2 Incident shock Mach number for a stoichiometric hydrogen/oxygen mixture at tailored conditions versus the concentration of helium or argon dilution.

Mach number by adding helium because the increased helium dilution can make the detonated gases lighter. However, the detonation temperature will become low because of the low chemical energy release. For the detonation driver operated with a hydrogen/oxygen mixture diluted with helium, the tailored Mach number that can be achieved is as high as 9. This Mach number can be applied to generate high-enthalpy test flows of total temperatures up to 8,000 K. The expansion tube can be used if higher enthalpies are needed. Jiang et al. reported the progress on the JF-16 expansion tunnel in which a hypervelocity flow at 10 km/s is achieved (Jiang et al. 2015).

6.2.3 Effects of Tailored Interface

The reflected incident shock interacts with the interface during shock tunnel operation, and three types of wave reflections can be classified as shown in Figure 6.3. In the tailored mode, as shown in Figure 6.3a, the condition of the driver and driven gases at the interface is matched with each other, as required by Eq. (6.4). Therefore, there are no additional wave reflections created by the interaction. The tailored mode can be achieved by carefully controlling the pressure and composition of the gases in the driver and driven sections, respectively. The tailored mode is always desirable as it can effectively prolong the test time (Yuan & Jiang 2021).

The stagnation pressure (P_5) in the shock tunnel reservoir is an ideal parameter that indicates the operation mode because the nozzle flow becomes uniform when the stagnation pressure is stable.

Figure 6.4 presents the measured data at the end of the driven section. These experiments were conducted at the stagnation temperature of 2,205 K and the following observations can be obtained:

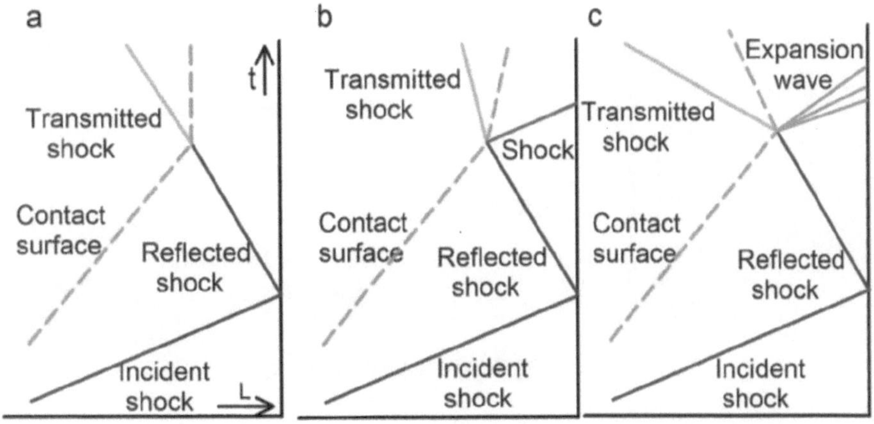

Figure 6.3 Interaction of the reflected shock with the interface: (a) Tailored mode. (b) Over-tailored mode. (c) Under-tailored mode.

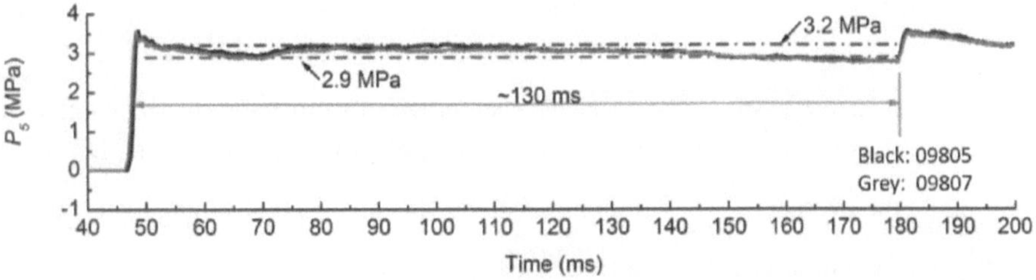

Figure 6.4 Stagnation pressure (P_5) measured at the end of the driven section.

1) The tailored-operation mode was successfully achieved in the JF-12 hypersonic flight-duplicated shock tunnel.
2) Before the reflected shock from the detonation front arrives at the end of the driven section, the uniform reservoir pressure was maintained for as long as 130 ms. This time duration is sufficiently long both for the flow pressure and heat transfer measurements, and even for the hypersonic combustion and aerodynamic force measurements.
3) The stagnation pressure histories measured in the two experiments coincide with each other, which indicates the good repeatability of the shock tunnel. The average pressure of the two experiments, from 50 ms to 175 ms, was 3.1 MPa.
4) The lower and upper dash-dot lines indicate constant values of 2.9 MPa and 3.2 MPa, respectively. From 50 ms to approximate150 ms, the stagnation pressure varied between the two lines, which means that the stagnation pressure was constant within ±5% difference only.

Figure 6.5 presents the normalized stagnation pressure change $\left(\frac{dP_5^*}{dt}\right)$ derived from the measured pressure variations. $P_5{}^*$ is the stagnation pressure normalized by

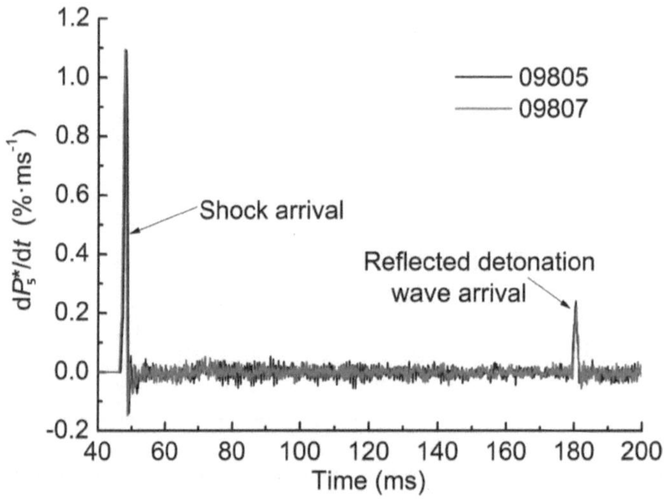

Figure 6.5 Normalized stagnation pressure change.

its average value. Two sharp peaks indicate the arrival of the incident shock wave and the reflected shock wave from the other end of the shock tunnel. During the test time, the average $\frac{dP_5^*}{dt}$ was about –0.048%/ms and –0.044%/ms, a very nice performance for a high-enthalpy shock tunnel.

Interface-tailoring is theoretically possible only for a given incident shock strength (Li et al. 2008a), and the fact imposes a severe limitation on shock tunnel operation. In practice, the over-tailored and under-tailored modes, as shown in Figure 6.6, are possible in the operation, but also acceptable if the reflected shock wave or expansion wave becomes successively weak within a limited time. After multiple reflections, the temperature or pressure changes do not occur significantly. The equilibrium temperature can be calculated by assuming that the reflection is isentropic, while the equilibrium pressure can be directly measured by a pressure sensor installed at the end of the driven section.

From the comparison between the measured stagnation pressures of different operation modes, as shown in Figure 6.6, where each pressure curve in the figure was normalized with its average value, the following remarks can be made:

1) In the under-tailored and over-tailored operation modes, it takes approximately 20 ms for the stagnation pressure to reach their equilibrium states, and approximately 100 ms of test time remains for experiments. This means that untailored operation modes are acceptable for the JF-12 hypersonic flight-duplicated shock tunnel.
2) The arrival time of the reflected shock wave from the detonation fronts in the three operation modes was 176.8 ms, 179.6 ms, and 183.5 ms, respectively. It seems that the operation mode has a minor effect on the reflected shock wave propagation.

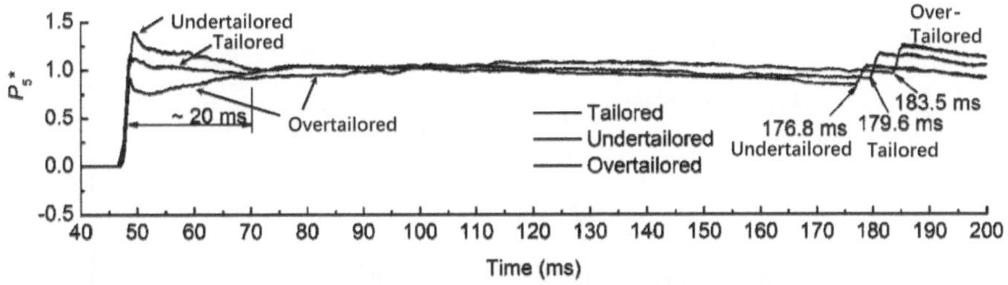

Figure 6.6 Comparison of the measured stagnation pressures in three different operation modes.

6.3 Gas Contamination Induced by Shock-Wave/Boundary-Layer Interaction

The test time of the reflected shock tunnel mainly depends on the time during which the test gas state in the reservoir chamber remains constant. If the influence of various wave systems on the gas state is neglected, the theoretical test time of the shock tunnels should be the time required for the test gas to flow completely through the nozzle throat. However, due to the influence of other factors, the actual test time of the shock tunnel is much shorter than the theoretical value. Among these factors, the test gas contamination induced by the interaction of the shock/boundary-layer/contact surface is an important one.

This is the second physical issue we plan to discuss in this chapter, being a typical kind of shock/boundary-layer interaction that takes place near the end of the driven section, as shown in Figure 6.7. The boundary layer develops while the incident shock propagates in the driven section to its end where the nozzle is attached. At the end of the driven tube is an end wall that has an opening to the nozzle. A thin diaphragm separates the driven section from the nozzle entrance. The shock reflection at the end wall generates the required total pressure and temperature, thus forming the reservoir condition in the shock tunnel. The shock/boundary-layer interaction occurs after the incident shock reflects back from the end wall. The test gas in the boundary layer flows towards the tube end while the flow in the tube center remains stationary. This flow motion causes the driver gas to mix with the test gas earlier than expected, resulting in test gas contamination that ruins the test flow quality and reduces the test time significantly (Li et al. 2008b).

The reflected shock wave bifurcation phenomenon was first reported by Mark (1958), and then Davies and Wilson applied it to the study of gas contamination problem, which has also been referred to as driver gas contamination in shock tubes and tunnels (Davies & Wilson 1969). Wilson and Chue had also carried out numerical calculations on the problem (Wilson 1993, Chue & Eitelberg 1998). Both theoretical predictions and numerical calculations showed that the phenomenon shortens the shock tunnel test time by the mixing of the test gas with driver gas and its effect is quite serious. In order to extend the effective test time of shock tunnels, the most direct method is to apply boundary layer suction technology. Burtschell had discussed this in detail (Burtschell & Cardoso 2001). However, for the detonation-driven high-enthalpy shock tunnel, due to its high operating pressure, boundary layer suction will bring potential dangers to the safe operation of the shock tunnel. So, it has been an open problem for decades.

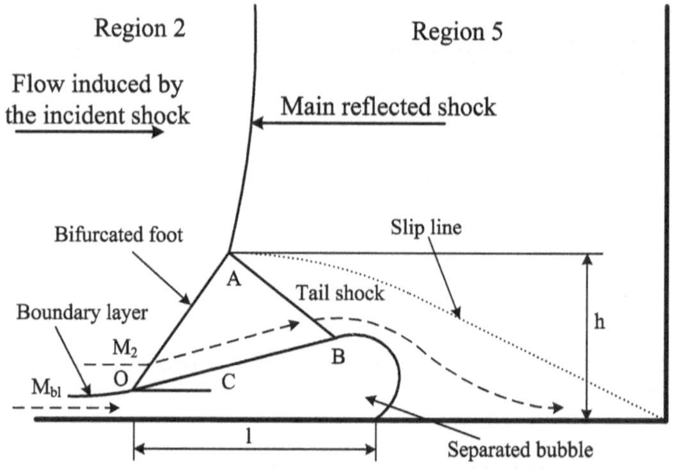

Figure 6.7 Schematic diagram of the interaction of the shock/boundary-layer at the end of the driven section.

6.3.1 Numerical Simulation of Shock-Wave/Boundary-Layer Interaction

Considering the gas contamination, Li and Jiang proposed a method of applying a circular sleeve in the shock tunnel to restrict the driver gas movement to avoid premature contamination of the test gas. Numerical calculations had been carried out to investigate the feasibility of this method and to demonstrate that the flow motion control method is indeed practical in engineering (Li et al. 2008b).

Taking a simple detonation-driven shock tunnel as an example, its configuration is schematically showed in Figure 6.8. The driven section is L_1 = 2,000 mm and its diameter D = 50 mm and D_{throat} = 9 mm. By solving the Navier–Stokes (N-S) equations with the DCD scheme, the numerical results are plotted in Figure 6.9. The figure shows the gas mass fraction distributions and pressure distributions at the end of the driven section at different times. Pressure contours are located on the left of the figure and the test/driver gas mass fractions on its right. The mass fractions of the driver gas are shown in white and the test gas in black.

The computation is initialized when the incident shock wave moves to 85 mm from the end wall. Since there is a continuously developing boundary layer after the incident shock wave, the reflected shock wave will interact with this boundary layer. Due to the influence of the boundary layer, at t = 0.0658 ms, the reflected shock wave near the wall forms a reflected shock wave system with a bifurcated structure. Because of the existence of the throat, the incident shock wave also undergoes a series of the complex processes such as reflection, diffraction, and convergence near the throat. However, the impact of these processes on the reflected shock wave only appears in the early stage of the shock reflection process, bending the reflected shock wave near the axis. As the reflected shock wave propagates upstream, the influence gradually weakens. From the pressure distribution at t = 0.0658 ms, it can be

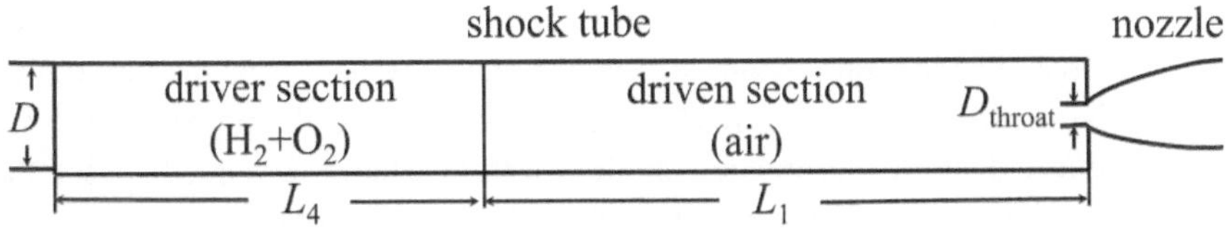

Figure 6.8 Schematic of the detonation-driven shock tunnel.

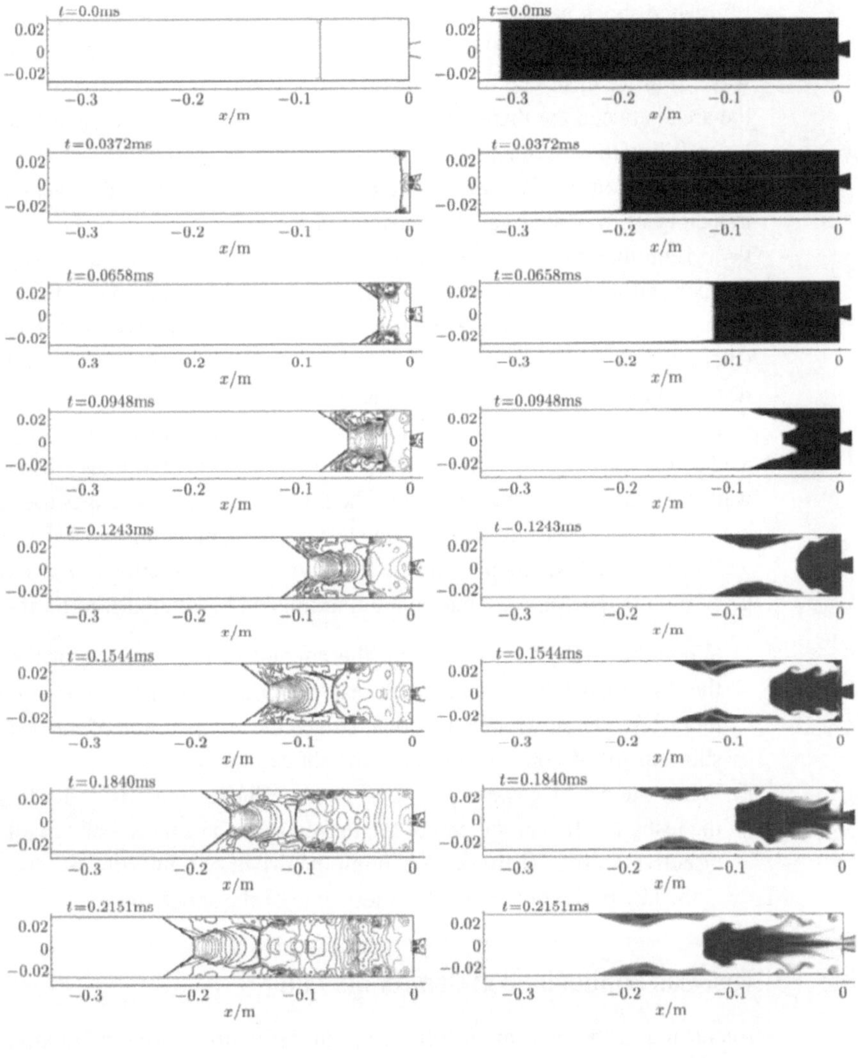

Figure 6.9 Pressure contours and driver/test gas mass fraction distributions in the reflected-shock region at successive time with the pressure distributions on the left and the driver/test gas mass fractions on the right (white: test gas; black: driver gas).

seen that the reflected shock wave near the axis had evolved into a normal shock wave, which is almost not affected by the flow field near the throat. The height of the triple-wave stem increases with the increase in the propagation distance of the reflected shock. Before meeting the contact surface, the angle between the front foot of the bifurcated shock and the wall is about 35 degrees.

When the reflected shock wave reaches the contact surface, the contact surface is deformed by the reflected shock wave with a bifurcated structure, and the driver gas penetrates into the test gas as shown by the mass fraction distribution at $t = 0.0948$ ms. The reason for this is that the contact surface near the tube wall first encounters the bifurcated shock wave, and its center part encounters the normal shock wave. The test gas wall jet induced by the reflected shock wave follows behind the reflected shock wave to move upstream, while the driver gas continues to move downstream through the area between the three wave points of the reflected shock wave and the boundary layer, filling the space left by the wall jet.

Starting from $t = 0.1243$ ms, the distribution of the gas mass fraction indicates the trajectory of the driver gas that moves to the nozzle. From the pressure distribution at $t = 0.1243$ ms, it can be seen that an upstream-moving shock wave is generated near the axis and obviously different from the ideal gas flow situation, indicating that the gas flow near the axis is experiencing two shock wave compression. After the driver gas enters into the test gas, the part that is close to the wall constitutes a wall jet, while the rest continues to move toward the end wall of the driven section and reaches the end wall at $t = 0.1544$ ms. Due to the obstruction of the end wall and the low pressure generated by the flow expansion of the nozzle, this part of the driver gas moves along the end wall and finally flows into the throat. The flow phenomenon changes the composition of the test flow in the nozzle, thus causing test gas contamination and a shortened test time.

From Figure 6.9, the mechanism underlying the interaction of the shock/boundary-layer and the resulting test gas contamination can be summarized briefly as follows:

1) When the wall jet interacts with the contact surface, mixing of the test gas with the driver gas takes place, which results in the driver gas entering the nozzle prematurely. The flow phenomenon changes the composition of the test gas, and thus shortens the effective test time of the shock tunnel.
2) When the driver gas enters into the nozzle throat, there is still a large amount of the test gas left near the end of the driven section of the shock tunnel, which is not effectively used. Therefore, delaying the driver gas entering into the nozzle throat is the key problem to extend the test time of the shock tunnel.

6.3.2 Test Gas Contamination Control Method

According to the flow motion from the investigation of the interaction of the shock wave/boundary layer/contact surface, as shown in Figure 6.9, a new concept proposed is to install a circular "sleeve" around the nozzle throat, as shown in Figure 6.10, to block the flow of driver gas along the tube wall into the nozzle, thereby increasing the effective test time of the shock tunnel.

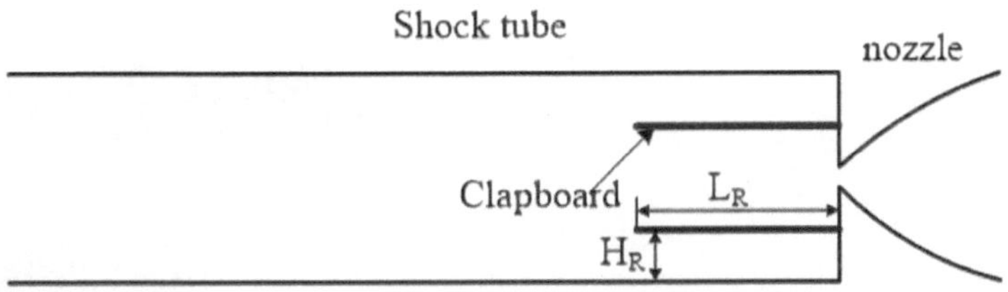

Figure 6.10 Reflected-shock region in the shock tunnel with a sleeve installation around the nozzle throat in the reservoir.

The installation position of the sleeve is shown in Figure 6.10. The length of the sleeve is L_R = 35 mm approximately equal to the distance that the reflected shock wave propagates when it encounters the contact surface. The distance between the sleeve and the tube wall is H_R = 11 mm, which is slightly larger than the thickness of the boundary layer. Other initial and boundary conditions are taken to be the same as those in the previous test case. The numerical simulations were carried out for concept verification. Pressure contours and driver/test gas mass fraction distributions in the reflected-shock region are presented in Figure 6.11 in a successive time order.

Figure 6.11 shows the pressure contours and gas mass fraction distributions in the reflected-shock region in the shock tunnel, in which a sleeve is installed around the nozzle throat. Due to the existence of the sleeve, the supersonic test gas flow after the incident shock wave forms a bow shock wave in front of the sleeve, as shown in the results at t = 0.0340 ms. When the incident shock wave moves near the front edge of the sleeve after its reflection from the end wall, the presence of the bow shock causes the reflected shock wave near the axis to protrude upstream, as shown in the results at t = 0.0904 ms. As the reflected shock wave propagates continuously to upstream, its curved part disappears and evolves into the same reflected shock structure as shown in Figure 6.9. The results at t = 0.12690 ms show that there is no significant change in the reflected shock wave structure due to the sleeve installation.

Because of the upstream motion of the reflected shock wave and the high pressure difference between the sleeve and the wall surface, the driver gas flow can move only across the front edge to the axis of the sleeve, and then into the nozzle, finally terminating the test time of the shock tunnel, as shown in the results at the t = 0.2758 ms. It can be also seen from the gas mass fraction distribution at t = 0.1269 ms when the wall jet of the test gas still exists, but the strength of the driver gas moving downstream is reduced due to the existence of the sleeve. The reason is that the spatial restriction blocks the driver gas movement and increases the pressure in the area between the sleeve and the tube wall. The driver gas flow that enters the region between the sleeve and the tube wall can no longer move toward the axis, as shown by the results from t = 0.1635 ms to 0.2387 ms.

Numerical simulations demonstrated that the test duration could be increased by 60% due to the application of the new concept. The length and height of the sleeve have important influences on prolonging the test time. If the sleeve is too long, the

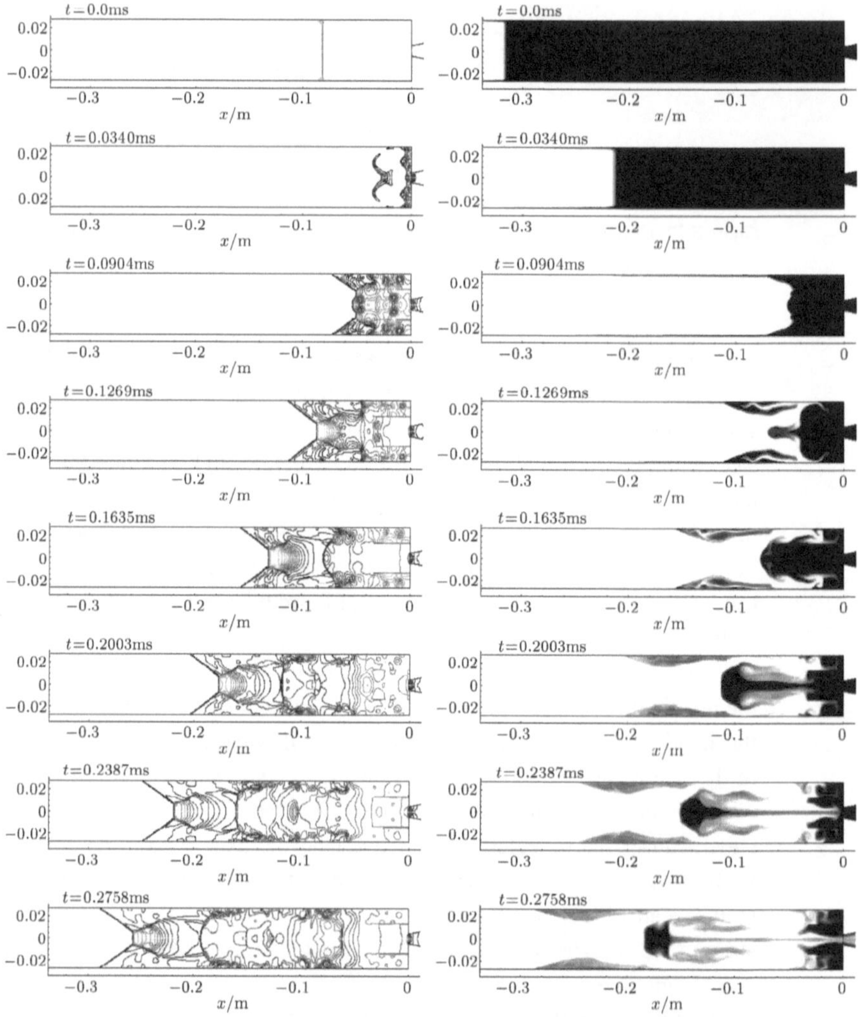

Figure 6.11 Pressure contours and driver/test gas mass fraction distributions in the reflected-shock region with a sleeve around the nozzle at successive time with the pressure distributions in left and the driver/test gas mass fractions in right (white: driver gas; black: test gas).

driver gas will enter the nozzle directly along it. If it is too short, the time when the driver gas enters the nozzle through the end of the sleeve cannot be effectively delayed. Similarly, if the sleeve is too far from the tube wall, the test gas cannot be fully utilized. If it is too close to the tube wall, the driving gas will enter the nozzle along the inner wall of the sleeve. A series of computational experiments shows that the radial distance between the tube wall and the sleeve should be slightly larger than the thickness of the boundary layer, and the length of the sleeve should be slightly less than the length of the test gas compressed by the reflected shock wave. This technique for gas contamination still needs further improvement.

6.3.3 Concluding Remarks

The new concept of adding a sleeve in the driven section to delay gas contamination is well verified with numerical simulations. The results show that a properly-designed sleeve can disrupt the flow of driver gas and effectively extend the test time of the shock tunnel. However, there is no proper experimental device that can accurately verify the accuracy of the existing theoretical and numerical simulations. Some researchers used the mass spectrometry to detect the driver gas in the test flow. Others proposed a test device consisting of a square groove and wedge to determine the composition of the test flow from the angle of shock reflection from the wedge. However, for the detonation-driven shock tunnel, these methods are of low accuracy because the difference between the specific heat ratios of the driver gas and the test gas is minor and the oblique shock angle is not very sensitive to gas composition. It seems that there is a long way to go for evaluating test gas contamination efficiently, but it is an interesting topic to try since the phenomenon is physically important and its application is also meaningful in engineering.

6.4 Shock Damping in a Vacuum Tank

The third issue related closely to the shock tunnel test time is precursor shock damping in the vacuum tank. The precursor shock wave comes from the nozzle flow starting process after the nozzle diaphragm ruptures, propagates into the test section, and then reflects back from the end wall of the vacuum tank that is applied to maintain a steady static pressure for the simulated flight altitude. The test flow will be disturbed once the reflected shock wave propagates into the test flow region, and the test time will be terminated because the test flow quality is disrupted. It would not be a problem at all for small shock tunnels because of their short test time, but becomes a severe issue for a large-scale long-test-duration shock tunnel.

Usually, the shock speed is higher than the sound speed in ambient air and may reach to about 400–500 m/s or more in the vacuum tank. If the test time is about 100 milliseconds, the shock-propagating distance is about 50 meters. To avoid reflected shock perturbations in the test flow, the length of the vacuum tank must be longer than half of the distance along which the shock wave propagates during the test duration. However, if one wants to develop a shock tunnel with a hundred of milliseconds of test duration, the length of the vacuum tank will become a serious problem for its engineering design and shock tunnel operation. For the JF-12 shock tunnel, it is necessary to provide a long enough distance for the precursor shock propagation in more than 200 ms to ensure a 100-ms test time. Assuming the precursor shock speed is 500 m/s, the length of the vacuum tank required is as long as 50 meters. It is not realizable in engineering design for a vacuum tank of 3.5 diameter because the tank stiffness will get weaker and weaker as its length becomes longer and longer. Therefore, the length reduction of the vacuum tank becomes a meaningful work to consider for the large-scale shock tunnel development (Jiang & Yu 2014).

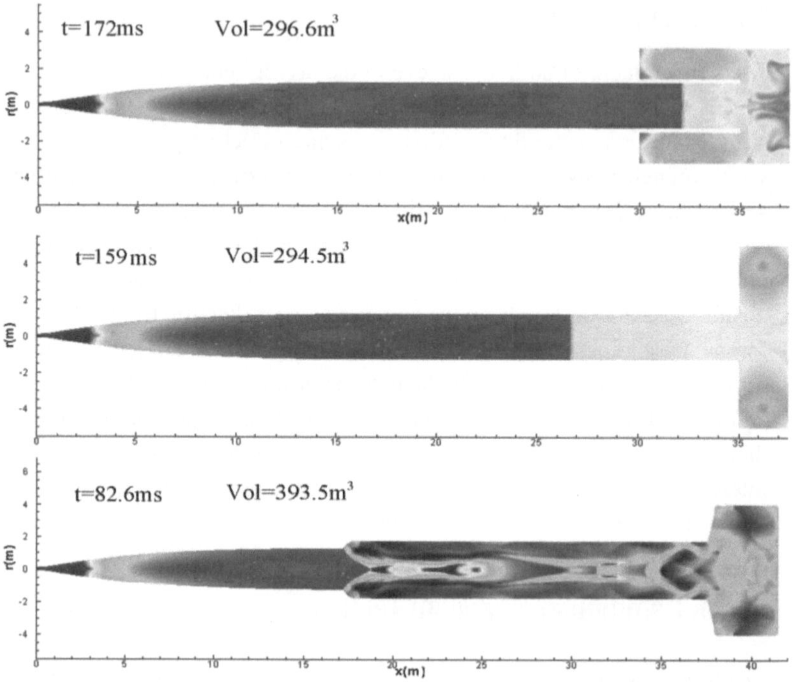

Figure 6.12 Precursor shock propagation in the vacuum tank and test section.

Several aerodynamic configurations were proposed based on shock dissipation concept for the vacuum tank and were investigated with numerical simulations. Results from two configurations are presented in Figure 6.12. The upper figure shows the pressure distribution in the E-shaped vacuum tank and the escaped time is about 172 ms calculated from the beginning of the nozzle-starting process. The lower figure shows the pressure distribution in the T-shaped vacuum tank and the escaped time is about 159 ms. The E-shaped vacuum tank demonstrates its good aerodynamic performance. For such an aerodynamic design, the delay time of more than 200 ms is achieved for a vacuum tank being 30 meters in length and 3.5 meters in diameter.

The physical process occurring in the vacuum tank is described as follows: the precursor shock that originates from the center portion of the E-shaped vacuum tank is reflected from a wedge installed at the end of the tank, thus splitting into two shocks that propagate laterally into two adjacent tanks that are arranged in parallel to the main tank forming an E-shaped configuration. This process results in strong shock dissipation so that the reflected shock is weakened. Furthermore, the longer shock propagation distance delays the reflection of the wave back into the test section. Consequently, the reflected shock wave gets weaker and weaker as it propagates back into the main vacuum tank and the disturbance on the test flow region is delayed greatly. As an additional benefit, the tank design is readily realizable in engineering and a lot of laboratory space is saved.

6.5 Other Issues Related with Test Time

The three key issues discussed in Sections 6.2–6.4 appear to be important for long-test-duration shock tunnels. All the phenomena are worthy of further investigations to improve performance. However, there are still other issues that cannot be ignored since they affect the test time more or less in different ways, such as ignition technique, diaphragm rupture, and pressure loss from diaphragm rupture. Experimental data will be presented to demonstrate their roles in long-test-duration shock tunnels and the related physical issues will be also discussed (Yuan & Jiang 2021).

6.5.1 Ignition Technique

When the shock tunnel adopts a detonation driver, the direct initiation of detonation, as opposed to the transition from deflagration to detonation (DDT), is always needed to achieve a perfect incident shock wave in the driven section. Figure 6.13 shows the reservoir pressure variations from two runs with direct initiation of detonation and DDT initiation, respectively. The experimental data show that the effect is quite obvious. The DDT process induces an unsteady stagnation flow with large pressure fluctuations, as seen in the black curve, and the direct initiation of detonation yields a uniform reservoir pressure profile as shown in the gray curve. The stagnation temperature cannot be obtained by solving the shock wave relationship because the shock wave structure in the driven section is too complex when the DDT process takes place. As a result, the freestream test flow parameters are unknown. Furthermore, the calculated results show that the DDT initiation should be avoided in shock tunnel operation.

The initiation of direct detonation is one of the key techniques for detonation-driven shock tunnels. Previous experimental data reveal a fact that the ignition energy must exceed a certain threshold value. The amount of energy is also related to the mixture ratio at the given pressure, temperature, and also the igniter. According to detonation theory, the strong blast wave generated by a powerful igniter upon the rapid deposition of its energy decays asymptotically to the C-J detonation. If the igniter energy is less than the critical value, the reaction zone will be progressively decoupled from the blast wave as the precursor shock wave decays, finally resulting in deflagration (Lee 2008, Jiang and Teng 2022).

For a large-diameter detonation driver filling with a high pressure mixture, the ignition power required for the igniter is very high. In the laboratory, the electrical exploding wire is one of the low-cost igniters to generate the required energy. Experiments were conducted to investigate the explosion process and Figure 6.14 presents the results captured by the high-speed camera. The imaging frame rate was 5,000 fps and the exposure time was 200 μs. Because the explosion was accompanied by radiation, we assumed that the gray scale of the image represents the energy release rate generated by the exploding wire. The normalized radiation intensity and radius are shown in Figure 6.15. It can be clearly seen that the energy-increasing process to the maximum ignition energy took approximately 2.6 ms. The long release time means that it is not feasible to initiate direct detonation by increasing the quantity of the electrical exploding wire.

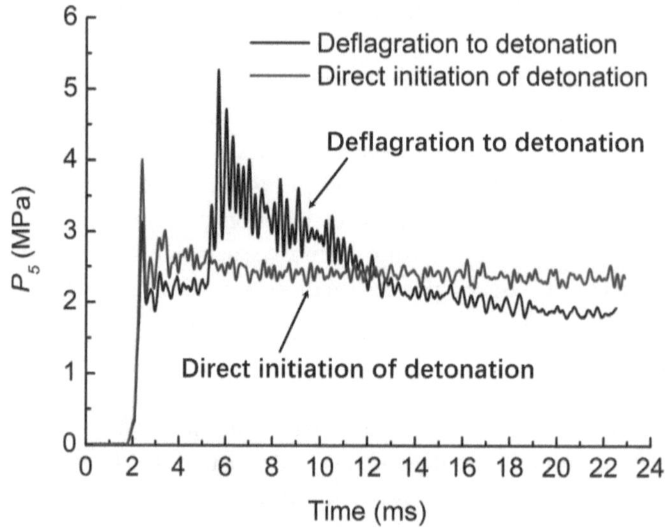

Figure 6.13 Effect of the transition from deflagration to detonation on the reservoir pressure variation.

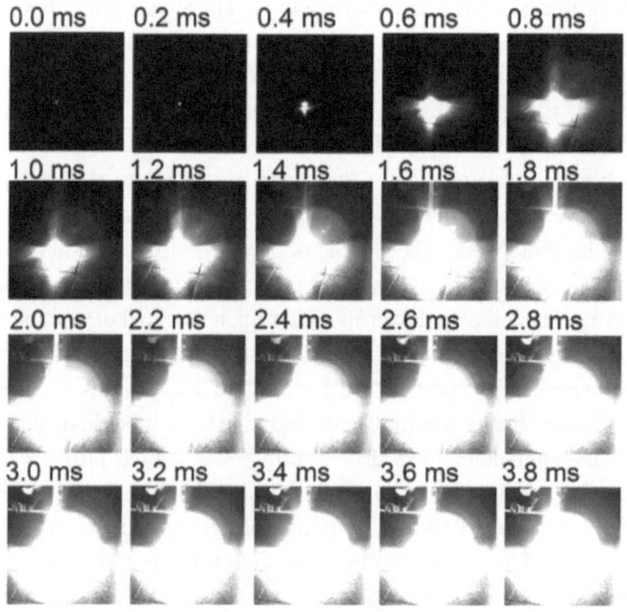

Figure 6.14 Measured explosion process of the electrical exploding wire.

To solve this problem, a special ignition tube with a length of 1,300 mm and an inner diameter of 60 mm was manufactured and installed perpendicularly in the driver section. A stoichiometric mixture of hydrogen and oxygen was filled into the ignition tube and ignited by the electrical exploding wire. The thus-produced high-temperature

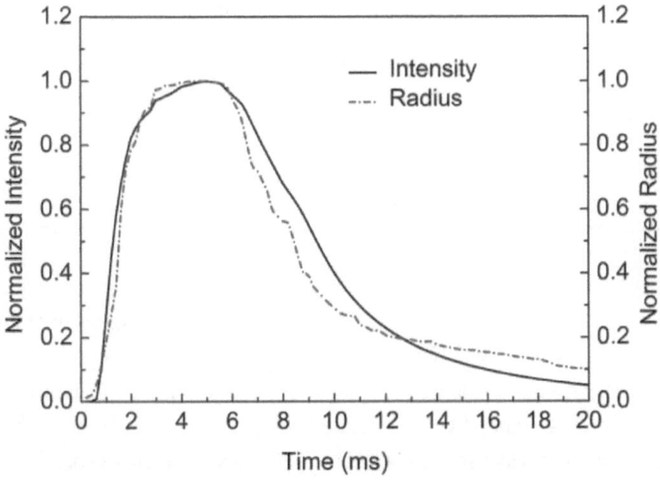

Figure 6.15 Derived normalized radiation intensity and radius.

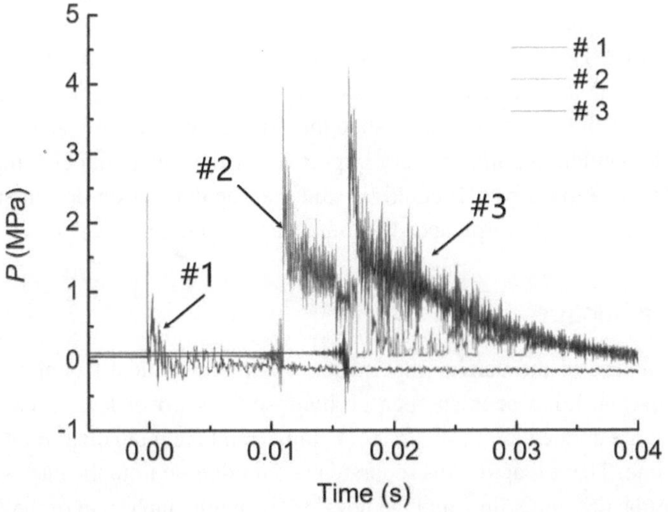

Figure 6.16 Typical outputs of the piezoelectric pressure sensors for measuring detonation wave velocity.

gases were accelerated and injected into the driver section, and then, direct detonation was triggered successfully.

To confirm the ignition performance, three piezoelectric pressure sensors, labelled #1, #2, and #3, respectively, were mounted along the driver section to measure pressure variations. The sensor distance measured from the right end of the driver section is 90, 25,522, and 37,772 mm, respectively. Experiments were conducted under different operating conditions and Figure 6.16 shows the typical sensor outputs.

Table 6.1 Measured speed of the detonation wave in the driver section

Mole ratios	V_{theory}	V_{12}	V_{23}
H2:O2:N2=2:1:2.0	2,296	2,499	2,501
H2:O2:N2=2:1:3.6	2,055	2,276	2,284
H2:O2:N2=2:1:3.7	2,048	2,252	2,279
CH4:O2:N2=1:1:2.7	2,148	2,531	2,574

The pressure profiles of sensors #2 and #3 clearly show the detonation front arrival and the following expansion waves. The peak value of sensor #1 is lower than that of sensors #2 and #3, mainly owing to the flow expansion generated after the diaphragm rupture.

The detonation wave speed was calculated using the time-of-flight method. Table 6.1 presents the detonation wave speeds, where V_{12} denotes the average speed between sensor #1 and sensor #2, and V_{23} denotes the speed between sensor #2 and sensor #3. The theoretical detonation speed was also calculated with the help of the Gaseq program (a chemical equilibrium program developed by Chris Morley, please refer to www.gasEq.co.uk for detailed introduction) for reference. From Table 6.1, it can be confirmed that the average speed V_{23} is almost the same as V_{12}, which indicates a quite stably-propagating detonation wave. Because sensor #1 was installed near the ignition position, the constant speed indicates that direct detonation was in fact initiated in the driver section. Hence, the design of the ignition tube has satisfied the experimental requirements.

Direct detonation initiation is very important for the shock tunnel equipped with a backward detonation driver. It could be said that the detonation-driven shock tunnel cannot provide good performances without direct detonation initiation.

6.5.2 Diaphragm Rupture

The manufacture of the diaphragm and its rupture are critical techniques for shock tunnel operation. It has been an open problem so far for decades. Three kinds of the diaphragms are applied in the JF-12 shock tunnel, and the main diaphragm is the most important one. These diaphragms separate the damping section, the backward detonation driver, the driven section, and the nozzle. The main diaphragm of the JF-12 shock tunnel has a diameter of 400 mm and is placed between the backward detonation driver and the driven section. The diaphragm function is to sustain the initial pressure before operation, to quickly break under the pulse load of the detonation wave, and to keep its mass without any fragment loss. This is because even small diaphragm pieces can cause great damage to the nozzle throat and the test model. Another requirement for the diaphragm is that its thickness must be as thin as possible to reduce the pressure loss generated from diaphragm rupture, and this imposes a special demand on the diaphragm material. Based on previous experiences, the scored groove can be used to control the rupture process and reduce diaphragm fragments into the test flow. Therefore, many experiments had been conducted to determine an appropriate material and the scoring methods in the JF-12 commissioning process.

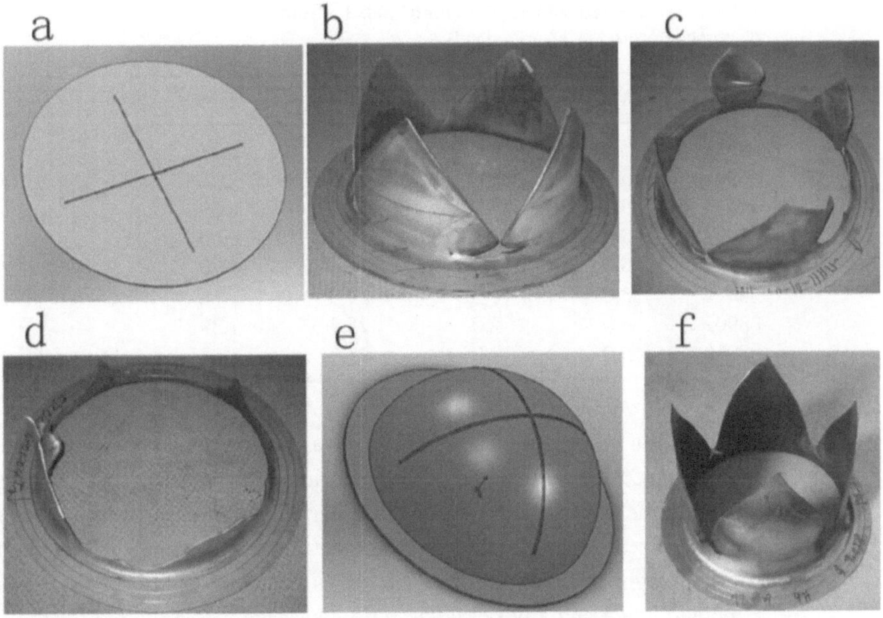

Figure 6.17 Main diaphragms: (a) Plain diaphragm; (b) Ruptured plain diaphragm (P_{4i} = 2.0 MPa); (c) Ruptured plain diaphragm (P_{4i} = 2.5 MPa); (d) Ruptured plain diaphragm (P_{4i} = 3.0 MPa); (e) Spherical diaphragm; (f) Ruptured spherical diaphragm (P_{4i} = 3.0 MPa).

For the JF-12 shock tunnel, ST14 cold-rolled steel was selected owing to its good tensile property. Four petals were scored on one side of the diaphragm for performance tests. The plain diaphragm, as shown in Figure 6.17(a), was used under a relatively low operating pressure. Figure 6.17(b) shows the ruptured diaphragm at an initial driver pressure of 2.0 MPa. From its weight, it is found that no fragments were produced during the rupture. When the initial driver pressure was increased to 2.5 MPa, it can be observed that the diaphragm petals were bent backward by the detonation wave. However, there are still no fragments broke off from the diaphragm. The initial driver pressure of 3.0 MPa is the limit for the plain diaphragm and ST14 cold-rolled steel is well demonstrated.

The rupture time of the plain diaphragm was determined from shock arrival at the piezoelectric pressure sensor downstream of the diaphragm. Table 6.2 shows the measured rupture time of the plain diaphragm. Experiments were conducted at the same incident shock Mach number, and the initial pressure of the driver section and the driven section can be varied. The average rupture time is 2.31 ms, 1.68 ms, and 1.385 ms, while the initial driver pressure is equal to 1.0 MPa, 1.5 MPa, and 2.0 MPa, respectively. The reason for doing this is that the detonation load increases when the initial pressure gets higher and higher.

With further increase of the initial driver pressure to 3.0 MPa, all four petals were detached from the diaphragm, as shown in Figure 6.17(d). This puts a limit on the possible operation capability of the shock tunnel with the plain diaphragm. A spherical diaphragm was designed by Zhao (2005), and manufactured into its configuration as

Table 6.2 Measured diaphragm rupture time, a 3.5-mm-thick diaphragm, scored to 1.2 mm depth

Run no.	*Ms*	P_{4i} (MPa)	P_1 (kPa)	Time (ms)
08920	5.0	1.0	12.6	2.32
08921	5.0	1.0	12.6	2.22
08923	5.0	1.0	12.6	2.39
07923	5.0	1.5	19.0	1.64
08929	5.0	1.5	19.0	1.73
08912	5.0	1.5	19.0	1.68
08902	5.0	2.0	25.3	1.39
08930	5.0	2.0	25.3	1.38

shown in Figure 6.17(e). The steel sheet was pre-stamped to be a spherical diaphragm with a dome. One advantage of the spherical diaphragm is that it can decrease significantly the stress at the root of the spherical diaphragm. Another advantage is the reduction in the rupture time for fragment generation control. The ruptured spherical diaphragm after its performance test is shown in Figure 6.17(f). The four petals remained intact and a possible large flow passage was provided for the reservoir gas expansion. The experiment indicated that the spherical diaphragm performed well under the high initial driver pressure.

6.5.3 Pressure Loss Caused by Diaphragm Rupture

To achieve a high-quality flow field in the test section, the initial parameters of the driven and driver sections should be determined. One-dimensional isentropic expansion, the normal shock-wave equations, and the detonation theory are helpful for obtaining the initial parameters of the driver and driven sections. However, the reservoir pressure loss caused by diaphragm rupture cannot be theoretically calculated. Thus, experiments have to be conducted to solve this problem. In the experiments, it is very difficult to measure effectively the reservoir pressure. An alternative method is to measure the incident shock Mach number (*Ms*), and then calculate the reservoir pressure (P_4) according to the shock-wave equations. The pressure loss (ΔP) is defined in Equation (6.5)

$$\Delta P = \frac{(P_4^{'} - P_4)}{P_4^{'}} \tag{6.5}$$

where $P_4^{'}$ is the detonation pressure, and the time-of-flight method was used to measure *Ms*. Three piezoelectric pressure sensors were installed along the driven section, with the distances from the main diaphragm being 1,952 mm, 8,032 mm, and 14,112 mm, respectively. The diaphragm was a 3.5-mm-thick plain diaphragm, with the nick depths of 1.2 mm, 1.0 mm, and 0.8 mm, respectively. The measured results are presented in Table 6.3. According to these data, the pressure loss is approximately 19% with a nick depth of 1.2 mm, and increased to 28% when the nick depth is 0.8 mm. This dataset provides a theoretical basis for determining the initial parameters of detonation-driven shock tunnels.

Table 6.3 Pressure loss due to diaphragm rupture

Run No.	Detonation H2:O2:N2	Nick depth (mm)	Ms	P_1 (kPa)	P_4 (MPa)	P'_4 (MPa)	Pressure loss%
08920	2:1:3.76	1.2	5.0	12.6	2.28	2.82	19.15
07923	2:1:3.76	1.2	5.0	19.0	3.46	4.26	18.78
08930	2:1:3.76	1.2	5.0	25.3	4.64	5.70	18.60
08407	2:1:2.75	1.0	7.1	6.0	4.56	6.07	24.88
08413	2:1:3.76	1.0	5.0	38.0	6.56	8.60	23.72
08420	2:1:3.76	1.0	5.3	18.0	4.31	5.70	24.39
06411	2:1:3.76	0.8	4.9	38.0	6.17	8.60	28.26
06409	2:1:4.00	0.8	4.7	41.0	6.10	8.46	27.90

6.6 Concluding Remarks

Theories and methods for developing the long-test-duration shock tunnel were introduced in this chapter. Besides the high-power detonation driver, the tailored-interface-matching condition for detonation-driven shock tunnels, the shock/boundary-layer interaction control concept at the end of the driven section, and the E-shaped aerodynamic configuration of the vacuum tank contribute much to the extension of the JF-12 test time. There are still other parameters that play important roles in test time extension, such as direct detonation initiation, diaphragm manufacture and rupture. It is believed that if one can make most parts of the shock tunnel work well, a long test duration may be realized. The goal may seem far away, but it is meaningful to try. Achieving long test duration opens the opportunity to many experiments which could not otherwise be feasible.

Nomenclature

a	sound speed
CH_4	Methane
D	diameter
H_2	hydrogen
L	length
Ms	shock Mach number
N_2	nitrogen
O_2	oxygen
P	pressure
P'	detonation pressure
P_5^*	normalized stagnation pressure
t	time
u	velocity

V	detonation wave speed
γ	ratio of specific heats
ΔP	pressure loss

Subscripts:

0	total or stagnation condition
1	initial state in driven gas
2	state behind incident or primary shock
3	state between the expansion wave and contact surface
4	initial state in driver gas
5	state behind reflected shock in test gas
7	state between the reflected shock wave and contact surface
i	initial
R	sleeve

References

Anderson, J. D. (1989). *Hypersonic and High Temperature Gas Dynamics*. New York: McGraw-Hill Book Company.

Bertin, J. J. (1994). *Hypersonic Aerothermodynamics*. Washington, DC: AIAA Education Series.

Bertin, J. J. and Cummings, R. M. (2003). Fifty Years of Hypersonics: Where We've Been, Where We're Going. *Progress in Aerospace Sciences*, **39**(6–7), 511–536.

Bertin, J. J. and Cummings, R. M. (2006). Critical Hypersonic Aerothermodynamic Phenomena. *Annual Review of Fluid Mechanics*, **38**, 129–157.

Burtschell, Y. and Cardoso, M. (2001). Numerical Analysis of Reducing Driver Gas Contamination in Impulse Shock Tunnels. *AIAA Journal*, **39**(12), 2357–2365.

Chue, R. S. M. and Eitelberg, G. (1998). Studies of the Transient Flows in High Enthalpy Shock Tunnels. *Experiments in Fluid*, **25**, 474–486.

Davies, L. and Wilson, J. L. (1969). Influence of a Reflected Shock and Boundary Interaction on Shock Tube Flows, *Physics of Fluid*, **12**(5), S37–S43.

Jiang, Z. and Yu, H. (2014). Experiments and Development of Long-Test-Duration Hypervelocity Detonation-driven Shock Tunnel (LHDst). *AIAA Paper 2014–1012* (Invited Lecture).

Jiang, Z. and Yu, H. (2017). Theories and Technologies for Duplicating Hypersonic Flight Conditions for Ground Testing. *National Science Review*, **4**(3), 290–296.

Jiang, Z., Hu, Z., Wang, Y., and Han, G. (2020a). Advances in Critical Technologies for Hypersonic and High-enthalpy Wind Tunnel. *Chinese Journal of Aeronautics*, **33**(12), 3027–3038.

Jiang, Z., Li, J., Hu, Z., Liu, Y., and Yu, H. (2020b). On Theory and Methods for Advanced Detonation-driven Hypervelocity Shock Tunnels. *National Science Review*, **7**(7), 1198–1207.

Jiang, Z. and Teng H. (2022). *Gaseous Detonation Physics and Its Universal Framework Theory*, Singapore: Springer.

Jiang, Z., Wu, B., Gao, Y., Zhao, W., and Hu, Z. (2015). Development of the Detonation-driven Expansion Tube for Orbital Speed Experiments. *Science China*, **58**(4), 695–700.

Jiang, Z., Zhang, Z., Liu, Y., Wan, C., and Luo, C. (2021) The Criteria for Hypersonic Airbreathing Propulsion and Its Experimental Verification. *Chinese Journal of Aeronautics*, **34**(3), 94–104.

Jiang, Z., Zhao, W., Wang, C., and Takayama, K. (2002). Forward-running Detonation Drivers for High-enthalpy Shock Tunnels. *AIAA Journal*, **40**(10), 2009–2016.

Lee, J. (2008). *The Detonation Phenomenon*. New York: Cambridge University Press.

Li, J., Feng, H., Jiang, Z., and Yu, Y. (2008a). Numerical Computation on the Tailored Shock Mach Numbers for a Hydrogen-oxygen Detonation Shock Tube. *Acta Aerodynamica Sinica*, **26**(3), 291–296 (In Chinese).

Li, J., Feng, H., and Jiang, Z. (2008b). Gas Contamination Induced by the Interaction of Shock/Boundary Layer in Shock Tunnel. *Chinese Journal of Theoretical and Applied Mechanics*, **40**(3), 289–296 (In Chinese).

Lu, F. K. and Marren, D. E. (2002). *Advanced Hypersonic Test Facility, Progress in Astronautics and Aeronautics*. Vol. 198, Reston, VA: AIAA.

Mark, H. (1958). The Interaction of a Reflected Shock with the Boundary Layer in a Shock Tube. *NACA TM No. 20050019632*.

Olivier, H., Jiang, Z., Yu, H., and Lu, F. K. (2002). *Detonation-driven Shock Tubes and Tunnels, Advanced Hypersonic Test Facilities*. Reston, VA: AIAA, 135–203.

Ozer, I. and Friedrich, S. (2016). *Experimental Methods of Shock Wave Research*. Heidelberg, New York, Dordrecht, London: Springer.

Park, C. (1990). *Nonequilibrium Hypersonic Aerothermodynamics*. New York: Wiley.

Wilson, G. J. (1993). Time-Dependent Simulation of Reflected-Shock/Boundary Layer Interaction. *AISAS-93-0480*.

Yuan, C. and Jiang, Z. (2021). Experimental Investigation of Hypersonic Flight-duplicated Shock Tunnel Characteristics. *Acta Mechanica Sinica*, **37**(3), 420–431

Zhao, W. (2005). Performance of a Detonation Driven Shock Tunnel. *Shock Waves*, **14**, 53–59.

7 Methods for Designing High-Enthalpy Flow Nozzles

Hypersonic high-enthalpy wind tunnels have been a challenge in ground testing facilities in aerospace research for decades and test flow uniformity is one of the most important parameters for evaluating wind tunnel performance. In regard to performance requirements, theories and methods for designing hypersonic flow nozzles at high enthalpy conditions are quite difficult, but very interesting, topics, especially when dissociation of air molecules takes place in wind tunnel test-gas reservoirs. In this chapter, fundamental theories and important methods for nozzle design are reviewed with an emphasis on two-dimensional axisymmetric nozzles for hypersonic high-enthalpy wind tunnels, including the method of characteristics (MOC), the graphic design method, the Sivells method, the theory for boundary layer correction, and computational fluid dynamics (CFD)-based design optimization methods. These theories and methods had been proposed based on several physical issues, respectively, which play important roles in nozzle flow expansion processes. These issues cover the expansion wave generation and reflection, boundary layer development, and real-gas effects of hypersonic high-enthalpy flows. Difficulties arising from applications of these methods in hypersonic high-enthalpy nozzle design are discussed in detail, and state-of-the-art nozzle design technologies that have been reached over decades are summarized with some brief comments. Finally, the prospects for hypersonic nozzle design methods and their numerical and experimental verifications are provided with the authors' viewpoint for readers' reference.

7.1 Introduction to Aerodynamic Nozzle Design

Wind tunnels are crucial to the development of aerospace technology and flight vehicles. Wind tunnel testing and the research on high-enthalpy flows for hypersonic vehicles impose higher requirements on the performance of hypersonic wind tunnels. To simulate the high-enthalpy flow in a wind tunnel, not only does the uniform flow produced from its nozzle need to reach a high Mach number (generally $Ma \geq 5$), but also the enthalpy of the test gas must be high enough to create the appropriate high-temperature effects. Some scholars often defined such a flow as "hypervelocity" (Stalker 1989) to distinguish it from the low-velocity high-Mach number flows produced by conventional hypersonic wind tunnels.

In the opinion of Stalker (1989), the reflected-shock tunnel is the best ground facility to simulate real-gas effects at high temperatures. The reflected tunnel mode makes the high-temperature air source "what you produce is what you use," which greatly reduces the heat load of the wind tunnel. To improve the performance of the shock tunnel, various powerful driver techniques have been developed in the world since the mid-1960s. Of those techniques, the free-piston driver and the detonation driver are the main techniques for high-enthalpy wind tunnels. With the free-piston driver technique proposed by Stalker (1967), the test gas can be compressed and heated to obtain a test flow with a velocity of 7–8 km/s and an enthalpy of up to 40 MJ/kg. However, the effective test duration is only 1–3 ms, which represents a huge challenge to meet test time requirements. Nevertheless, the free-piston shock tunnel has become one of the most commonly utilized facilities in researching real-gas effects, because it can generate high Reynolds number and high Mach number flows simultaneously and can reproduce the high-stagnation-temperature flight conditions of hypersonic vehicles in Earth's atmosphere. Bird (1957) was the first to propose using a detonation-driven shock tube to generate a high-enthalpy gas source, and conducted computations and analysis on the detonation-driver techniques that are applied at the upstream end of the driver section and near the main diaphragm, respectively. Yu (1989) proposed to add a damping tube at the upstream end of the driver section to reduce the risk of high loading arising from the ultra-high-pressure reflected detonation, so that the detonation-driven technique can be used to produce high-enthalpy flows. Jiang and Yu (2017) presented a long-test-duration shock tunnel theory that extended the test duration of the reflected-detonation-driven shock tunnel to 130 ms. The JF-12 shock tunnel developed based on this theory was capable of reproducing air flows at Mach numbers from 5 to 9 and for an altitude of 25–50 km. But compared with conventional hypersonic tunnels, shock tunnels have much shorter effective test durations. Therefore, the produced high-quality test flows have become a key factor to evaluate the success of the shock tunnel. Furthermore, the impact of real-gas effects on the nozzle contour should be considered in the design of high-enthalpy shock tunnels. In order to obtain high-quality test flows, contour corrections must be made to hypersonic tunnel nozzles which are designed with conventional methods.

In the 1920s, MOC proposed by Prandtl and Busemann (1929) laid a theoretical foundation for the design of modern wind tunnel nozzles. Many MOC-based nozzles have been designed for supersonic and hypersonic wind tunnels. In the 1940s and 1950s, a variety of graphic design methods (modified MOC) have been proposed for designing high-velocity nozzles (Busemann 1931, Preiswerk 1940, Puckett 1946). However, with the increased requirement on flow quality, these nozzles designed by the graphic method could no longer work well. On the basis of MOC, various analytic design methods have also been developed for nozzle design (Foelsch 1946, 1949; Cresci 1958; Sivells 1969, 1970, 1978). Based on radial flow assumptions, the nozzle contour can be created through the MOC method using theoretical and empirical formulas. Proposed by Sivells, the axial-Mach-number-distribution-based approach has greatly improved the quality of nozzle flow. This classic design method for high-velocity nozzles is still in use today.

With the in-depth exploration of hypersonic flight, it is found that real-gas effects have significant impacts on the aerodynamic characteristics of the flight vehicle, bringing high-enthalpy flow to attention. Since the 1940s and 1950s, some scholars have started exploring the design methods of high-enthalpy nozzles to obtain high-quality test flows for hypersonic flight. The earliest methods only focus on the impact of the ratio of specific heats γ (Guentert et al. 1959, Enkenhus & Maher 1962), and adopt MOC for the nozzle contour computation. In practice, it is found that certain design errors are presented if only the change of γ is considered. But such errors are unacceptable for hypervelocity and high-enthalpy nozzles. Therefore, CFD is used for nozzle contour computation and optimization in order to meet the quality requirements for the test flow. In the early 1990s, a better understanding of the limitations of the MOC in the design of hypersonic nozzles has been gained. MOC is effective for the design of nozzles with a Mach number of 8 or less. To design a nozzle with Mach numbers more than 8, it is recommended to perform the contour design by solving the parabolized Navier–Stokes (PNS) equations (Benton 1990). Many scholars (Korte 1992, Korte, Hedlund, Anandakrishnan 1992a, Korte, Kumar, Singh, Grossman 1992b, Korte, Kumar, Singh, White 1992c, 1993, Keeling 1993, Korte & Hodge 1994, 1995, 2000, Tolle 1997, Shope 2006, Chan et al. 2018) have applied CFD to optimize the design of hypersonic and high-enthalpy nozzles. They couple different optimization algorithms into the CFD solver, and then automatically iterate to solve the equations and get optimized nozzle contours. The CFD solver coupled with optimization methods improves the flow quality of hypersonic nozzles significantly, but the computational program and the optimization process of these methods are extremely complicated. The robustness of the solvers is also poor. Then, Gaffney (2006) and Tang et al. (2019) decoupled the computational process of the CFD solver and the optimization method, which greatly reduces the complexity of the design process without any impact on nozzle design efficiency and flow quality. Thus, such methods are convenient for engineering applications.

Most of the literature on design theories and methods of wind tunnel nozzles have been published before the 1970s. Limited by computer technology, the studies at that time tend to use analytic and semi-analytic methods. With the rapid development of the computer and CFD technologies, the computation and simulation of nozzle flows have become more accurate. The design method using CFD with computer-aided optimization has therefore drawn considerable attention. Accordingly, design methods for high-enthalpy nozzles have been improved. Several improved design theories and correction methods for high-temperature gas effects are brought up. Nevertheless, designing a high-enthalpy nozzle with superior flow quality remains a challenge due to the complexity of high-temperature reacting gas flow and limited by the equations for physicochemical characteristics of nonequilibrium flow. The higher the Mach number and the enthalpy of the gas flow, the larger the deviation between the actual flow parameters and the design results. Hypersonic flight technology is the core technology of the future aerospace industry. Wind tunnel experiments are its main research means. Therefore, research on design theories and methods of hypersonic nozzles is of great significance.

7.2 Fundamentals of Nozzle Flows and Hypersonic Nozzles

In a reflected-shock tunnel, the nozzle is placed between the driven section and the test section. The air is heated in the driven section through the shock wave. Then, the generated high-temperature and high-pressure air flows through the expanded nozzle and reaches the test section quickly to form a uniform test flow (see Figure 7.1). The nozzle contour design is closely related to the geometric layout and thermodynamic state in the driven and test sections. This chapter focuses on the design of two-dimensional axisymmetric nozzles and the high-enthalpy flows produced from such nozzles, which can be considered as the basis for designing other types of nozzles.

The reservoir condition of the driven section is determined by the state of the reflected shock wave, which is filled with a high-temperature reacting gas. The expansion and accelerated air motion in the high-enthalpy nozzle have complicated physiochemical reaction mechanisms and high-temperature gas effects, making it difficult to accurately evaluate them with either computational simulation or experimental measurements. Therefore, early nozzle design methods adopt a quasi-one-dimensional chemical equilibrium flow model for flow analysis. Suppose that the high-enthalpy flow is inviscid, heat insulated, and the nozzle flow under chemical equilibrium is isentropic. In the nozzle flow analysis, the conclusion regarding isentropic flow can be used to supersede the momentum equation or the energy equation. Generally, the condition of chemical equilibrium does not cause the system to be irreversible. Namely, if a deviation occurs in a reaction-equilibrium system with the initial pressure p and initial temperature T due to the change of certain conditions, and the values return to initial p and T at the final state, the chemical compositions will also go back to the initial composition. If chemical equilibrium is reversible, any equilibrium chemically reacting nozzle flow that is heat insulated without shock waves and viscosity will be isentropic (Tong et al. 1990). For an equilibrium chemically reacting gas, the two-dimensional axisymmetric nozzle flow can be simplified as a quasi-one-dimensional flow. The cross-sectional area is a variable $A = A(x)$, but any flow parameter is constant across any axial section. The flow state and the area ratio A/A^* at any given location depend on three parameters, that is, p, T, and the local speed u, and satisfy the equations below (Anderson 2006):

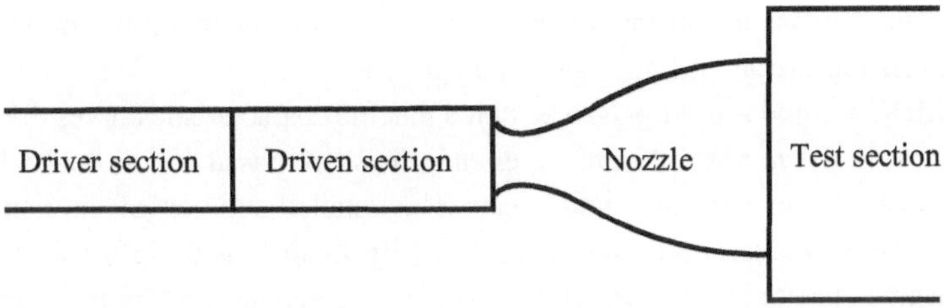

Figure 7.1 Basic configuration of the reflected-shock tunnel.

$$\frac{A}{A^*} = g_1\left(p_0, T_0, u\right) \tag{7.1}$$

$$\frac{T}{T_0} = g_2\left(p_0, T_0, u\right) \tag{7.2}$$

$$\frac{p}{p_0} = g_3\left(p_0, T_0, u\right) \tag{7.3}$$

For a calorically perfect gas (specific heat ratio $\gamma = 1.4$), the nozzle flow characteristics are governed by the local Mach number only, namely:

$$\frac{A}{A^*} = f_1\left(Ma\right) \tag{7.4}$$

$$\frac{T}{T_0} = f_2\left(Ma\right) \tag{7.5}$$

$$\frac{p}{p_0} = f_3\left(Ma\right) \tag{7.6}$$

Thus, the equilibrium flow through the nozzle can be simplified as an isentropic quasi-one-dimensional flow, based on which the expansion flow characteristics can be explored. This has been demonstrated in detail in the book by Anderson (2006). However, the simplification does not apply to nonequilibrium flows.

The nonequilibrium flow phenomenon adds to uncertainties in the nozzle design for the high-enthalpy wind tunnel. Considering the complexity of nonequilibrium flows, Hall and Treanor (1968) have limited the flow analysis to the steady state. In 1969, a new solution for nonequilibrium flow, which was proposed by Anderson (1970a, 1970b), involves the unsteady state. By solving the equations with a time-advancing finite-difference method, this solution avoids the problems with the steady state analysis (Anderson 2003). In many research papers and books, the analyses for nonequilibrium nozzle flow assume that local equilibrium is reached at the throat; downstream of the throat, nonequilibrium calculation starts. This can avoid dealing with the problems of the saddle-point singularity and the unknown mass flow. Erickson (1963) and Harris and Albacete (1964) have conducted detailed analysis and research on this. In the research of nozzle design theories and methods, the equations for two-dimensional axisymmetric nonequilibrium flows were generally solved using MOC (Prandtl & Busemann 1929) or CFD. The increase in entropy along the streamline in a nonequilibrium flow is a result of irreversible factors in finite-rate chemical reaction processes. When the nonequilibrium MOC method is used, it is to be noted that the characteristic curves that go through any point in the nonequilibrium supersonic flow are based on the Mach line and the streamline with the frozen sound speed. The detailed solution of the nonequilibrium flow using MOC can be found in the book by Zucrow and Hoffman (1977). In the design of hypersonic nozzles, MOC presents theoretical limitations when the Mach number is increased. The CFD method for finding a solution to the whole flow has been widely used in nozzle design. But with the increasing enthalpy in hypervelocity flows, especially for high-enthalpy flows with very high Mach numbers, limitations in the accuracy of the current chemically reacting model and the CFD technique are also observed. Therefore, there are still difficulties and problems in the nozzle design for high-enthalpy hypersonic tunnels that have not been effectively solved, as listed below:

(1) At present, the hypersonic nozzle design still follows the conventional methods of decades ago. With the development of computational technology, the computation cost of some methods has been reduced. But the assumptions in most of these methods are not applicable to hypersonic and high-enthalpy flows. There is an urgent need to develop new effective theories and methods for designing high-enthalpy flow nozzles.
(2) In high-enthalpy flows, the specific heat capacity ratio of the gas is a variable. High-temperature, chemical reactions, nonequilibrium, and other issues also have a greater impact on high-enthalpy flows. In addition, with the increase in the flow speed, high-temperature gas effects bring an even greater impact.
(3) Designed with conventional methods, hypersonic nozzles, especially hypervelocity nozzles, are very long. Indeed, a long uniform flow area (generally called the diamond-shaped area) results at the nozzle exit, which cannot be reduced using the current shortening technology. The extra-long nozzle causes a much thicker boundary layer at the exit, even exceeding 50% of the cross-sectional area of the nozzle exit. In addition, with the current computation technology, it is difficult to accurately evaluate the development of the boundary layer (mainly turbulent) of the high-enthalpy flow, which makes it impossible to accurately control the thickness of the boundary layer.
(4) Moreover, it is difficult to accurately predict and evaluate the complex high-enthalpy flow with current CFD numerical simulation. At the same time, its reliability can be hardly verified using the current experimental measurement technology. Problems in the research methods restrict the development of design theories and methods for high-performance and high-enthalpy nozzles.

7.3 Design Methodology of Hypersonic Nozzles

As high-velocity flight vehicles pose higher requirements on hypersonic tunnel tests, the aerodynamic tests require that the flows, in particular high-enthalpy flows, have higher quality. This highlights the importance and engineering value of high-enthalpy nozzles which have good performance. Therefore, this section focuses on the main theories and conventional methods of designing wind tunnel nozzles, which have always played an important role in the design of hypersonic nozzles.

7.3.1 Design Principles for Wind Tunnel Nozzles

Hypersonic nozzles are divided into axisymmetric, two-dimensional, three-dimensional, and asymmetric nozzles by shape. Generally, when the Mach number is high, the total

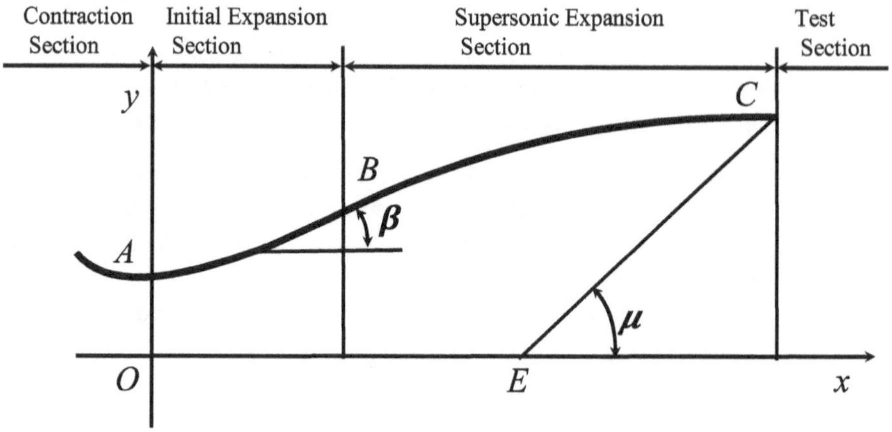

Figure 7.2 Structure of axisymmetric nozzles.

temperature of the flow must be increased in order to prevent liquefaction of the test gas. For a two-dimensional nozzle, maintaining the stability at the throat section is difficult and the expansion efficiency is low. It therefore has a longer length than the axisymmetric nozzle under the same design conditions. Due to this fact, most hypersonic nozzles are axisymmetric.

In early times, hypersonic nozzles were designed based on the supersonic MOC so that the expansion curve from the throat to the exit can ensure that the high-velocity flow at the nozzle exit is straight and uniform. The expansion curve can be divided into a front section (from A to B, initial expansion section) and a back section (from B to C, supersonic expansion section), with point B being the inflection point (see Figure 7.2). As shown in Figure 7.2, the section where OA is located is the smallest cross section at the throat; the line from point A, through B, to C represents the contour of the nozzle; β is an oblique angle on the contour; and tan β_B is the slope of the wall at point B. β_B is the maximum expansion angle which can be obtained from the Prandtl–Meyer angle (ν_1) corresponding to the nozzle exit Mach number.

$$\beta_B \leq \frac{\nu_1}{2} \tag{7.7}$$

The axial section between A and B is called the initial expansion section. In this section, after being expanded and accelerated, the flow gradually deflects outward along wall AB. A series of expansion waves are generated in the flow before it reaches the inflection point B. The section from B to C is called the supersonic expansion section or damping section. The flow continues to expand in the damping section but mainly expands inward. The slope of the curve BC gradually decreases and becomes zero at point C. The flow generates a series of compression waves along the back section, which can offset the reflected waves generated by the expansion waves at the front section. The Mach angle corresponding to the Mach number at the test section is μ_1, namely the included angle between the straight line EC and the axis. When reaching the EC line, the flow finishes expansion and acceleration, moves in parallel to the axis, and achieves the Mach number required for the design.

In the conventional design of hypersonic nozzles, the size of the nozzle exit at the test section is generally a constant. The Mach number at the exit depends on the throat size according to the area ratio formula

$$\frac{A}{A^*} = \frac{r^2}{r^{*2}} = \frac{1}{Ma}\left(\frac{\gamma-1}{\gamma+1}Ma^2 + \frac{2}{\gamma+1}\right)^{\frac{\gamma+1}{2(\gamma-1)}} \tag{7.8}$$

where A is the cross-sectional area of the nozzle, A^* is the cross-sectional area at the throat, r is the nozzle radius, r^* is the radius at the throat section, Ma is the local Mach number at the throat, and γ is the specific heat capacity ratio of the gas, which is 1.4 for air as an ideal gas.

Equation (7.8) is effective for designing hypersonic nozzles with a low total temperature. Since it does not take into account the change of the specific heat capacity ratio due to high temperatures, a large error could be produced in designing high-enthalpy nozzles. Limited by technical conditions and development at the early stage, conventional methods are still used to design the contour of hypersonic nozzles and even high-enthalpy nozzles. During the design process, the impact of real-gas effects needs to be appropriately corrected, which will be discussed in detail later.

Based on the above design principles, conventional methods for the design of nozzles, including low-velocity and high-velocity nozzles, basically include three steps:

(1) Suppose that the flow in the nozzle is ideal, design the potential flow (inviscid) contour according to the isentropic relationship;
(2) Estimate the development of the boundary layer on the wall, which is actually the estimated displacement thickness of the boundary layer;
(3) Superimpose the estimated displacement thickness onto the potential flow contour in the normal direction.

In addition, the design approach discussed mainly focuses on the expansion section of hypersonic and high-enthalpy nozzles. As for the contraction section of the nozzle, its design (the polynomial fitting method is mainly used) is relatively simple and has no direct impact on the design of the supersonic section, so it is not discussed here.

7.3.2 Conventional Design Methods for Hypersonic Nozzles

As mentioned above, MOC has laid a theoretical foundation for the design of modern nozzles. The nozzles for early supersonic and hypersonic wind tunnels were mostly designed based on MOC. Since MOC has been successfully applied in designing the nozzles of supersonic wind tunnels, the nozzle contours of hypersonic wind tunnels at that time were also designed based on this method. Although many special problems were encountered in hypersonic wind tunnels (such as real-gas effects, heat dissipation and deformation of the throat, and the thick boundary layer), a considerable amount of design work produced satisfactory results after necessary corrections were made.

A variety of theoretical methods have been derived from MOC, including the graphic method and the analytic method. Nozzle contours designed with these methods can be basically divided into three types:

(1) Curved throat contours (with a large "radius of curvature") designed with the maximum expansion angle $\beta_B = 0.5\nu_1$;
(2) Curved throat contours (with a large "radius of curvature") designed with the maximum expansion angle $\beta_B < 0.5\nu_1$;
(3) "Abrupt change" throat contours (with a small "radius of curvature") designed with the maximum expansion angle $\beta_B = 0.5\nu_1$.

Among these methods, the third one produced the shortest nozzle, followed by the first, and then the second. The nozzle designed with the third method had an abrupt expansion near the throat and a large axial pressure gradient, which often caused an abrupt change in the boundary layer thickness on the nozzle wall, and even flow separation. The flow was actually not uniform. The second method used a smaller maximum expansion angle. The nozzle therefore was too long to produce a thin boundary layer for hypersonic flows. Thus, in contrast, the first method was commonly used.

With the conventional MOC design method, the boundary layer of the nozzle will become too thick to ignore when the hypersonic flow moves at a high Mach number (in particular, in modern nozzles that are generally designed for large wind tunnels). So, the boundary layer must be corrected to meet the requirement of the design Mach number at the nozzle exit.

7.3.2.1 Graphic Design Method

Scholars have proposed various graphic methods for designing the nozzles in the 1940s and 1950s. The Busemann method (Busemann, 1931, Preiswerk, 1940) is the most fundamental one among method-of-characteristics mesh methods. It assumes that the flow at the throat is uniform. Based on the fundamental nozzle design theory and experience, the method sets an initial curve AB where the slope gradually increases and $\beta_B = 0.5\nu_1$ is satisfied at point B.

As shown in Figure 7.3, curve AB is replaced with polylines $AA_1A_2 \ldots A_NB$, where A_i is the inflection point. Each polyline turns outward at the same (or variable) angle $\Delta\beta$. The smaller $\Delta\beta$ is, the more accurate the result will be. According to the characteristics theory, there are expansion waves emanating from point A_i, reflected waves from the axis or wall, and the generated expansion waves are intersected with the reflected waves. All these waves can be obtained by computation. All the areas surrounded by the wave lines are uniform flow areas. Finally, points A_i are connected in sequence to form a continuous smooth nozzle contour. The supersonic flow travels around the outer obtuse angle to produce the expansion wave system which only approximates to a single wave. This method is very useful for designing nonconventional nozzles, while the analytic method developed later is more convenient and computationally accurate for conventional hypersonic nozzles.

Puckett (1946) modified the conventional Busemann method. The modified method has many advantages. First, using the simplified initial design curve can save almost

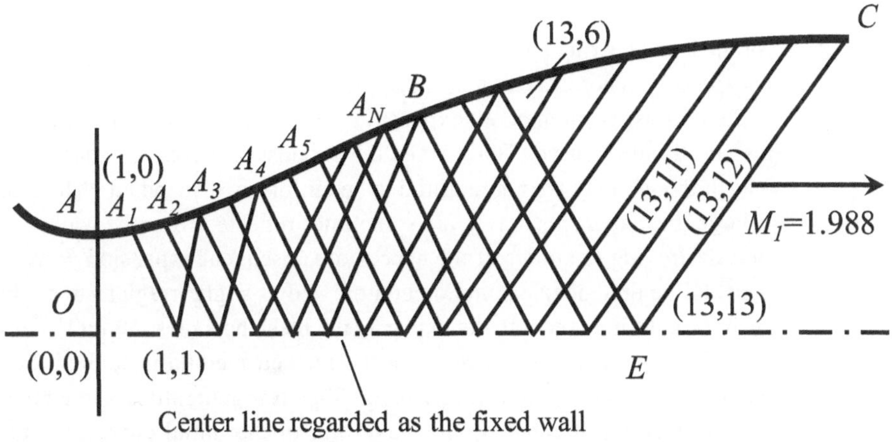

Figure 7.3 Characteristics mesh for nozzle design.

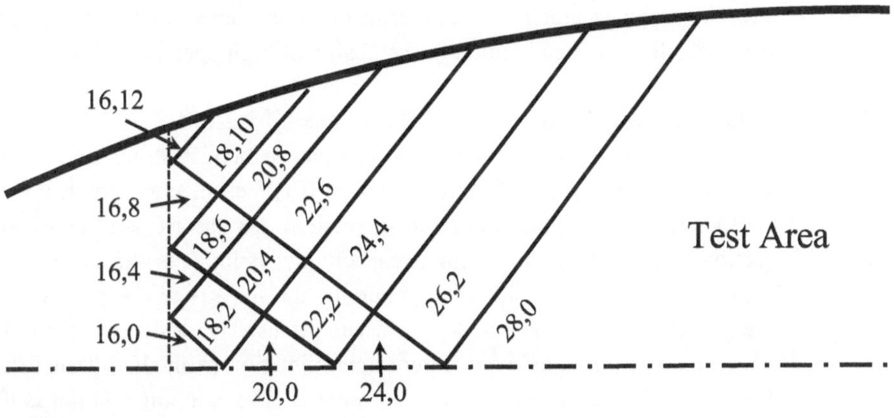

Figure 7.4 Puckett method for nozzle design.

half the time or efforts for the nozzle design. The characteristic curves are calculated from the middle of the nozzle to both ends. It is assumed that the flow passing through the nozzle at the maximum expansion (the inflection point) has a uniform speed and a uniformly variable flow direction (see Figure 7.4). Based on experience, setting these conditions is reasonable. With these values, the exit (damping section) of the nozzle can be determined in the same way as the Busemann method. Similarly, the initial expansion section can be constructed. In addition, if β is less than β_B, one or multiple inflection points must be added to reflect the expansion waves. Since the reflected wave can be selected, several initial curves AB can be selected, so that the design Mach number can be reached at the contour of the corresponding damping section. If the mesh size can be infinitely small, there will be an infinite number of initial curves for selection, in theory. Likewise, there will be countless damping curves corresponding to specific initial expansion curves. Despite this, it does not imply that any contour satisfying the area ratio requirement (Equation 7.8) is appropriate. On the other hand, for most engineering applications, the error arising from the use of an arbitrary initial curve can be ignored as long as the appropriate correction has been made.

In the Busemann method, expansion waves are generated at certain points along a smooth initial curve. That means the expansion waves are spaced in an orderly way, although they are not required to be uniformly distributed. When the mesh size is large, the expansion waves are sometimes reflected from the wall at some points, thus destroying the order of the spaces of subsequent expansion wave systems. As a result, the final nozzle damping contour shows slight irregularities. Theoretically, such irregularities seem to disappear as the mesh becomes infinitely dense, but they cannot be avoided in practice when the Puckett method is applied. So in order to obtain a smooth contour as much as possible, it is generally assumed that the damping curve is not affected by the waveform of the initial curve; the mesh is made denser to ignore the error. However, limited by the computational technology at that time, the efficiency is decreased with the increased computing workload when the mesh becomes denser. Also, the analytic method then became available and is far superior to the graphic method in terms of the produced flow, so the graphic method has basically not been applied to the design of high-speed nozzles.

7.3.2.2 Analytic Design Method and the Foelsch Method

The analytic method is similar to the graphic method. With reference to the contour structure of the nozzle in Figure 7.2, point B is the inflection point; the slope of the contour ABC is the largest at point B; sections AB and BC are the initial expansion section and the damping section, respectively; point A is the starting point of the contour, while C is the end point; generally, the slope is zero at points A and C. This method generally adopts radial flow assumptions. The design principle of the initial expansion section is to use different approximation methods to transform the sonic flow at the throat into a radial flow at point B. The damping contour is to transform the supersonic or hypersonic radial flow into a uniform flow with the design Mach number and traveling parallel to the axis. In addition, all expansion waves on the wall must be damped.

Based on the MOC theory and methodology, early scholars have developed various analytic methods for the nozzle design. The common feature of such analytic methods is that the flow at inflection point B is assumed to be a supersonic radial flow (see Figure 7.2). On this basis, the curve after point B is obtained through computation with the analytic method. The design principle for this curve is to damp all the expansion waves reflected along the curve. Though different methods are used for processing the contour before point B, there is a shared idea that the curve at the front section should be able to transform the uniform sonic flow at the throat into a supersonic radial flow at the inflection point.

In the design of nozzle contours, an inviscid potential flow contour is first designed based on a rigorous theoretical method, then boundary layer correction is conducted. However, one big problem here is that it is difficult to evaluate and calculate the boundary layer with sufficient accuracy. Especially for the boundary layers of hypersonic and high-enthalpy flows, neither theoretical nor numerical methods can hardly ensure an accurate evaluation. So, a very accurate contour design seems to be completely unnecessary. In addition, the theoretically calculated wave system may be completely different from the actual flow conditions. So far, most potential flow contours have therefore been designed using the simple approximate method, which greatly reduces the design workload.

For the design of nozzles (especially for low-velocity nozzles) without consideration of the impact of high-temperature effects in the early stage, the idea of radial flow assumption of the analytic method is reasonable to a certain extent. The design results basically meet usage requirements. For the design of nozzles with a high Mach number, various optimization techniques have to be applied after the design in order to obtain a high-quality flow. This section reviews and analyzes several classical analytic methods for nozzle design (including methods that are still in use today).

In 1948, Foelsch (1946, 1949) proposed an approximate method. In this method, a certain area of the nozzle is assumed to have a radial flow, the initial expansion section is designed using an empirical formula, and the flow is transitioned to the uniform area using MOC (see Figure 7.5). However, due to the limitation of the radial flow assumption, nozzles designed with this method are generally very long. The direct connection between the radial flow area and the uniform flow area results in a discontinuous axial velocity gradient, which then leads to an unsatisfactory flow quality. The Foelsch method does not determine a contour based on theory and does not guarantee that the radial flow assumption can be met at the inflection point. Instead, it uses an empirical curve with a curvature that changes monotonically and the area ratio formula is satisfied at point B. The curve equation is as below:

$$y = y^* + x^2 \frac{\tan \beta_B}{x_B}\left(1 - \frac{x}{3x_B}\right) \tag{7.9}$$

where

$$x_B = \frac{3}{2}\left(y_B - y^*\right)\cot \beta_B \tag{7.10}$$

In the initial expansion contour, the actual flow at the inflection point differs from a radial flow, which has certain impacts on the uniformity of the flow in the test section. The damping contour connects the starting and end points (B, C) with a discontinuous curvature, but has no impact on the hypersonic nozzle with a fixed wall.

Crown (1948) improved the Foelsch method and gave a new curve (see Eq. 7.11) as the initial expansion contour. In general, however, there is no substantial improvement over the original Foelsch method.

$$y = y^* + x_B \log_{10} \beta_B \left(\frac{x}{x_B}\right)^3 \left(1 - \frac{x}{2x_B}\right) \tag{7.11}$$

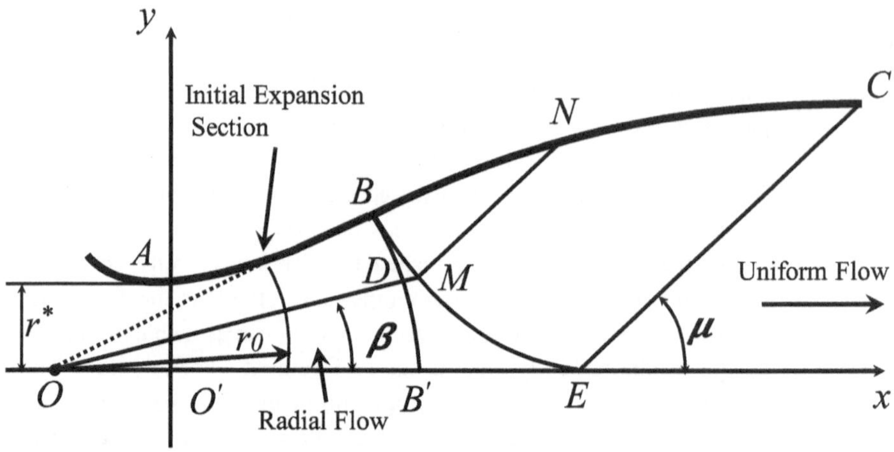

Figure 7.5 Foelsch method for nozzle design.

where

$$x_B = 2\left(y_B - y^*\right)\cot\beta_B \tag{7.12}$$

The Foelsch method is of great significance because it "reasonably" processes the flow near the nozzle throat using the radial flow assumption, and greatly simplifies the design of the wall contour at the throat which is the most complicated in the nozzle design. As a result, a series of nozzle design methods have been derived based on the radial flow assumption and applied to the design of nozzles for most high-velocity and low-velocity wind tunnels that have been constructed in the last century.

7.3.2.3 Cresci Method

In 1958 the Foelsch methods were modified by Cresci (1958). A damping area was added in the contour, aiming at reducing the discontinuity of the axial velocity gradient due to the algorithms of the Foelsch method. In the Cresci method, the flow in the nozzle is divided into three areas (Figure 7.6): the area OABD is the throat expansion area which is considered to be a radial flow from the origin; the area BDEC is the transition area which rectifies and straightens the radial flow into a parallel flow; the area that is labeled as uniform flow in Figure 7.6 is a region with uniform and constant Mach number parallel flow.

The initial expansion section AB is designed with the empirical curve from the Foelsch method. It is assumed that the axial velocity distribution satisfies a polynomial in the damping area BDEF. Based on experience, the Mach number at point D is 0.2 less than the Mach number at point E, that is, the design Mach number. In this case, suppose that the velocity distribution of the DE section satisfies a cubic polynomial:

$$F = c_0 + c_1\overline{x} + c_2\overline{x}^2 + c_3\overline{x}^3 \tag{7.13}$$

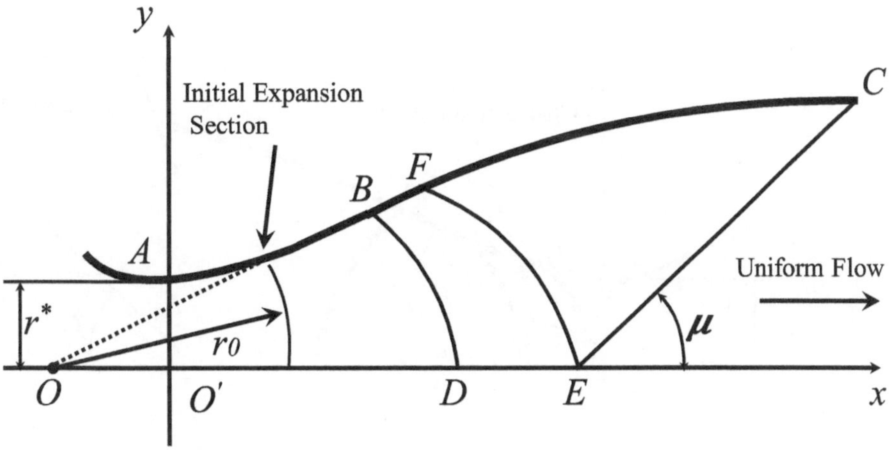

Figure 7.6 Cresci method for nozzle design.

where $\bar{x} = (x - x_D)/(x_E - x_D)$, the cubic polynomial satisfies the following boundary conditions:

If $\bar{x} = 0,\ F = F_D$, then

$$\left[\frac{dF}{dx}\right]_D = (x_E - x_D)\left(\frac{dF}{dx}\right)_D \tag{7.14}$$

If $\bar{x} = 1,\ F = F_E$, then

$$\left[\frac{dF}{dx}\right]_D = 0 \tag{7.15}$$

According to

$$3(F_E - F_D) = (x_E - x_D)\left(\frac{dF}{dx}\right)_D \tag{7.16}$$

the value of x_E is selected. The positions of points D and E can be determined by substituting the boundary conditions into the cubic polynomial. The leftward characteristic curves through point E is a straight line with the design Mach number, so the endpoint C of the characteristic curve can be determined. With BD, DE, and EC as the boundary conditions, the characteristics mesh throughout the flow and the nozzle contour of the BC section can be obtained based on the characteristics theory and the mass conservation law. Since the empirical curve from the Foelsch method is still used in designing the initial expansion section in the Cresci method, it is difficult to optimize the nozzle contour obtained on this basis. However, compared with the Foelsch method, the Cresci method is a more mature technology in nozzle contour design.

7.3.2.4 Sivells Method

The previous nozzle design methods consider almost entirely the contour, which easily leads to phenomena such as disturbance accumulation at the axis and flow separation on the wall. In the late 1960s and the 1970s, Sivells (1969, 1970, 1978)

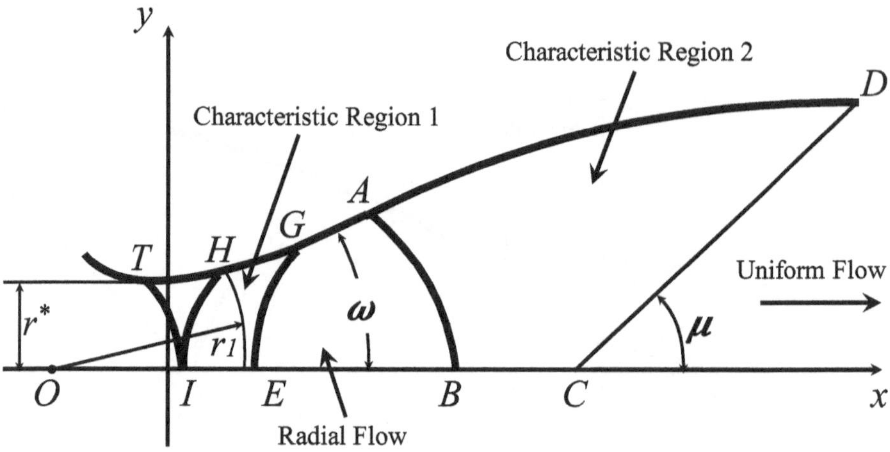

Figure 7.7 Sivells method for nozzle design.

proposed a nozzle design method based on the axial velocity and Mach number distribution according to the design idea of the Cresci method and the Hall transonic theory (Hall 1962). With the Sivells method, the nozzle contour with continuous curvature can be obtained by setting the axial Mach number distribution. As a result, there is no concentrated compression wave inside the nozzle, and the Mach numbers at the exit are uniformly distributed, which greatly improves the quality of the flow.

The design contour of the nozzle in the Sivells method is determined by the characteristics mesh among the known boundary points, the given axial Mach number distribution, and the approximate analytic solution at the sonic throat. As shown in Figure 7.7, the nozzle is divided into three areas: a throat area TIEG, a radial area GEBA, and a downstream area ABCD. In the radial flow area, the area between the characteristic curves EG and AB is a radial flow; the contour from point G to point A is a straight line inclined to the axis at an angle ω. Therefore, determining the contour includes two steps. That is to say, two sets of characteristic equations are required to be solved. First, the throat contour from point H to point G on the branch line is determined. Point H is located on the leftward characteristic curve starting from point I with the sonic speed. The flow in the throat area is calculated using the Hall transonic flow solution. Next, the downstream contour from point A to point D is determined. In Figure 7.7, the characteristics CD is a straight line where point D is the theoretical endpoint of the nozzle contour, namely, the position of the nozzle exit. According to these assumed boundary conditions, the flow enclosed in the IHGE and BADC areas can be solved by combining the characteristics mesh with the mass conservation law, thereby obtaining the nozzle contours along HG and AD. The contour of section TH is obtained through the transonic solution. In this method, the velocity distribution from point I to point C is first assumed. Then, the velocity distribution matches the theoretical transonic conditions at the throat, radial flow conditions through an intermediate region, and design flow conditions at the nozzle exit. The axial velocity distribution from

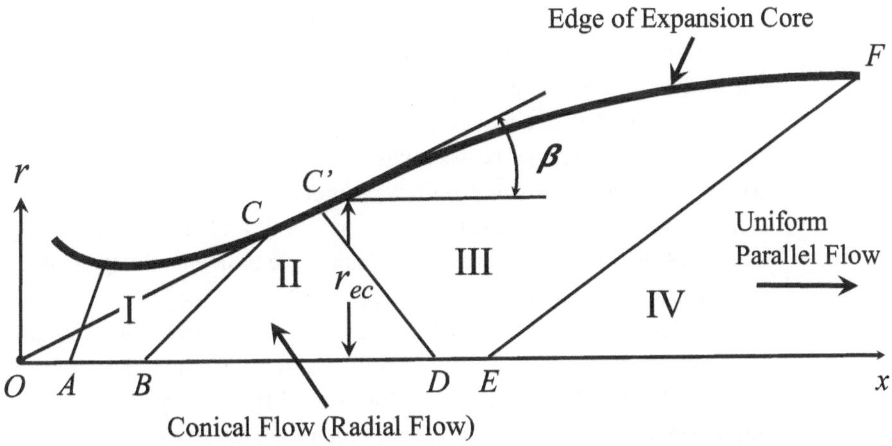

Figure 7.8 Schematic diagram for the design of inviscid nozzle contour.

point I to point C can be divided into three parts. The first part is from point I with the sonic speed in the throat area to point E where the radial flow is assumed to start. The velocity distribution of this part is defined by a quartic polynomial. The second part is from point E to point B where the radial flow ends. Its velocity distribution is given by the conditions generating the radial flow. The third part is from point B to point C where a uniform flow starts. Its Mach number distribution is defined by a quintic polynomial. The coefficients of the polynomial are selected based on the requirements that the second derivatives of the axial velocity are continuous all the time and the derivative at point C is zero. For the detailed solving process and nozzle design process of the Sivells method, see the references by Sivells (1969, 1970, 1978).

As proposed by Potter and Carden (1968), the conventional MOC was used to design two nozzles (one with *Ma* = 9 and the other with *Ma* = 10) for the AEDC-VKF (The von Kármán Gas Dynamics Facility at Arnold Engineering Development Center) low-density hypersonic wind tunnels. The Cresci method (1958) and the Sivells method (1963) were combined to design the inviscid potential flow contour. As shown in Figure 7.8, the Cresci method based on the radial flow assumption was used for the design in the damping area (Area III). According to the Sivells design process, the Mach number distribution for the axis was given to complete the contour design. Because the impact of variable specific heat capacity ratio was ignored and that the nozzle flow was mistaken for being fully laminar due to low density, certain design errors emerged. The test results showed that the actual Mach numbers at the exits of the two nozzles were 10.15 and 9.30, respectively, instead of the design Mach numbers of 9 and 10. It was seen from the numerical results that the Sivells method can produce better nozzle flow than the Foelsch and Cresci methods. Its flow quality can even be further improved with the increase of Mach number. Therefore, the Sivells method is more applicable for designing hypersonic nozzles that need to meet the requirements of high flow quality.

Restricted by the computational technology at the time when the Sivells method was proposed, the nozzles, especially high-velocity nozzles, have very large lengths due to their conservative design. In order to solve this problem, many scholars have proposed a number of design methods to shorten the nozzles. For example, the most common one is based on the assumption of a straight sonic line at the throat (Ali et al. 2012). These conventional nozzle shortening methods can meet the design requirements of lower Mach number nozzles, but they can hardly ensure the quality of the flow when the Mach number is high. So they are basically not applicable to designing the nozzles for hypersonic wind tunnels.

In practice, designed with MOC, the inviscid potential flow contour can basically meet the design requirements of hypersonic nozzles which do not need to take into account high-temperature effects. Moreover, different improvement methods can be implemented to meet several special conditions involved in the nozzle design. To some extent, the Sivells method is a milestone that brings about the maturity of conventional MOC nozzle design technology. So far, the Sivells method is one of the internationally recognized design methods for high-velocity nozzles that feature high accuracy and have gained wide usage in engineering applications.

7.3.2.5 Theory and Method for Boundary Layer Correction

Due to the impact of gas viscosity, the flow inside the nozzle is different from the inviscid flow. The gas viscosity has the following impact on the nozzle flow: (1) the flow near the nozzle wall forms a boundary layer. The distribution of the boundary layer thickness along the nozzle wall affects the core flow in the nozzle. (2) The wave system in the nozzle and the boundary layer on the nozzle wall interfere with each other. Such interference not only affects the flow in the boundary layer, but also changes the occurrence and reflection of the wave system on the nozzle wall. Therefore, the requirement of completely eliminating the expansion wave reflection in the potential flow design cannot be met.

If the contour coordinates of the potential flow design are directly used for the nozzle contour, two problems will occur. One is that the wave system shape is changed because a viscous wall boundary layer exists and its thickness is increasing. As a result, the flow reaching the exit of the nozzle is not uniform. The other problem is that the thicker boundary layer reduces the effective potential flow area, thereby changing the area ratio. As a result, the actual Mach number at the exit is lower than the design Mach number. Therefore, the resulting flow in the test section cannot meet the design requirements. These two problems are more prominent and serious in hypersonic flows, especially in high-enthalpy nozzle flows. Therefore, boundary layer correction must be performed on the designed potential flow contour and its coordinates, which generally means correcting the effect of the boundary layer on the nozzle flow rate. Specifically speaking, the displacement thickness of the boundary layer at each point of the potential flow contour is calculated and the potential flow contour is displaced in the normal direction. Namely, the theoretical contour is moved outward by the equivalent displacement thickness. For low-velocity nozzles with thin boundary layers, the displacement thickness is generally displaced in the direction of the vertical axis in the calculation.

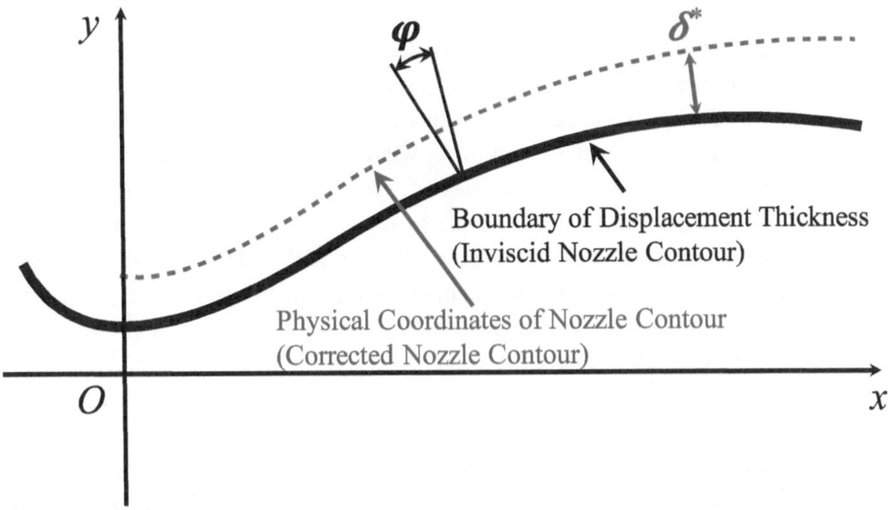

Figure 7.9 Schematic diagram for boundary layer and displacement thickness line.

The boundary layer thickness at the wall leads to the change of the cross section of the nozzle. Such change can be denoted by the distribution of the displacement thickness δ^*. The displacement thickness shown in Figure 7.9 is defined as:

$$\delta^* = \int_0^{\delta} \left[1 - \frac{\rho u}{\rho_1 u_1} \right] dy, \tag{7.17}$$

where ρ_1 and u_1 are respectively the density and the velocity in the boundary layer. A nozzle contour that is designed for a potential flow does not account for the impact of δ^* on the nozzle area. Therefore, the inviscid nozzle contour is corrected by displacing the wall contour outward in the normal direction by δ^*. The boundary layer correction of the nozzle is generally performed on the contour wall of the nozzle according to the following formulas:

$$\begin{cases} x = x_{\text{potential flow}} - \delta^* \sin\theta_{\text{potential flow}} \\ y = y_{\text{potential flow}} + \delta^* \cos\ {}_{\text{potential flow}} \end{cases} \tag{7.18}$$

where $x_{\text{potential flow}}$ and $y_{\text{potential flow}}$ are the coordinates for the potential flow contour and θ is the inclination angle of the potential flow wall relative to the x-axis. The growth law of δ^* on the wall is correlated to the characteristics of the viscous flow. When the flow in the nozzle is accelerated in the axial direction, the boundary layer on the wall is generally in a favorable pressure gradient. The distribution of the boundary layer thickness δ^* can be determined by theoretical, empirical, and numerical methods. In the nozzle design for low-velocity wind tunnels, the computation and evaluation of the displacement thickness of the boundary layer are relatively imprecise. A theoretical approximation or an empirical formula is often used. Because the nozzle is short with a thin boundary layer, this simple correction method will not cause a large design deviation at the exit of the nozzle. Generally, quite satisfactory results can be obtained. But the deviation cannot be ignored for hypersonic as well as high-enthalpy nozzles, because they are longer with thicker boundary layers. Therefore, no matter which method is used, the effect of boundary layer correction actually depends on the accurate evaluation of the boundary layer.

The boundary layer on the nozzle walls in hypersonic wind tunnels (except for low-density wind tunnels) is generally a turbulent one. In order to prevent the liquefaction of gas composition in the wind tunnel, the test gas is generally heated to several hundred to several thousand K. Some nozzle walls also require cooling (e.g., in electric-arc wind tunnels). Because of wall cooling, there is heat exchange on the wall, which makes the computation of the nozzle flow more complicated. For hypersonic boundary layers with heat exchange and pressure gradients, especially high-enthalpy gas boundary layers, the computation and evaluation methods are currently not yet mature. In most cases, the computation and evaluation methods of low-velocity turbulent boundary layers are used for that of hypersonic boundary layers. The Sivells–Payne method (Sivells & Payne 1959) is more commonly used. This method is relatively simple in computation, but does not take into account the impact of real-gas effects on the development of the boundary layer. The Persh–Lee method (Persh & Lee 1956), originally proposed by Preiswerk (1940), computes the axisymmetric momentum equation and accounts for high-temperature effects.

On the whole, conventional nozzle design methods basically use different techniques to evaluate the thickness of the boundary layer and calculate the displacement thickness, before making corrections on the designed potential flow contours. They are known as the method of characteristics and boundary layer approach (MOC/BL).

7.3.2.6 Limitations of MOC

The above nozzle design methods are conventional MOC-based analytic methods. The development of contour design for hypersonic nozzles with ideal gas has been relatively mature since the late 1940s. By the 1980s, with the development of CFD computational technology, the flow from the MOC/BL-designed nozzles has been more easily verified and evaluated through CFD numerical computation (Benton 1989). Benton et al. (1990) has analyzed the limitations of conventional MOC applications in the design of axisymmetric hypersonic nozzles in detail. He believes that the conventional MOC/BL nozzle design method is unreliable for the design of hypersonic nozzles. MOC/BL is quite effective for designing the nozzles with a Mach number less than or equal to 8 at the exit. With the increase in Mach number, especially for hypervelocity and high-enthalpy flow, the boundary layer becomes very thick compared to the diameter of the uniform flow area. In this case, nozzles that are designed based on MOC/BL hardly meet the design requirements. After solving the Navier–Stokes (N-S) equations for two MOC/BL-based nozzles with Mach numbers of 13.5 and 17, respectively, Benton (1989) discovered that the CFD results were less consistent with the experimental results. Even when the impact of the specific heat capacity ratio was considered, the results of numerical evaluation were not satisfactorily improved. So, when the quality of the flow produced by the designed nozzle is not ideal or the Mach number distribution does not meet the design parameters, the thin boundary layer assumption is considered to be unreasonable to a large extent. Compression waves are present in all the hypersonic nozzles analyzed

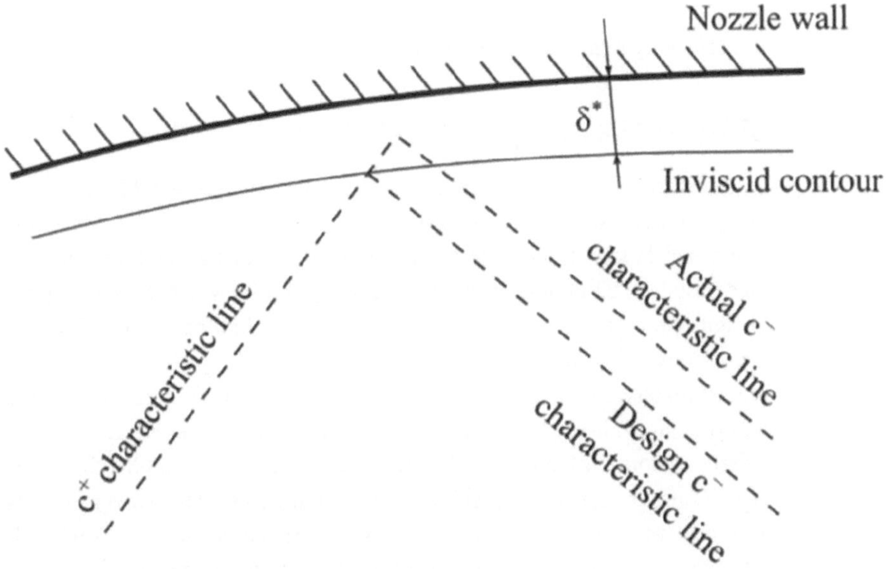

Figure 7.10 Lagging between actual characteristic curves reflection and design characteristic curves reflection.

by Benton, which result in irregular flow rates. These waves always occur near the inflection point. He recommends that a design method should be developed using PNS equations. Candler and Perkins (1991) have analyzed the reflection of characteristic curves in the thick boundary layer of hypersonic nozzles. The basic assumption behind the MOC/BL method is that the boundary layer flow and the nozzle core flow are not coupled. For a low-Mach-number nozzle with a thin boundary layer and a short geometrical shape, this assumption is reasonable. The reflection locations of the flow characteristic curves are basically on the inviscid contour. However, after the reflection through a thick boundary layer, there is a certain lag between the actual reflected characteristic curves and the calculated ones that are reflected through the potential flow boundary. That means the reflection of characteristic curves may actually occur inside the thick boundary layer instead of at the potential flow boundary line (see Figure 7.10), which results in the actual reflected characteristic curves lagging behind the design reflected characteristic curves, thereby reducing the flow quality of MOC/BL-based nozzles (Craddock 2000, Chan et al. 2018). Therefore, a nozzle designed with the classic MOC/ BL has incurred deviations in the reflected characteristics of the flow, which makes the expansion waves in the damping section incompletely eliminated, thus resulting in a nonuniform flow at the exit. Benton (1990) believes that this impact actually starts when the Mach number is approximately 7 if the MOC/BL method is used to design hypersonic nozzles. Once the turbulent boundary layer grows and reaches the exit of the nozzle, the proportion of the boundary layer thickness to the nozzle radius rises with the increase of the Mach number. Therefore, in order to improve the flow quality of hypersonic nozzles, more accurate nozzle design methods need to be developed. Such methods should be able to accurately evaluate the boundary layer.

7.3.3 CFD-Based Design-by-Analysis Method

In order to overcome the limitations of the conventional MOC/BL method in the design of hypersonic nozzles, Korte (1992, 1993), Korte et al. (1992a, 1992b, 1992d), Korte and Hodge (1994, 1995) proposed a new design method for hypersonic nozzles based on CFD numerical techniques in the early 1990s (see Figure 7.11 for the design optimization process). Obviously, this technology cannot be realized without a vigorous development of computational technology. By making the optimization algorithm based on a nonlinear least squares method (Huddleston 1989) coupled with an efficient high-accuracy CFD solver (Korte 1991), the Korte method can create a nozzle contour with the least changes in flow angularity and Mach number at the exit. A CFD program is used to solve the PNS equations. This solver ensures the optimization of the flow at the exit of the MOC/BL-based nozzle. This method is called "design-by-analysis" by Shope (2006). The main advantage of the Korte method is that it uses a solver based on the N-S equations, making it possible to have the correct modeling method to handle the interaction between the core flow (inviscid flow) and the boundary layer in a high-Mach-number nozzle. During the design process, the computational results of the nozzle flow and the nozzle contour are substituted into the optimization algorithm. The algorithm iterates until a uniform flow is obtained at the exit. The accuracy of the Korte method depends on the accuracy of the CFD solver and the performance of the optimal solution that the optimization algorithm can achieve. In addition, this design method can be used to design not only axisymmetric nozzles but also 3D nozzles by simply using a three-dimensional flow solver.

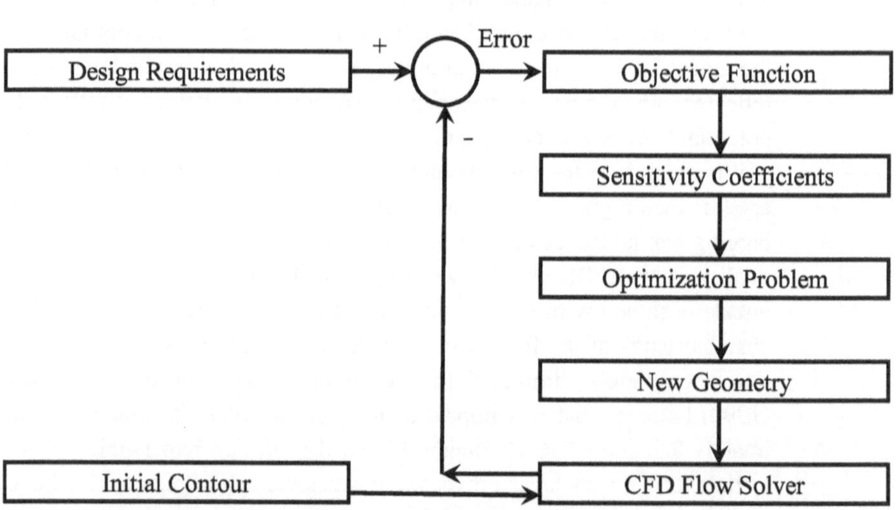

Figure 7.11 CFD-based design optimization process.

Subsequently, Korte et al. (1992b) and Korte (1993) improved the CFD-based design method for hypersonic nozzles. The improved method makes it possible to complete the entire design process for hypersonic flow by using CFD technology, without the use of the MOC/BL-based initial nozzle contour. In the design process of the improved Korte method, the Mach number distribution along the axis is optimized according to the parameters such as the Mach number and the flow angularity at the exit. The Mach number distribution along the axis in the optimization objective function continues to be used as in the Sivells method (Sivells 1970). The Korte method basically removes the constraints of the conventional MOC. The designed nozzle flow is more accurate than that generated by the theoretical method. It is, however, obvious that this "CFD-based design-by-analysis" method highly depends on the numerical simulation accuracy. Therefore, for high-enthalpy nozzle flows with nonequilibrium chemical reactions, the CFD-based numerical method needs to be further tested. The impact of real-gas effects will be discussed in detail later.

In 2000, Korte (2000) developed a set of codes for processing high-temperature flows (where the specific heat capacity ratio could change with temperature) through Sivells' irrotational flow MOC (Sivells 1970). The codes were written in the FORTRAN 66 language and therefore became less versatile later. This method is also used by Gaffney (2006) to design the nozzles for the HYPULSE shock-expansion tunnel that can accommodate nonuniform inlet flows. The HYPULSE tunnel is a high-enthalpy ground testing facility owned by NASA Langley Research Center in the United States. The tunnel nozzles can generate a non-irrotational flow with an enthalpy of $Ma = 15$. Such nozzle flow conditions are useful for evaluating the performance of a scramjet engine as the flow entering the nozzle from the acceleration section of the tunnel is always highly nonuniform. This method combines the rotational flow MOC with the solution of Euler equations. Its advantages include: it is not limited by the structure (or programming language) of the existing codes; it does not need to go through the program line by line for possible modification; it runs relatively fast; the number of points defining the nozzle contour basically depends on the specified number of initial points. In the Gaffney method, viscous effects are not directly considered, but the nonuniform distribution of the flow properties of the HYPULSE nozzles is a result of viscous effects and obtained by solving the rotational equations. Final viscous-corrected contours can be generated by iterating the MOC solution with a full N-S solution (Gaffney & Korte 2004) (usually with only one or two iterations).

Theoretically, the Korte method allows any combination of CFD solver and optimization algorithm during the design process. Some scholars (Keeling 1993, Tolle 1997, Shope 2006, Chan et al. 2018) have made some improvements based on the Korte method, including the use of different CFD solvers and optimization algorithms. During the early stage, Korte developed a nozzle design program CAN-DO (Korte 1992c). This program uses a cubic spline interpolation for contour design, explores the optimal exit flow by adjusting the spline interpolation points, and analyzes the nozzle flow with the PNS solver. In recent years, Chan et al. (2018) have adopted an open-source Reynolds-averaged Navier–Stokes CFD solver. The Eilmer code (Gollan & Jacobs 2013) combines it with the simplex optimization method

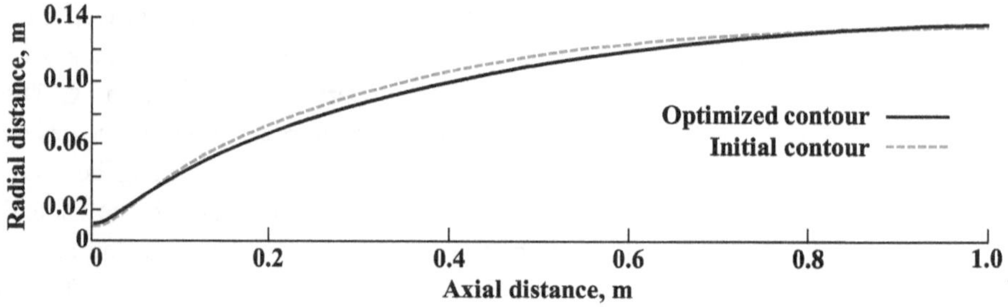

Figure 7.12 Comparison of the initial contour and the final contour of the nozzle optimized with the method proposed by Chan et al. ($Ma = 7$).

(Nelder & Mead 1965) for contour design. The optimization goal is to obtain a flow with minimum flow angularity and minimum deviation from the design Mach number at the exit. With this program, Chan et al. (2018) designed the nozzles (Ma = 4, 7, 10) for the T4 shock tunnel (Hannemann et al. 2016) at the University of Queensland, Australia, and carried out the numerical and experimental comparison. The numerical simulation results of the nozzle flow show that the core flow area has good uniformity; the change in the Mach number at the exit is less than 0.5%; the change in flow angularity is less than 0.05°; the changes in the static temperature and flow rate are less than 1%; and the changes in the pitot and the static pressure are less than 2%. In addition, by conducting experiments and comparing the measured pitot pressure on several sections at the downstream of the three optimized nozzle exits, the good matches with CFD numerical simulations proved that the proposed "CFD-based design-by-analysis" method is effective for the nozzle design. Figure 7.12 shows the computation and optimization results of Chan et al. (2018) for the nozzle of $Ma = 7$.

The "CFD-based design-by-analysis" method has achieved good results in the design of hypersonic or hypervelocity nozzles, but its design process is complicated. In particular, for a design that adopts a CFD-coupled optimization algorithm, the method needs to select a suitable optimization algorithm since different N-S solvers may be used, making it difficult to be popularized and applied in engineering applications. Tang et al. (2019) have optimized a series of hypervelocity and high-enthalpy nozzles using the CFD-based MOC design. Different from the Korte and Gaffney methods, this method does not couple the optimization algorithm with the N-S solver to optimize the contour of the hypersonic nozzle. Instead, based on Sivells' irrotational MOC method, it corrects and optimizes the inviscid contour (potential flow contour) and the high-temperature boundary layer simultaneously by using the CFD solutions and solving the N-S equations set (which takes into account high-temperature chemical nonequilibrium). It is the separate solving of the CFD solution and the conventional matured MOC, as well as the combined optimization for the contour design that make this design method simple and convenient to use. The boundary layer displacement thickness, which is obtained based on the boundary layer definition and by using the CFD numerical flow solution from each iteration step, is used as the basis for the boundary layer correction in the next iteration step. In addition, the flow information at each boundary will also be applied to the optimization computation of the next iteration

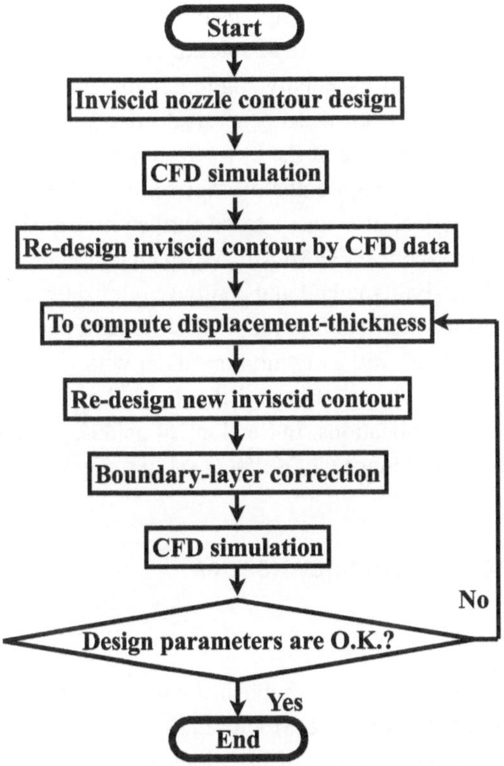

Figure 7.13 Flow chart for CFD-based iterative optimization in Tang's work.

step. All the initial and boundary conditions in the MOC are derived from the nozzle flow data of the CFD numerical method. Generally, a satisfactory exit flow can be created through two or three iterations. The evaluation of whether the nozzle flow meets the design requirements is to examine whether the flow angularity and the Mach number distribution uniformity in the nozzle exit core area meet the test conditions. Figure 7.13 shows the process and steps of Tang's optimization method.

In general, the "CFD-based design-by-analysis" methods for nozzles are roughly divided into two types. One is the computational fluid dynamics/optimization method (CFD/OM) where a CFD numerical solver is tightly coupled with an optimization algorithm. This method tries to abandon the MOC analytic design method or only uses the MOC design contour as the initial input condition. Compared with the conventional MOC/BL theoretical analytic design method, the CFD/OM for hypersonic nozzles includes the impact of viscous effects into the optimization process, so boundary layer correction is unnecessary. With this method, a more uniform flow can be created from the nozzle core area and the quality of the flow has been improved to a certain extent. This is especially true for a nozzle in which the boundary layer is thick and the Mach number is large. Of course, there are several criticisms of the method. First, limited by the number of control points used in the optimization process, the method has a much higher computation cost than the MOC/BL method. Second, the codes must be modified for every given situation, which would cause much inconvenience for other users. Third, the accuracy of the design results of this method still entirely depends on the accuracy of the CFD numerical solution and the optimization algorithm. For hypervelocity flows and especially high-enthalpy flows, therefore, the associated chemical reactions and even thermochemical nonequilibrium will occur with the increase in the total temperature of the flow. Currently, it is still challenging to accurately simulate flows in the high-enthalpy regime. The effectiveness of the nozzle design optimization method also needs to be further evaluated according to the actual situation. The other type is CFD/MOC/OM which combines a CFD solver with the conventional MOC analytic method and contour optimization. It can greatly reduce the computation workload. By solving the CFD solution separately and using a mature MOC design method, as well as combining them with an optimization of the nozzle contour, it could reduce the technical difficulty in the design process and provide high robustness in the applications. In Section 7.4 on design technologies of high-enthalpy nozzles, we will introduce CFD/MOC/OM in more detail.

7.4 Design Technologies of High-Enthalpy Nozzles

7.4.1 Real-Gas Effects in Nozzle Design

In the 1950s and 1970s, a number of scholars published a series of reports (Armstrong & Smith 1951, Ruptash 1952, Beckwith & Moore 1955, Johnson et al. 1963) and books (e.g., Lukasiewicz 1973) on the design of supersonic and hypersonic nozzles for ideal gas flow (Shapiro 1953, Zucrow & Hoffman 1977). When the nozzle exit radius is selected as the characteristic length, the boundary layer thickness can be ignored. Under this assumption, the nozzle flow can be regarded as an inviscid flow. Once the potential flow contour of the nozzle is determined, the design of the nozzle contour can be completed by estimating the displacement thickness to correct for the boundary layer. For the hypersonic nozzle with a relatively low Mach number, ideal results can be achieved with conventional design methods. However, with the increase in Mach number, total temperature, and total enthalpy, the impact of real-gas effects has to be considered. Hypervelocity shock tunnels have the characteristics of high total temperature, high total pressure, and high enthalpy. At the entrance (contraction section) of the wind tunnel nozzle and in the initial expansion area near the throat, the flow is of high temperature and high pressure in most of the working conditions. In general, as the gas temperature increases, as in a high-enthalpy flow, a point will be reached when chemical reactions start to occur. If the temperature exceeds 800 K, the vibrational energy of the gas molecules will be excited, and the specific heat capacity ratio of the gas γ will become a function of temperature rather than a constant. If the temperature reaches 2,000 K, the oxygen will begin to dissociate. At 4,000 K, the oxygen will be almost completely dissociated into oxygen atoms, while the nitrogen begins to dissociate into nitrogen atoms. For high-temperature and high-pressure flows, if the ideal gas assumption is still used with the nozzle design problem, a large error will occur and the conventional optimization methods will no longer be effective.

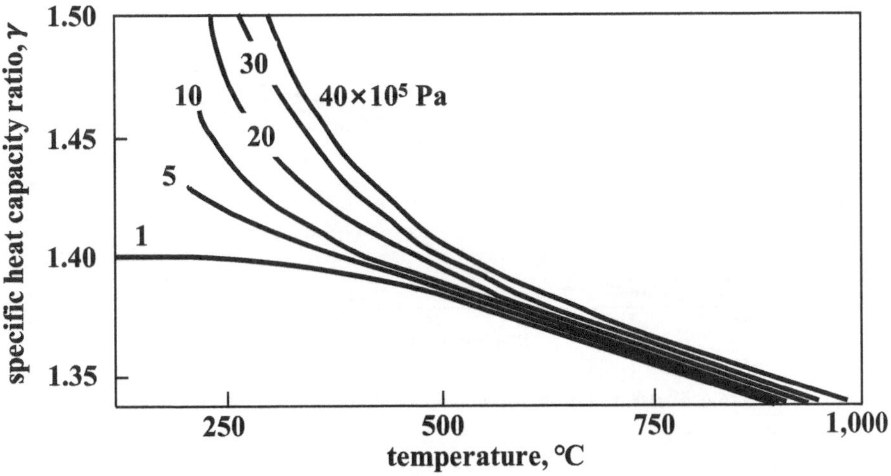

Figure 7.14 Effects of temperature and pressure changes on specific heat capacity ratio.

The specific heat capacity ratio γ reflects the ratio of the enthalpy to the internal energy, which changes with the gas temperature. For air at normal temperature and normal pressure, $\gamma = 1.4$, and γ can remain unchanged for an ideal gas. A gas can be considered as an ideal gas as long as the temperature and pressure do not change much. Under hypersonic and high-enthalpy conditions, the nozzle flow will be affected by real-gas effects. In the design of high-enthalpy nozzles, the impact of the variable specific heat capacity ratio should be considered. Early research works on the design of high-enthalpy nozzles mainly consider the impact from the aspects of varying temperature and pressure, but the impact from other high-enthalpy flow phenomena, such as vibration, dissociation, and ionization, have been selectively ignored. They assume that the flowing gas of the nozzle is a semi-perfect gas, which is still suitable for the state equation, while taking into account the conditions that the thermodynamic properties (such as the specific heat capacity ratio) do vary with temperature.

To explain the effect of variable γ on nozzle design, Figure 7.14 shows how temperature and pressure changes affect the specific heat capacity ratio. In a compressible flow, the changes in many properties of the gas are associated with γ, so a wind tunnel test is generally required to maintain the same γ as that in the real flight condition. The variation of γ can be regarded as a problem mainly for hypersonic or supersonic wind tunnels. According to the research on high-temperature real-gas effects, γ variation generally must be considered when the Mach number reaches 7. When the flight velocity or the test flow velocity reaches $Ma = 10$, the value of γ will range from 1.15 to 1.2 in the gas behind the shock wave on the flight vehicle. Large γ variation will affect not only the test results but also the nozzle design for hypersonic wind tunnels.

The research by Johnson and Boney (1975) can clearly explain the effect of γ on the nozzle contour. Figure 7.15 presents a comparison of the nozzle contours obtained for different exit Mach numbers and with different specific heat capacity ratios. Obviously, with different specific heat capacity ratios, even for nozzles with the same area

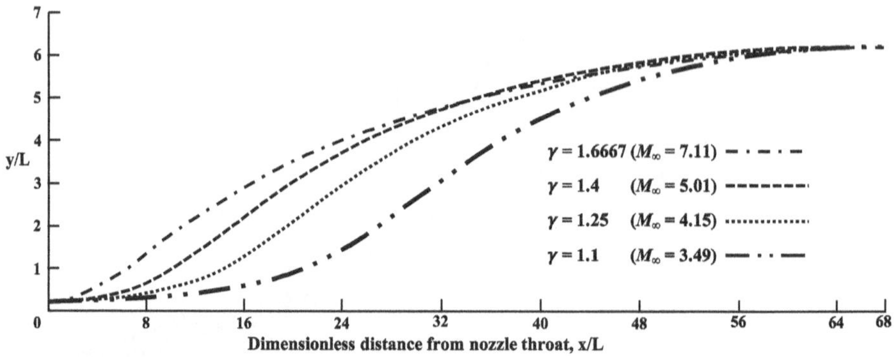

Figure 7.15 Effect of γ on nozzle contour in the case of the same area ratio.

ratio, the Mach numbers at the nozzle exit are quite different, which greatly differ from the results for an ideal gas. However, the work by Johnson and Boney considers a fixed area ratio in order to compare the change in the nozzle contour, which is a simplification in itself, because the area ratio of high-enthalpy nozzles is also affected by real-gas effects, which therefore can lead to large differences. This point has been explained in Tang's work (Tang et al. 2019).

7.4.2 MOC with Variable Specific Heat Capacity Ratio

As early as in the 1950s, some scholars conducted a series of studies on the design of hypersonic nozzles under the influence of real-gas effects (Guentert & Neumann 1959, Erickson & Creekmore 1960, Enkenhus & Maher 1962, Johnson et al. 1963, Johnson & Boney 1975). The inviscid contours of early supersonic nozzles with real-gas flow were still designed and computed based on MOC, but the actual changes in the thermodynamic properties of the gas have been taken into account. For example, the variable γ is taken as a main design parameter. Then, the contour of the physical wall of the nozzle is finally determined by superimposing the calculated displacement thickness onto the inviscid nozzle contour. Only the variable γ is considered while other influencing factors are ignored, which is indeed a lot of simplifications in itself. However, it is a realistic and practical method for early studies of high-enthalpy nozzle design to only consider the effect γ on the thermodynamic properties of the gas.

Guentert et al. (1959) proposed a design method for the inviscid contours of high-enthalpy axisymmetric nozzles. This method only calculates the gas with chemical reaction by applying the MOC with a variable isentropic index. Subsequently, Enkenhus and Maher (1962) proposed a similar method for calculating axisymmetric nozzle contours for high-temperature gases, but independently developed a computational program for the design of nozzles with real-gas effects. Johnson et al. (1963) improved the above methods by directly applying the thermodynamic data of the isentropic expansion to the MOC computational procedure and designed a set of high-enthalpy axisymmetric nozzles for a Mach number of 17 and a total temperature of 2,800 K.

Korte (2000) has developed a simplified design method with variable γ. The inviscid potential flow contour is designed by the Sivells method based on the MOC approach. As shown in Figure 7.7, in the design process of a nozzle with variable γ, the flow in the area HIEG is first calculated. The specific heat capacity ratio γ_1 is the average value in this area, based on which the initial expansion contour HG can be created. Then, the flow in the area ABCD is calculated. The specific heat capacity ratio γ_2 is the average value in this area, based on which the damping contour AD is created through computation. Then, the area ratio is calculated with a modified γ, so that the length of the straight line GA can be determined. The initial expansion contour HG, the damping contour AD, and the linear segment GA are made to match with each other in order to meet the requirements of the following area ratio formula. In the formula, the equilibrium flow assumption is adopted, so the area ratio formula is derived based on the isentropic relation (Eggers 1949):

$$\frac{A^*}{A} = \left(\frac{A^*}{A}\right)_i F\left(T_0, Ma\right) \tag{7.19}$$

$$\left(\frac{A^*}{A}\right)_i = Ma\left(\frac{\frac{\gamma_i+1}{2}}{1+\frac{\gamma_i-1}{2}Ma^2}\right)^{\frac{\gamma_i+1}{2(\gamma_i-1)}} \tag{7.20}$$

where A^* and A are the nozzle throat area and the nozzle exit area, respectively, $\left(\frac{A^*}{A}\right)_i$ is the area ratio of the ideal gas, the subscript i indicates an ideal gas, and Ma is the design Mach number. Without taking the variable γ into account, we have

$$F\left(T_0, Ma\right) = 1. \tag{7.21}$$

When γ is considered a variable, $F\left(T_0, Ma\right)$ is expressed by the formula below (Eggers, 1949):

$$F\left(T_0, Ma\right) = 1 + B_4\left(Ma\right)b_{\rho_0} + C_4\left(Ma\right)\frac{c_{\rho_0}}{RT_0^{\,2}} + D_4\left(\frac{\theta}{T_0}, Ma\right)\left(\frac{\theta}{T_0}\right)^{-\left(\frac{\theta}{T_0}\right)} \tag{7.22}$$

where

$$B_4\left(Ma\right) = \frac{\gamma_i+1}{2\left(\gamma_i-1\right)}\left[B_1\left(Ma\right) - B_1\left(Ma^*\right)\right] \tag{7.23}$$

$$C_4\left(Ma\right) = \frac{\gamma_i+1}{2\left(\gamma_i-1\right)}\left[C_1\left(Ma\right) - C_1\left(Ma^*\right)\right] + \frac{\Gamma_i-\gamma_i}{\gamma_i}\left[\left(\frac{\rho}{\rho_0}\right)_i^{3-2\gamma} - \left(\frac{\rho^*}{\rho_0}\right)_i^{3-2\gamma_i}\right] \tag{7.24}$$

$$D_4\left(\frac{\theta}{T_0}, Ma\right) = \frac{\gamma_i+1}{2\left(\gamma_i-1\right)}\left[D_1\left(\frac{\theta}{T_0}, Ma\right) - D_1\left(\frac{\theta}{T_0}, Ma^*\right)\right]$$
$$+ e^{-\left(\frac{\theta}{T_0}\right)\left[\left(\frac{T_0}{T}\right)_i - 1\right]}\left\{\left(\frac{T_0}{T}\right)_i + \frac{T_0}{\theta}\left[1 - \frac{\left(\gamma_i-1\right)^2}{\gamma_i}\left(\frac{\theta}{T_0}\right)^2\left(\frac{T_0}{T}\right)_i^2\right]\right\}$$

$$-e^{-\left(\frac{\theta}{T_0}\right)\left[\left(\frac{T_0}{T^*}\right)_i-1\right]}\left\{\left(\frac{T_0}{T^*}\right)_i+\frac{T_0}{\theta}\left[1-\frac{(\gamma_i-1)^2}{2\gamma_i}\left(\frac{}{T_0}\right)^2\left(\frac{T_0}{T^*}\right)_i^2\right]\right\} \quad (7.25)$$

$$B_1(Ma)=2\left[\left(\frac{P}{P_0}\right)_i-\left(\frac{\rho}{\rho_0}\right)_i\right]+\frac{\gamma_i-1}{\gamma_i}\left[1-\left(\frac{P}{P_0}\right)_i\right] \quad (7.26)$$

$$C_1(Ma)=2\frac{\Gamma_i}{\gamma_i}\left(\frac{\rho}{\rho_0}\right)_i^{2-\gamma_i}\left[1-\left(\frac{\rho}{\rho_0}\right)_i^{1-\gamma_i}\right]-3\frac{\gamma_i-1}{\gamma_i}\left[\;-\left(\frac{\rho}{\rho_0}\right)_i^{2-\gamma_i}\right] \quad (7.27)$$

$$D_1\left(\frac{\theta}{T_0},Ma\right)=\frac{\gamma_i-1}{\gamma_i}\left[1-e^{-\left(\frac{\theta}{T_0}\right)\left[\left(\frac{T_0}{T}\right)_i-1\right]}\left\{1-(\gamma_i-1)\left(\frac{\theta}{T_0}\right)\left(\frac{T_0}{T}\right)_i\left[\left(\frac{T_0}{T}\right)_i-1\right]\right\}\right] \quad (7.28)$$

$$\Gamma_i=\gamma_i^2-3\gamma_i+3 \quad (7.29)$$

where θ is the characteristic vibrational temperature, which is affected by the gas components. $\frac{P}{P_0}$, $\frac{\rho}{\rho_0}$, and $\frac{T}{T_0}$ are obtained from a series of one-dimensional flow isentropic relations. See the reference by Eggers (1949) for the detailed derivation process.

In fact, the effect of a variable γ on the upstream flow of the nozzle (especially the flow near the throat) is much larger than that on the downstream flow (the area from the damping section to the nozzle exit). Because the downstream flow has expanded to approximately an ideal gas, it is more necessary to focus on the high-temperature effects in the radial flow area. In fact, a great deal of simplification has been made for high-enthalpy flow, especially hypervelocity flow, because the downstream gas is basically frozen while the gas near the throat has a complicated composition and is basically in a chemically nonequilibrium state. Zonars (1967) provides a comparison between the theoretical evaluation and the test measurements of nozzle wall pressure with equilibrium and nonequilibrium flows. Notably, as the total temperature increases, the equilibrium effects become more obvious. As a result, the evaluation and analysis of nonequilibrium nozzle flow becomes more difficult. For the design of high-enthalpy nozzles, using the MOC technique that only considers a variable γ will also cause certain design errors. The design error for hypervelocity nozzles will become very large. It is imperative to develop new design methods or use an optimization of the nozzle contour based on CFD numerical techniques in order to meet nozzle design requirements.

7.4.3 Research on CFD-Based Design of High-Enthalpy Nozzles

In high-enthalpy gas flows, the specific heat capacity ratio γ is variable. Several fluid dynamic equations derived under the perfect gas assumption are no longer valid in high-temperature gas dynamics. For such flows, analytical solutions may not exist but some form of numerical solution can be obtained. In addition, considering merely the variation of γ and ignoring the effects of the other factors have been a significant simplification for high-temperature gas flows that are extremely complicated, so large errors will inevitably occur. With the rapid development of computational technology and to solve this problem, several scholars have conducted feasibility verification and flow confirmation on the high-enthalpy nozzles designed by using CFD numerical simulation. To some extent, the evaluation of real-gas effects is easier to implement by using CFD techniques, which is an advantage of CFD-based nozzle design methods. This is because if the computation cost allows, we can conduct the evaluation and computation of the overall flow of the high-enthalpy nozzle by applying the appropriate chemically reacting model and test-gas reaction components in the CFD solver or commercial numerical simulation software.

Several scholars, represented by Korte, first applied CFD simulation to the optimization design of hypersonic and high-enthalpy nozzles. As mentioned above, Korte's CFD-based analysis and optimization method allows any combination of CFD solver and optimization algorithm in the design process. Therefore, the design process for high-enthalpy flow nozzles becomes easier. What we need to do is to include in the CFD solver a nonequilibrium model of high-temperature chemical reactions. Then, high-accuracy contour design can be achieved under the corresponding high-enthalpy conditions. Korte designed a set of helium-driven Mach 15 nozzles using the nozzle design program CAN-DO (Korte et al. 1992c). But in the PNS solver used for the analysis of the nozzle flow, real-gas effects are ignored. Subsequently, Korte et al. (1992b) and Hollis (1992) designed a 22-inch helium tunnel nozzle at the NASA Langley Research Center using the CAN-DO software. However, they simply modified the state equation in the CFD solver and the results obtained basically met the application requirements at that time. In 2000, Korte (2000) improved the area ratio of the real-gas nozzle flow in his design method to address real-gas effects (see Figure 7.16). The main idea is to solve a steady one-dimensional flow at a given Mach number, and iterate until the desired nozzle exit condition can be obtained.

Researchers at AEDC (Shope 2005, Shope & Tatum 2005) have applied CFD-based analyses and optimization design methods to high-enthalpy nozzles with

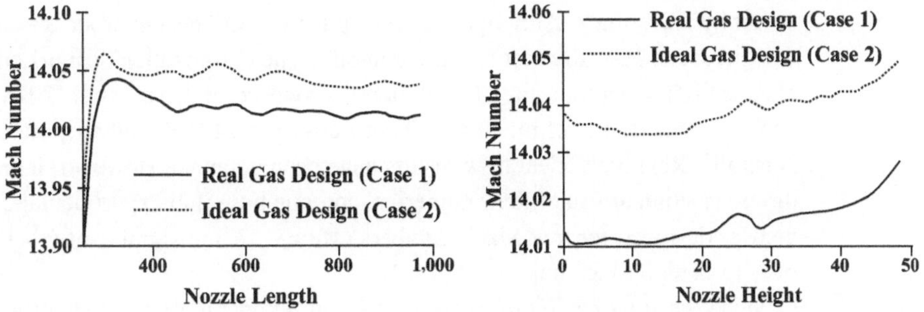

Figure 7.16 Numerical comparison for the nozzle design with real gas and with ideal gas (Left: Mach number distributions along the axial line; Right: Mach number distributions at the cross-section of nozzle exit).

chemically reacting flows and developed a new characteristics tracking code. Multiple optimization algorithms, such as the least-squares optimization (Scales 1985) and the spline-interpolation optimization (Conte & Boor 1972) are integrated, controlled, and implemented through programming scripts. Shope's nozzle design code allows the expansion of chemically reacting nozzle flows and can modify the spline points based on the code of Korte et al. Its CFD solver uses a computation method called data-parallel line relaxation to solve the N-S equations with chemical reactions (Wright & Candler 1998, Candler 2004, 2005). The solver computes an effective thermally and calorically perfect gas model with frozen gas components that include CO_2, H_2O, N_2, O_2, and NO. The percentage of each component is obtained using the CEA96 program (Gordon & McBride 1996) of the NASA Glen Research Center.

Shope's optimization method is roughly divided into two steps. First, the MOC-based Sivells method is adopted to design the initial nozzle contour. In the characteristics computation, it is assumed that the gas flowing through the nozzle is limited to thermally and calorically perfect gas. The nozzle exit design parameters (such as Mach number and velocity) that are approximately in thermodynamic equilibrium can be obtained by setting γ and the gas constant. In the second step, the obtained initial nozzle contour is further optimized in order to remove the limitations of MOC. The main design process is to correct the spline interpolation of the nozzle contour coordinates. The spline curve only goes through a few defined points along the initial nozzle contour, rather than interpolating all the coordinate points of the nozzle contour. The spline interpolation correction aims to allow a sufficient number of coordinate points to be used for defining the high-accuracy nozzle contour, while only a few nodes (design parameters) are used in the correction. Then, this set of design parameters was used in the repeated optimization process (namely, the MOC design process) to further improve the quality of the high-enthalpy flow. In the end, Shope has designed the Aerodynamic and Propulsion Test Unit (APTU) Mach 6 nozzle of the Arnold Engineering Development Center (AEDC) using this code and obtained good results. However, this method has not been further verified for high-enthalpy flows at higher Mach numbers.

Gaffney (2006, 2007) has designed the nozzle for NASA's HYPULSE shock-expansion tunnel (SET). He designed a set of $Ma = 15$ nozzles with an Euler equation solver by adjusting γ. Although the conventional MOC design process is adopted, the design results basically met the requirements. The optimization design code based on chemically nonequilibrium CFD, which is proposed by Chan et al. (2018), has been applied in the design of the $Ma = 10$ nozzle for the T4 high-enthalpy shock tunnel in Australia. Relatively ideal flow results have been obtained. However, it can be found through a comparison of their numerical computations and experimental data that the nozzles designed for low Mach numbers ($Ma = 4, 7$) are significantly better than the ones for a high Mach number ($Ma = 10$).

Since the flow has high pressure and high temperature, real-gas effects must be considered in the design of the nozzles for reflective impulse high-enthalpy ground facilities. The combined use of CFD codes and optimization techniques (CFD/OM), as well as the conventional MOC with boundary layer correction (MOC/BL), are the two main methods for designing nozzles for hypersonic wind tunnels. When the boundary layer of the nozzle wall is very thick, more accurate results can be obtained with a CFD-based nozzle design. However, for the reflective impulse facility, the nozzle has a cold wall with extremely high pressure, which generally results in a thinner boundary layer. Therefore, satisfactory design results can also be achieved using the classic MOC/BL. Since a finer grid resolution is required to accurately capture the thin boundary layer, it is more time consuming and expensive to embed the CFD solution in the optimization procedure as compared with the conventional MOC design method.

Tang et al. (2019) have improved the classic MOC/BL method to design hypervelocity and high-enthalpy nozzles that takes real-gas effects into account. The nozzle flow is simulated through CFD numerical simulation. Then, the numerical flow data is extracted and substituted into MOC/BL for iterative optimization. Therefore, this method (CFD/MOC/OM) is different from the conventional MOC/BL and does not use the more complex CFD-embedded-in-optimization method in CFD/OM. It combines CFD numerical simulation, MOC/BL, and optimization method in the design process, which is a simpler combined optimization solution. Tang et al. (2019) carried out a design and iterative optimization using CFD/MOC/OM for a high-enthalpy nozzle at a Mach number of 17. During the correction of high-temperature gas effects, not only a variable γ but also the species change of the nonequilibrium gas are considered. The average value of the area ABCD (see Figure 7.7) is selected for computation since the specific heat capacity ratio does not change significantly in this area. In the area TIEG, the variation of γ with temperature is considered in the CFD simulations. The computations are performed using the NASA polynomial curve fits (McBride et al. 1993, 2002)

$$\gamma = a_1 T^{-2} + a_2 T^{-1} + a_3 + a_4 T^1 + a_5 T^2 + a_6 T^3 + a_7 T^4. \tag{7.30}$$

The coefficients of each species in the polynomial (see Equation 7.30) can be obtained through a table lookup. The information on gas species is extracted by high-accuracy nonequilibrium CFD simulations. Calculating the area ratio requires the characteristic vibrational temperature, which is also affected by the change of gas species. The species composition and its concentrations in the corresponding area are extracted by CFD techniques to calculate the characteristic vibrational temperature. Then, a new area ratio can be worked out. The initial flowfield data and computational conditions are extracted from the simulated nozzle flowfield obtained by CFD computation using the inviscid nozzle contour. In the iterative computation, the flow parameters will gradually converge. Generally, a satisfactory result can be achieved in two or three iterations. For hypervelocity nozzles with a very high enthalpy, the number of iterations may be increased accordingly.

In Tang's work (2019) on the design of large-scale high-enthalpy nozzles, two key design techniques are adopted. One is the fitting of the area ratio of hypervelocity and high-enthalpy nozzles based on CFD data. Since the conventional theoretical formula of the nozzle area ratio is only suitable for the case of equilibrium flow, the actual

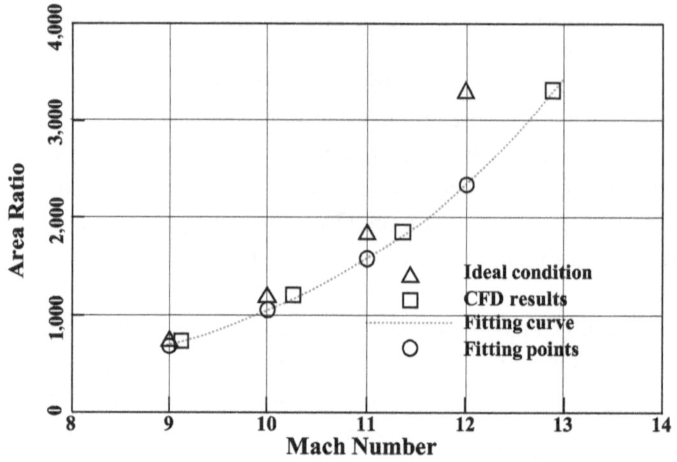

Figure 7.17 Comparison of area ratios obtained by theoretical and CFD calculations (Tang et al. 2019).

Mach number calculated by nonequilibrium CFD simulations can have a large deviation from the design Mach number. In order to obtain a more accurate inviscid flow solution, the theoretical area ratio obtained from an equilibrium flow is corrected by using a fitting method. The triangles in Figure 7.17 are the area ratios obtained using Equation (7.19) for different Mach numbers. The squares are for the actual Mach numbers at the nozzle exit to which the area ratios correspond in the nonequilibrium CFD simulation. A curve-fitting procedure is carried out based on the relationship between the theoretical and simulated Mach numbers at the nozzle exit. A new area ratio (circles in Figure 7.17) corresponding to each Mach number is extracted from the fitted curve, and then substituted into the code to update the nozzle contour. After several iterations, the final area ratio is obtained when the actual Mach number is consistent with the design Mach number. In fact, the fitting process of the area ratio is a relatively more accurate evaluation of the boundary layer of the nonequilibrium nozzle flow.

The other key design technique is that the desired nozzle contour optimization is realized using the endpoint slope-controllable Hermitian spline interpolation. On the whole, the obtained spline curve is very similar to the reference points given by the original design. The optimized curve solves the problem of local fluctuation at the reference points. In fact, in the design process of high-enthalpy nozzles, in order to ensure the flow quality at the nozzle exit, the nozzle contour must meet the following conditions: (1) monotonically increase in area, (2) smooth first-order derivatives (continuous second-order derivatives), (3) the nozzle contour goes through the defined coordinate points and the first-order and second-order derivatives at the defined points are zero or controllable, and (4) a refined connection control at the throat section and the segmented connection points can be achieved to prevent the occurrence of any interruptions along the nozzle contour.

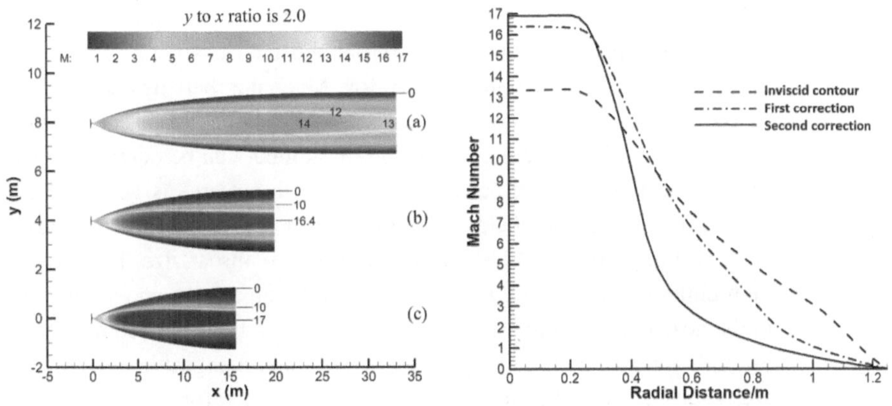

Figure 7.18 Iterative optimization for $Ma = 17$ nozzles. Left: (a) initial flow calculated from the inviscid contour, (b) result by the first iteration, and (c) result by the second iteration. Right: Mach number distributions at the nozzle exit (Tang et al. 2019).

The CFD/MOC/OM iterative optimization of the Mach 17 nozzle designed by Tang et al. (2019) can be summarized in Figure 7.18. The top image on the left (Figure 7.18 (a)) shows the Mach number contours calculated using nonequilibrium CFD for the initial inviscid nozzle contour. With each correction for viscous and nonequilibrium effects, the nozzle exit Mach number approaches the target design value of 17, while the exit area is increased to accommodate the boundary layer (Figure 7.18(b), (c) on the left). Note also that the nozzle length is decreased with each iteration. The radial Mach number distributions across the nozzle exit for the different iterations are also shown in Figure 7.18 (right), demonstrating that the core flow achieved the target value of 17 after two iterations.

7.4.4 Nozzle Design with Interchangeable Throats

It is highly desirable to design an axisymmetric hypersonic nozzle that is capable of generating multiple exit Mach numbers, as the need to replace the entire nozzle for each discrete Mach number would require a high cost and a large workload. This can be accomplished by designing a throat section that is interchangeable. The interchangeable-throat technique is to design the nozzle throat block that allows the cross-sectional area at the throat to be changed, by adding a straight segment at the inflection point of the nozzle expansion section or by redesigning the initial expansion section, while retaining the downstream supersonic expansion section of the hypersonic nozzle. In this way, the area ratio of the nozzle for a given design Mach number is changed accordingly, thereby changing the Mach number at the nozzle exit. In practical engineering applications, a nozzle with interchangeable throats can be designed in advance and the Mach number of the test flow can be varied by directly replacing the throat component before a wind tunnel test.

To obtain different Mach numbers at the nozzle exit, the ratio of the exit cross-sectional area to that at the throat must be changed. Using interchangeable throats achieves different nozzle exit flow Mach numbers by changing the throat area on the premise that the nozzle exit area is kept unchanged. For nozzles at different enthalpy conditions, the nozzle exit Mach number can be determined using different area ratios. For a low-velocity ideal gas, the throat area is generally determined by Eq. (7.8). For a high-enthalpy flow, the impact of real-gas effects needs to be considered, so Eq. (7.19) can be used to determine the throat size. However, in the actual high-enthalpy flow, with the increase of total temperature and the flow rate, the error of the nozzle flow parameter determined by Eq. (7.19) will become increasingly large. For the design of high-enthalpy nozzles with interchangeable throats, a more accurate area ratio becomes the most critical parameter. Therefore, a more ideal design result can be obtained by fitting the area ratio with CFD data.

A nozzle with interchangeable throats shares the same damping section (supersonic expansion section), so we only need to design a series of initial expansion sections for the throat part to meet the design requirements at different Mach numbers. An interchangeable throat changes the size of the throat radius by adding or reducing a straight segment at the "turning point." Therefore, the nozzle contour determined by a certain design method should have a straight segment near the inflection point (the "turning point" of the contour). From the hypersonic nozzle design methods above, it can be found that the straight segment is short by these design methods and the short length sometimes cannot meet the required modified throat area, especially in the case of changing a high Mach number design to a low Mach number one (namely to obtain a throat with a large size by reducing the length of the straight segment). On the contrary, the effect of the short straight segment on the designed nozzle is relatively small, which can be ignored in most cases.

After completing the design of the initial nozzle contour, if we wish to change the throat cross-sectional area, the inflection point (the turning point, namely point E or A in Figure 7.19) or the straight segment (such as line EA in Figure 7.19) of the nozzle

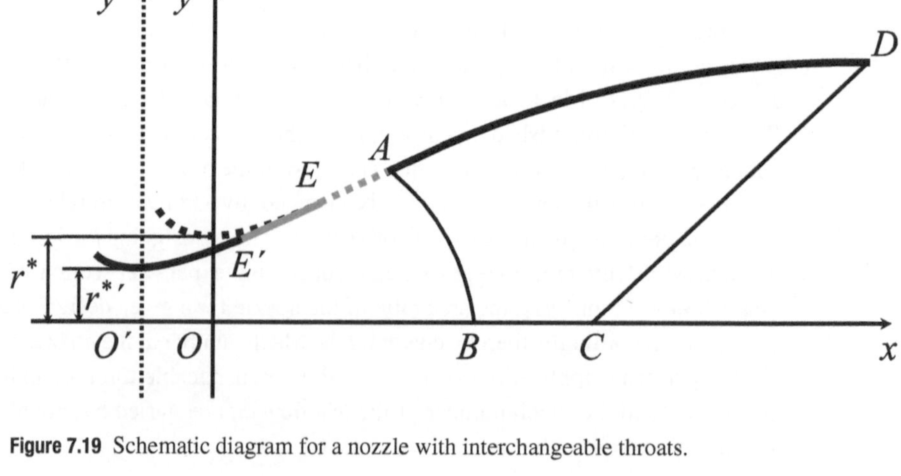

Figure 7.19 Schematic diagram for a nozzle with interchangeable throats.

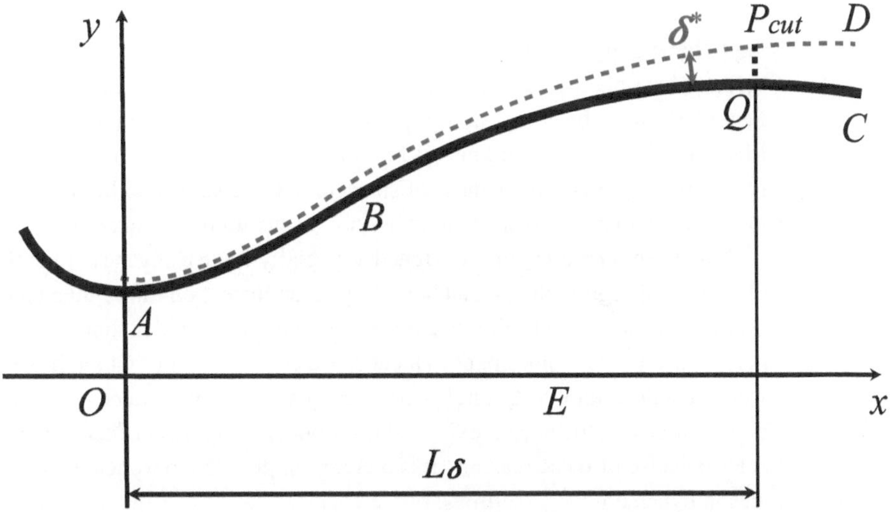

Figure 7.20 Schematic diagram for nozzle truncation method.

contour should first be found. Then, the slope of the inflection point or straight segment should be calculated. In addition, a spline interpolation can also be performed on the nozzle contour to obtain the slope at each point. Alternatively, the first-order derivatives of the nozzle contour are solved. Generally, the inflection point or the straight segment has the most extreme slope. After the modified point is determined, a new throat radius for producing the required design Mach number is obtained using the area ratio computation, without changing the nozzle exit area. Then, a new nozzle is designed using the same expansion angle parameter. Next, the inflection point of the new nozzle contour is found. At this time, the slope of the inflection point should be consistent with that of the original nozzle contour. Furthermore, the length of the line EE' is adjusted so that the inflection point of the initial expansion section of the new nozzle is coincident with point E'.

In practical engineering applications, the interchangeable-throat nozzle design is generally used in hypervelocity (high Mach number) or high-enthalpy (high total temperature and total pressure) nozzles. In these cases, real-gas effects in the nozzle flow should not be ignored. The variable γ of the gas needs to be considered during the nozzle design. For the specific design method, refer to the above descriptions on high-enthalpy nozzle design.

7.4.5 Truncation of the Designed Nozzle

Hypersonic and high-enthalpy nozzles designed with conventional methods are generally very long, which is caused by different nozzle design methods on the one hand, and determined by the flow characteristics of hypersonic nozzles on the other hand. In particular, the diamond-shaped area of uniform nozzle flow is very long. From the perspectives of nozzle design and engineering application, it is desirable to make the nozzle as short as possible without sacrificing flow uniformity. Aiming at the shortening of hypersonic nozzles, several scholars have made attempts in terms of the design method, but only the cases with low Mach numbers have been designed and analyzed. Few works have been reported or published in the literature on nozzle shortening for Mach numbers of 10 or higher. While there have been a lot of discussions about the necessity and cost-effectiveness of shortening hypersonic and high-enthalpy nozzles, the methods to perform shortening are not discussed in the literature.

After the nozzle contour is designed, especially when using the conventional method of inviscid design with viscous boundary layer correction rather than using optimization with detailed CFD solutions, it is sometimes necessary to shorten the nozzle. This may be due to the requirement to meet space constraints, or that the full-length nozzle would result in an overly thickened boundary layer that reduces the uniformity and core flow size at the nozzle exit. A shorter nozzle would also decrease the fabrication cost. Because of these reasons, it is a common practice to truncate the designed nozzles in hypersonic test facilities.

The location on the contour to truncate a high-enthalpy nozzle is determined with the conditions that the cross-sectional Mach number distribution should be kept uniform at the truncation position and the radial velocity component should meet design index requirements. The steps are as follows:

(1) Carry out a CFD numerical simulation of the flowfield with the designed nozzle contour (dashed line in Figure 7.20). From the solution, compute the displacement thickness and determine the boundary line (solid line in Figure 7.20) that separates the potential core flow from the displacement thickness, according to the definition of the boundary layer and the displacement thickness.
(2) Compute the first-order derivative (i.e., slope) of the potential flow boundary line. If the derivative at point C of the nozzle exit (at the end of the boundary line, see Figure 7.20) is negative, it implies that the boundary layer has thickened so much that the cross-sectional area of the potential core flow is decreasing in the axial direction near the nozzle exit. The area contraction not only would reduce the core flow area at the nozzle exit, but would also lead to the generation of compression waves and nonparallel flow, thus destroying the flow uniformity at the nozzle exit. If the derivative at point C is indeed negative, determine the location (point Q) at which the derivative vanishes. With the horizontal coordinate x_Q of point Q as the x-coordinate of the truncation point, the truncated position P_{cut} of the final nozzle contour can be determined.
(3) With the truncated nozzle contour, a CFD simulation of the nozzle flowfield is again performed to determine whether it meets design requirements. If the derivative at point C of the truncated nozzle is zero, the truncation is deemed acceptable on the premise that the test flow quality is met. If not, repeat step (2) until all the design requirements are met.

In theory, if there is no error in the initial design's boundary layer correction, the derivative at point C should be zero. But in reality, errors in the evaluation of the boundary layer may exist due to real-gas effects. As a result, the potential flow/

displacement thickness boundary line obtained by CFD will not coincide with that obtained from the less-accurate design procedure. This is especially true for the case where the derivative at point C is negative, making it necessary to perform truncation in order to obtain a shorter nozzle and a larger core area of uniform flow.

7.5 Summary and Prospects

With the rapid development of CFD techniques, many achievements have been made on high-velocity nozzle design. Associated theories and methods can be divided into three categories: the first is the MOC/BL method combining the conventional MOC and boundary layer correction. This method has been evolving since the MOC was proposed in 1929 and is still in use today. With this method, good design results can be obtained for conventional hypersonic wind tunnel nozzles which need not take high-temperature effects into account. However, the conventional MOC has limitations on the design of high-enthalpy nozzles. Its improved methods also hardly meet the increasingly higher evaluation standard of flow quality. Therefore, the solving method of the nozzle flowfield is shifted from MOC to CFD numerical simulation with higher accuracy. The second one is the CFD-based optimization method (CFD/OM) in which the optimization method is embedded in the CFD solver. Since its first appearance in the early 1990s, the CFD/OM method has been evolving with the improvement of computational technology and developed various coupling modes between the CFD solver and the optimization algorithm. CFD/OM is a tight coupling of the optimization algorithms and the CFD solvers, which removes the constraints of conventional methods. Boundary layer and real-gas effects are also reduced through the CFD solver, eliminating the corrections of the boundary layer and high-temperature effects. However, a criticism of this method is that the demanding coupling of the CFD solver and the optimization algorithm, together with the high computation cost, make it difficult to apply in the engineering field. The third one is CFD/MOC/OM, which is the MOC optimization design based on CFD numerical techniques. This method has been developed for solving the problem of the complex design process in the second category to keep the design simple. The initial contour of the inviscid potential flow of high-enthalpy nozzles is still designed with the simple and mature MOC technique. The CFD numerical simulation of chemically nonequilibrium flow is then used to optimize the high-enthalpy nozzle contour. This method does not require a tight coupling between the optimization algorithm and the CFD solver. The CFD data obtained from each iteration step is taken as the initial and boundary conditions of the MOC design, and substituted into the solver for the new solution. The method is simple, easy to use, and robust. Satisfactory optimization results can generally be obtained in two or three iterations.

As hypersonic nozzle design evolves from theoretical analysis, the semi-empirical formula approximation, the boundary layer correction (MOC/BL), to the CFD-based optimization methods (CFD/OM and CFD/MOC/OM), the design accuracy has been getting higher. The CFD-based optimization offers prominent advantages, especially for high-enthalpy nozzle design. Seen from the development history of nozzle design theories and methods, they have been evolving with the development of computer technology and computational methods, as well as driven by the demand for aerospace technology. Since the beginning of the twenty-first century, especially in the past ten years, hypersonic flight has become a key technology for the future aerospace industry. Therefore, an increasingly high requirement has been put on the quality of hypervelocity and high-enthalpy test flows. This not only gives rise to the need for performance improvement in most hypersonic and high-enthalpy ground facilities developed in the last century, but also motivates the research and development of new hypervelocity wind tunnels. Hence, theoretical and methodological research on hypersonic nozzle design is becoming more important. By reviewing the research progress over the years, it can be concluded that the following aspects need to be further explored in the field of hypersonic nozzle design.

(1) The MOC/BL design method alone can no longer produce a nozzle that can meet the requirements of hypervelocity and high-enthalpy wind tunnels, but the optimization design method combining it with CFD numerical simulation can still obtain high-quality high-enthalpy flows, so it is valuable to further improve the MOC/BL method. For example, more accurate test data on high-enthalpy flows can be used to provide more reasonable MOC setting conditions, thereby improving or correcting the design deviation brought by real-gas effects. In addition, the method in which the axial velocity or Mach number distribution is set based on the Sivells method can still obtain relatively satisfactory results. In the definition of Bezier control points for nozzle contour optimization, it is recommended that the radial coordinates of the last 2 to 3 Bezier control points be preferably the same and are kept constant during the optimization process to ensure that the flow at the nozzle exit always stays parallel to the nozzle axis, thereby ensuring that the radial component of the exit flow is zero.
(2) The CFD/OM method has the disadvantages of being a complex process and incurs a high computational cost, but with the development of computer technology, it probably represents the development trend in the near future. The major optimization objective functions of the CFD/OM algorithm are the Mach number distribution and flow angularity at the nozzle exit. Further optimization and improvement can be made by adjusting the objective functions. More parameters of the exit flow (such as static pressure and static temperature) can be taken as the optimization parameters to design high-quality nozzles.
(3) Because the CFD/OM method must be bound to the accuracy of the CFD simulation, a high-accuracy CFD technique that can handle thermochemical nonequilibrium conditions also needs to be developed simultaneously in view of real-gas effects. This actually requires that the high-temperature chemically reacting model used in the CFD solver must be further improved based on more accurate experimental data obtained in high-enthalpy ground facilities.
(4) CFD techniques have developed rapidly. However, for quite a long time, it will still be a great challenge to achieve the high-accuracy simulation capability for

hypersonic and high-enthalpy flows. In particular, most thermochemical reaction models have limitations for hypervelocity flows (with a total temperature of 5,000–10,000 K). Therefore, a method worth exploring is to take high-accuracy high-enthalpy test data as the basis for nozzle optimization design or the set boundary conditions. This mainly involves two ways: one is to correct the contour at the throat based on the experimental data of the designed nozzle. This way is simple and straightforward but not cost-efficient. Partial "redesign and re-correction" is necessary for the machined nozzle. The other is to take the high-enthalpy flow-field data as design reference conditions and make correction and optimization with CFD techniques. The measured test parameters, such as flow enthalpy and Mach number, can then be used as reference conditions to optimize the nozzle design.

The theory and method for hypersonic nozzle design and their evaluation cover a broad range of knowledge, which requires combining effective methods and techniques that involve theory, computation, and experiment. It is one of the most difficult research subjects and is still challenging to date. Improving the test flow quality through hypersonic nozzle design is of great significance in enhancing the accuracy and reliability of hypersonic test data. It is also important for supporting the development of advanced aerospace vehicles. Therefore, the research on the theory and methods of hypersonic nozzle design is fundamental and should receive great interest.

Nomenclature

a_i	NASA polynomial curve-fit coefficients
A	cross-sectional area
A^*	nozzle throat area
c_i	polynomial coefficients
F	function variable
L	nondimensionalizing scale factor for nozzle contours
Ma	Mach number
p	pressure
r	radius
T	temperature
u	velocity
u_1	velocity in the boundary layer
x	horizontal coordinate
y	vertical coordinate
β	oblique angle of supersonic expansion section
δ	boundary layer thickness
δ^*	boundary layer displacement thickness
γ	ratio of specific heats

μ	Mach angle
θ	inclination angle of the potential flow or characteristic vibrational temperature
ρ_1	density in the boundary layer
ν_1	Prandtl–Meyer angle

Subscript

0	total or stagnation condition
B	nozzle inflection point
D,E	nozzle centerline design point
I	ideal gas

Superscript

*	throat

References

Ali, M. H., Mashud, M., Bari, A. A., and Islam, M. M. U. (2012). Numerical Solution for the Design of Minimum Length Supersonic Nozzle. *ARPN Journal of Engineering and Applied Sciences*, **7**(5), 605–612.

Anderson John, D. Jr. (1970a). A True-Dependent Analysis for Vibrational and Chemical Nonequilibrium Nozzle Flows. *AIAA Journal*, **8**(3), 545–550.

Anderson John, D. Jr. (1970b). Time-Dependent Solutions of Nonequilibrium Nozzle Flows – A Sequel. *AIAA Journal*, **8**(12), 2280–2282.

Anderson John, D. Jr. (2003). *Modern Compressible Flow: With Historical Perspective*, 3rd ed. New York: McGraw–Hill.

Anderson John, D. Jr. (2006). *Hypersonic and High-Temperature Gas Dynamic*, 2nd ed. Reston, VA: American Institute of Aeronautics and Astronautics.

Armstrong, A. H. and Smith, M. G. (1951). Two-Dimensional Supersonic: Nozzle Design, Part I: Theory, Armament Research Establishment. *Report No. 5/51*.

Beckwith, I. E. and Moore, J. A. (1995). An Accurate and Rapid Method for the Design of Supersonic Nozzles. *NACA TN 3322*.

Benton, J. (1989). Design and Navier-Stokes Analysis of Hypersonic Wind Tunnel Nozzles. M.S. Thesis. Raleigh, NC: North Carolina State University.

Benton, J., Edwards, A., and Perkins, J. (1990). Limitations of the Method of Characteristics When Applied to Axisymmetric Hypersonic Nozzle Design. *AIAA Paper 90-0192*.

Bird, G. A. (1957). A Note on Combustion Driven Shock Tubes. In *Hypersonic Facilities in the Aerodynamics Department, Royal Aircraft Establishment*, ed. P.A. Hufton, AGARD Report 146.

Busemann, A. (1931). Gasdynamik. Handbuch der Experimental-physik, Akademische Verlagsgesellschaft. *Leipzig*, **4**(1), 341–460.

Candler, G. (2004). *APTU Nozzle Code Manual*, Version 3.0 (included with DPLR software distribution).

Candler, G. (2005). *Hypersonic Nozzle Analysis Using an Excluded Volume Equation of State*. AIAA-2005-5202, 35th AIAA Fluid Dynamics Conference and Exhibit, Toronto, Ontario, Canada.

Candler, G. and Perkins, J. (1991). Effects of Vibrational Nonequilibrium on Axisymmetric Hypersonic Nozzle Design. *AIAA Paper 91-0297.*

Chan, W. Y. K., Jacobs, P. A., Smart, M. K., Grieve, S., Craddock, C. S., and Doherty, L. J. (2018). Aerodynamic Design of Nozzles with Uniform Outflow for Hypervelocity Ground-Test Facilities. *Journal of Propulsion and Power*, **34**(6), 1467–1478.

Conte, S. D. and Boor, C. (1972). *Elementary Numerical Analysis*, 2nd ed. New York: McGraw-Hill.

Craddock, C. S. (2000). *Design of the Axisymmetric HyShot Nozzle for T4.* Department of Mechanical Engineering, University of Queensland Report 2/00, Brisbane, QLD, Australia.

Cresci, R. J. (1958). *Tabulation of Coordinates for Hypersonic Axisymmetric Nozzles.* WADC Technical Note 58–300, Pts. I and II. Wright Air Development Div.

Crown, J C. (1948). *Supersonic Nozzle Design.* NACA Technical Note No. 1651. Moffett Field, California: Ames Aeronautical Laboratory.

Eggers, A. J. Jr. (1949). *One-Dimensional Flows of an Imperfect Diatomic Gas.* NACA TN 1861. Ames Aeronautical Laboratory, Moffett Field.

Enkenhus, K. R. and Maher, E. F. (1962). *The Aerodynamic Design of Axisymmetric Nozzles for High-Temperature Air.* NAVWEPS Rep.7395, U.S. Naval Ord. Lab. (White Oak, Md.).

Erickson, W. D. (1963). Vibrational Nonequilibrium Flow of Nitrogen in Hypersonic Nozzles. *NASA TN D-1810.*

Erickson, W. D. and Creekmore, H. S. (1960). A Study of Equilibrium Real-Gas Effects in Hypersonic Air Nozzles, Including Charts of Thermodynamic Properties for Equilibrium Air. *NASA TN D-231.*

Foelsch, K. (1946). A New Method of Designing Two-Dimensional Laval Nozzles for a Parallel and Uniform Jet. *North American Aviation Report No. NA-46-235.*

Foelsch, K. (1949). The Analytical Design of an Axially Symmetric Laval Nozzle for a Parallel and Uniform Jet. *International Journal of Orthodontics*, **20**, 39–40.

Gaffney, R. L. J. (2006). Design of a Pulse-Facility Nozzle Using the Rotational Method of Characteristics. *Journal of Spacecraft and Rockets*, **43**(6), 1216–1224.

Gaffney, R. L. J. (2007). Erratum on design of a Pulse-Facility Nozzle Using the Rotational Method of Characteristics. *Journal of Spacecraft and Rockets*, **44**(2), 496–504.

Gaffney, R. L. J. and Korte, J. J. (2004). Analysis and Design of Rectangular-Cross-Section Nozzles for Scramjet Engine Testing. *AIAA Paper 2004-1137.*

Gollan, R. J. and Jacobs, P. A. (2013). About the Formulation, Verification and Validation of the Hypersonic Flow Solver Eilmer. *International Journal for Numerical Methods in Fluids*, **73**(1), 19–57.

Gordon, S. and McBride, B. J. (1996). Computer Program for Calculation of Complex Chemical Equilibrium Compositions and Applications, I. Analysis, II. User's Manual and Program Description. *NASA Reference Publication 1311.*

Guentert, E. C. and Neumann, H. E. (1959). Design of Axisymmetric Exhaust Nozzles by Method of Characteristics Incorporating a Variable Isentropic Exponent. *NASA TR R-33.*

Hall, I. M. (1962). Transonic Flow in Two-dimensional and Axially-symmetric Nozzles. *Quarterly Journal of Mechanics and Applied Mathematics XV*, **4**, 487–508.

Hall, J. G. and Treanor, C. E. (1968). Nonequilibrium Effects in Supersonic Nozzle Flows. *AGARDograph No. 124.*

Hannemann, K., Itoh, K., Mee, D. J., and Hornung, H. G. (2016). Free Piston Shock Tunnels HEG, HIEST, T4 and T5. In *Experimental Methods of Shock Wave Research*, eds. O. Igra and F. Seiler. New York: Springer, pp. 181–264.

Harris, E. L. and Albacete, L. M. (1964). *Vibrational Relaxation of Nitrogen in the NOL Hypersonic Tunnel No. 4*. Naval Ordnance Lab., TR 63–221, White Oak, MD.

Hollis, B. R. (1992). Real Gas Flow Parameters for NASA Langley 22-Inch Mach 20 Helium Tunnel. *NASA CR-4462*.

Huddleston, D. H. (1989). *Aerodynamic Design Optimization Using Computational Fluid Dynamics*. Ph.D. Dissertation. Knoxville, TN: University of Tennessee.

Jiang, Z. and Yu, H. (2017). Theories and Technologies for Duplicating Hypersonic Flight Conditions for Ground Testing. *National Science Review*, **4**(3), 290–296.

Johnson, C. B., Boney, L. R., Ellison, J. C., and Erickson, W. D. (1963). Real-Gas Effects on Hypersonic Nozzle Contours with a Method of Calculation. *NASA TN D-1622*.

Johnson, C. B, and Boney, L. R. (1975). A Method for Calculating a Real-Gas Two-Dimensional Nozzle Contour Including the Effects of Gamma. *NASA TM X–3243*.

Keeling, S. L. (1993). A Strategy for the Optimal Design of Nozzle Contours. *The 28th AIAA Thermophysics Conference. AIAA Paper 1993–2720*.

Korte, J. J. (1991). An Explicit Upwind Algorithm for Solving the Parabolized Navier-Stokes Equations. *NASA TP 3050*.

Korte, J. J. (1992). Aerodynamic Design of Axisymmetric Hypersonic Wind-Tunnel Nozzles Using Least-Squares/Parabolized Navier-Stokes Procedure. *Journal of Spacecraft and Rockets*, **29**(5), 685–691; also *AIAA Paper 92-0332*.

Korte, J. J. (1993). A CFD-Based Aerodynamic Design Procedure of Hypersonic Wind-Tunnel Nozzles. *AGARD Conference Proceedings*, **514**: 36.1–36.10.

Korte, J. J. (2000). Inviscid Design of Hypersonic Wind Tunnel Nozzles for a Real Gas. *AIAA Paper 2000-0677*.

Korte, J. J. and Hodge, J. S. (1994). Prediction/Measurement of Flow Quality in Hypersonic Wind-tunnel Nozzles Designed Using CFD. The 18th AIAA Aerospace Ground Testing Conference. *AIAA-94-2544*.

Korte, J. J. and Hodge, J. S. (1995). Flow Quality of Hypersonic Wind-Tunnel Nozzles Designed Using Computational Fluid-Dynamics. *Journal of Spacecraft and Rockets*, **32**(4), 569–580.

Korte, J. J., Hedlund, E., and Anandakrishnan, S. (1992a). A Comparison of Experimental Data with CFD for the NSWC Hypervelocity Wind Tunnel Mach14 Nozzle. *AIAA Paper 92-4010*.

Korte, J. J., Kumar, A., Singh, D. J., and Grossman, B. (1992b). Least-Squares Parabolized Navier-Stokes Procedure for Optimizing Hypersonic Wind-Tunnel Nozzles. *Journal of Propulsion and Power*, **8**(5), 1057–1063.

Korte, J., Kumar, A., Singh, D., and White, J. (1992c). *CAN-DO, CFD-Based Aerodynamic Nozzle Design and Optimization Program for Supersonic/Hypersonic Wind Tunnels*. Presented at the 17th Aerospace Ground Testing Conference, Reston, Virigina: American Institute of Aeronautics and Astronautics. *AIAA Paper 92-4009*.

Korte, J. J., Hedlund, E., and Anandakrishnan, S. (1992d). A Comparison of Experimental Data with CFD for the NSWC HypervelocityHypervelocity Wind Tunnel Mach14 Nozzle. *AIAA Paper 92–4010*.

Lukasiewicz, J. (1973). *Experimental Methods of Hypersonics, Gasdynamics, Part I*. Vol. 3. New York: Dekker, pp. 80–86.

McBride, B. J., Gordon, S., and Reno, M. A. (1993). Coefficients for Calculating Thermodynamic and Transport Properties of Individual Species. *NASA Technical Memorandum 4513*.

McBride, B. J., Zehe, M. J., and Gordon, S. (2002). *NASA Glenn Coefficients for Calculating Thermodynamic Properties of Individual Species. NASA/TP-2002-211556*. Cleveland, Ohio: Glenn Research Center.

Nelder, J. A., and Mead, R. (1965). A Simplex Method for Function Minimization. *Computer Journal*, **7**(4), 308–313.

Potter, J. L and Carden, W. H. (1968). Design of Axisymmetric Contoured Nozzles for Laminar Hypersonic Flow. *Journal of Spacecraft and Rockets*, **5**(9), 1095–1100.

Prandtl, L. and Busemann, A. (1929). *Naherungsverfahren zur Zeichnerischen Ermittlung von Ebenen Strömungen mit Übershllgeschwindigkeit*. Zurich, Switzerland: Füssli Verlag.

Preiswerk, E. (1940). Application of the Methods of Gas Dynamics to Water Flows with Free Surface, Part I. Flows with no energy dissipation. *NACA TM 934*.

Puckett, A. E. (1946). Supersonic Nozzle Design. *Journal of Applied Mechanics*, **13**(4), A265–A270.

Ruptash, J. (1952). *Supersonic Wind Tunnels Theory, Design and Performance*. University of Toronto, Institute of Aerophysics: Toronto, Canada.

Scales, L. S. (1985). *Introduction to Non-Linear Optimization*. New York: Springer-Verlag, Chap. 4.

Shapiro, A. H. (1953). *The Dynamics and Thermodynamics of Compressible Fluid Flow*. Vol. 1. New York: The Ronald Press Company.

Shope, F. L. (2005). Design Optimization of Hypersonic Test Facility Nozzle Contours Using Splined Corrections. *AEDC-TR-04-2*.

Shope, F. L. (2006). Contour Design Techniques for Super/Hypersonic Wind Tunnel Nozzles. The 24th AIAA Applied Aerodynamics Conference, *AIAA Paper 2006-3665*.

Shope, F. L. and Tatum, K. E. (2005). Reacting Navier-Stokes Nozzle Contour Design Optimization. The 23rd AIAA Applied Aerodynamics Conference, *AIAA Paper 2005-5081*.

Sivells, J. C. and Payne, R. G. (1959). A Method of Calculating Turbulent Boundary Growth at Hypersonic Mach Numbers. *AEDC-TR-59-3*.

Sivells, J. C. (1969). Aerodynamic Design of Axisymmetric Hypersonic Wind Tunnel Nozzles. AIAA 4th Aerodynamic Testing Conference. *AAIA Paper 69-337*.

Sivells, J. C. (1970). Aerodynamic Design of Axisymmetric Hypersonic Wind-Tunnel Nozzles. *Journal of Spacecraft and Rockets*, **7**(11), 1292–1299.

Sivells, J. C. (1978). A Computer Program for the Aerodynamic Design of Axisymmetric and Planar Nozzles for Supersonic and Hypersonic Wind Tunnels. *AEDC-TR-78-63*.

Stalker, R. J. (1967). A Study of the Free-piston Shock Tunnel. *AIAA Journal*, **5**(12), 2160–2165.

Stalker, R. J. (1989). Hypervelocity Aerodynamics with Chemical Nonequilibrium. *Annual Review of Fluid Mechanics*, **21**, 37–60.

Tang, B., Wang, Y. P., and Jiang, Z. L. (2019). Nozzle Design for a Large-scale High-enthalpy Shock Tunnel. *Science China Physics, Mechanics & Astronomy*, **49**, 074701. (in Chinese)

Tolle, R. (1997). A New Optimum Design Code for Hypersonic Nozzles, Utilizing Response Surface Methodology. 35th AIAA Aerospace Sciences Meeting and Exhibit. *AIAA Paper 1997-0519*.

Tong, B. G., Kong, X. Y., and Deng, G. H. (1990). *Aerodynamics*. Beijing: Higher Education Press. (in Chinese)

Yu, H. (1989). Recent Developments in Shock Tube Application. In *Proceedings of the 1989 Japanese National Symposium in Shock Wave Phenomena*, ed. K. Takayama. Sendai, Japan: Tohoku University, 1–10.

Zonars, D. (1967). Nonequilibrium Regime of Airflows in Contoured Nozzles: Theory and Experiments. *AIAA Journal*, **5**(1), 57–63.

Zucrow, M. J., Hoffman, J. D. (1977). Gas Dynamics. *Volume. 2 – Multidimensional Flow*. New York: John Wiley & Sons, Inc.

8 Aerothermodynamic Testing and Hypersonic Physics

Hypersonic vehicles encounter an aerothermodynamic environment characterized by strong shocks, high temperatures, and chemical reactions. The surrounding air flow around such a vehicle in the hypersonic regime behaves quite differently from supersonic flows. The kinetic energy associated with hypersonic flight is converted into internal energy which increases the air flow temperature and induces endothermic reactions, such as molecular vibrational excitation, dissociation, and ionization of the air flow near the vehicle surface. Thus, the air flow becomes a chemically reacting gas medium. It is a challenge to develop flow diagnostic and aerodynamic measurement technologies with high-precision for high-enthalpy wind tunnel tests. There are, generally, three types of measurement technologies widely used in exploring high-enthalpy flows, including heat-transfer measurement, aerodynamic balance, and optical diagnostic techniques. These technologies are based on different physical principles and have been developed with great success during the recent decades and contributed a lot to high-enthalpy tunnel data acquisition. In this chapter, hypersonic testing with these measurement technologies is reported to demonstrate the progress on high-enthalpy flow experiments. Four kinds of experiments are included in this chapter and these topics are aerodynamic forces and moments tests, aerothermal heating measurements, hypersonic boundary-layer development diagnostics, and supersonic combustion and scramjet engine tests. Actually, there are a lot of interesting topics to investigate, but these four are important not only to understand aerothermal dynamic physics but also to support the development of hypersonic vehicles.

8.1 Fundamentals for High-Enthalpy Flow Experiments

A high-enthalpy flow is referred to as a gas flow with high kinetic energy in which the velocity may reach approximately 5 km/s or above. In such a flow, the total enthalpy H_0 can be primarily measured by the kinetic energy $\frac{1}{2}U^2$ as the static enthalpy h is negligible as compared with the kinetic enthalpy. There may exist thermochemical reactions behind the shock or within the boundary layer around a hypervelocity vehicle during the transition of kinetic energy into internal energy. In other words, the primary features of a high-enthalpy flow are the high kinetic energy and the high stagnation temperature. The key aerodynamic phenomena and their effects on aircraft

Table 8.1 Examples of high-enthalpy shock tunnels in operation over the world

Tunnel	Driver mode	Total length (m)	Exit nozzle diameter (m)	Test time (ms)	Mach number
JF-12	Detonation driver	~275	1.5 or 2.5	100–130	5–9
LENS-II	Heated H2 or He	~65	1.07, 1.52 or 1.83	20–100	7–15
HEG	Free-piston driver	~62	0.88	3–6	6–21

performance were first discovered during the atmospheric reentry of space vehicles such as space capsules or space shuttles. Such vehicles encounter extremely strong nose shock waves and viscous friction along the surfaces that can heat the surrounding air to a temperature up to thousands of K. Molecular vibrational excitation, gas dissociation, and atom ionization may occur successively as the gas temperature increases. In such a situation, the air will no longer be an ideal gaseous mixture but a chemically reacting medium. The substantial change in the flow medium results in changes to the constitutive relation of the high-enthalpy flow in which the energy transition takes place within chemically reacting gases and flow motion. This is the fundamental issue of high-enthalpy flow and boosts fundamental studies on it. The aforementioned physicochemical processes occurring in high-enthalpy flow at micro scales may have significant effects on the aerodynamic performance of hypersonic vehicles. Being different from subsonic and supersonic flows, the chemically reacting gas flow features nonlinearity, nonequilibrium, multi-scales, and multi-physics properties.

After more than sixty years of research work, hypersonic ground test facilities suitable for exploring aerothermochemistry still rely on high-enthalpy shock tunnels. Many shock tunnels were built over the world, for example, the Large Energy National Shock (LENS) tunnels in the United States, the High Enthalpy Shock Tunnel (HIEST) in Japan, the High Enthalpy Shock Tunnel Göttingen (HEG) in Germany, the JF-10 high-enthalpy shock tunnel, the JF-12 hypersonic flight-duplicated shock tunnel, and the JF-16 hypervelocity expansion tunnel in China. Some examples of high-enthalpy shock tunnels running in the world and their main features are listed in Table 8.1. Valuable data have been produced with these shock tunnels for both hypersonic gas dynamics and hypersonic vehicle developments.

The challenge is obvious for developing measurement techniques for the diagnosis of high-enthalpy flow and aerodynamic measurements under such a severe thermo-environment. A series of measurement techniques have been developed and applied to high-enthalpy flow tests during the recent decades. These techniques can be classified into three types, that is, the balance for measuring aerodynamic forces and moments, thermal sensors for aerodynamic heating, and optical techniques for flow diagnosis. When developing these techniques, one of the critical issues is the trade-off between measurement accuracy and device survivability under the chemically reacting environment of high-enthalpy flows. The test duration is another critical issue. The short test time of high-enthalpy tunnels brings forth severe constraints to the technical development of aerodynamic balances. The third issue is the high-temperature flow environment which can lead to surface erosion, washout, oxidization, and charged ions to the surface-mounted sensors, which affect their lifespan and experimental data accuracy. Optical techniques for flow visualization and diagnosis work in a nonintrusive mode, and can keep high-enthalpy flow phenomena free from disturbances. One can obtain flow structures, temperature, gas species, and flow velocity using special optical techniques. Image quantification and data accuracy are the key issues for improving optical measurement techniques.

8.2 Aerodynamic Forces and Moments of Hypersonic Vehicles

Traditional hypersonic wind tunnels usually produce test flows with low total temperatures and at low sound speeds, and therefore, thermochemical reactions, one of the key mechanisms in hypersonic flow, are ignored in the experiments. As a result, real-gas effects on aerodynamic force and moment measurements become a very difficult problem to solve in hypersonic ground tests, and it was once identified as an "unknown unknown" (Bertin & Cummings 2006). During the high Mach number and high-altitude segment of the first entry flight of the Space Shuttle Orbiter, the vehicle exhibited a nose-up pitching moment increment relative to pre-flight prediction.

Woods et al. (1983) noted that the Space Shuttle Orbiter required a much larger body flap deflection than the preflight prediction of 7.5° in the Aerodynamic Design Data Book to maintain trim at the proper angle of attack ($\alpha = 40°$). The actual deflection ($\delta \approx 16°$) was close to the limit, and any further deflection could have resulted in a catastrophic accident. A comparison of equilibrium-air computations with the perfect gas model indicates that the real-gas effects at very high Mach numbers are responsible for at least the main part of the so-called hypersonic anomaly (Maus et al. 1984).

In order to improve the accuracy of ground experimental data, various advanced hypersonic test facilities were developed (Lu & Marren). One of the main concerns was related to the aerodynamics forces and moments during hypersonic vehicle reentry. Unfortunately, there is no single ground-based facility capable of duplicating the hypersonic flight environment, and so far different facilities are used to address various aspects of design problems associated with hypersonic flight (Bertin & Cummings 2003, 2006). Considering the thermochemistry in hypersonic flow, high-enthalpy shock tunnels are capable of generating high-temperature flows but their effective test duration is too short for force and moment measurements. Moreover, for hypersonic experiments, there is no reliable scaling criterion available as that in low-speed wind tunnel tests, which makes the aerodynamic force and moment measurements for hypersonic vehicles even more difficult.

Under the support of the National Major Project of Scientific Instrumentation R&D, a long-test-duration hypersonic detonation-driven shock tunnel was developed based on the backward-running detonation driver and was named the JF-12

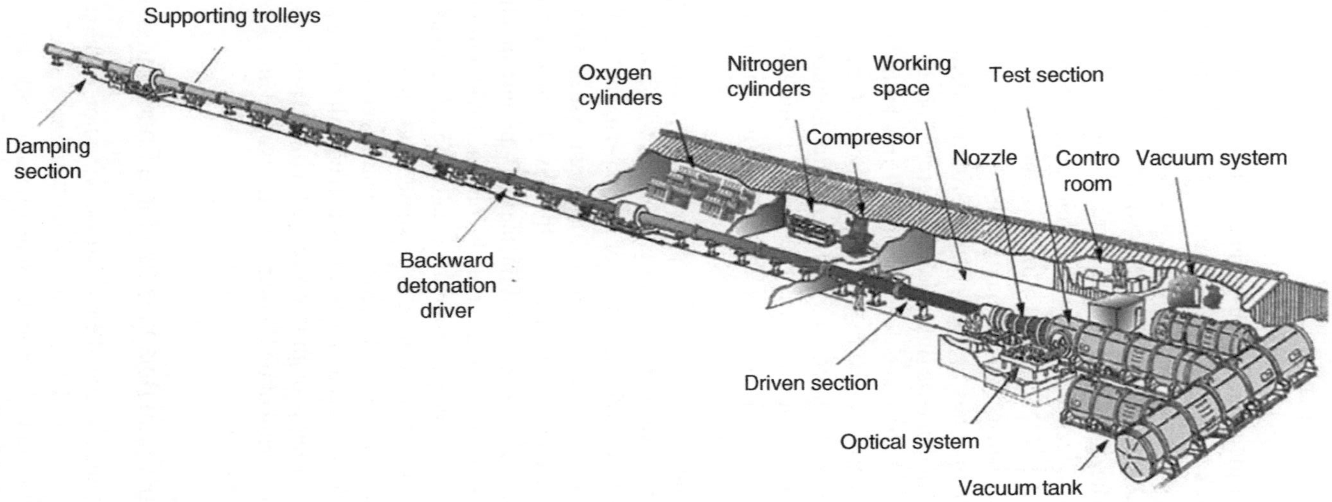

Figure 8.1 Schematic of the long-test-duration hypersonic detonation-driven shock tunnel.

hypersonic shock tunnel under the serial number in the State Key Laboratory of High Temperature Gas Dynamics (LHD) laboratory. Its performance tests demonstrated that the facility is capable of reproducing the pure-air flow with Mach numbers from 5 to 9 at an altitude of 25–50 km with 100 ms of test duration. A description of the JF-12 facility is provided in Section 5.3.2.2, but it is briefly restated here for the reader's convenience. Figure 8.1 shows a sketch of the JF-12 hypersonic shock tunnel facility. From right to left, the first part is the E-shaped vacuum tank used for damping the wave reflection during the nozzle-starting process. The second part is the test section which is 15 m long and 3.5 m in diameter at the outlet. The contoured nozzle is 15 m long and 2.5 m in diameter, while another smaller nozzle with an exit diameter of 1.5 m is also available for the Mach number of 5–7. Next to the nozzle is the driven section which is 89 m long and 720 mm in diameter. The detonation driver is 99 m long and 400 mm in diameter. The driver operates in the backward-running detonation mode, that is, the detonation is initiated at its right end. The detonation driver and the driven section are connected with the transition component in which the inner tube diameter is gradually reduced from 720 to 400 mm. Between the detonation driver and the transition component is a diaphragm rig used to produce the incident shock wave in the shock tunnel after the direct detonation initiation. The damping section is located at the far-left end of the facility and is 19 m in length and 400 mm in diameter (Jiang & Yu 2014).

Taking advantage of the JF-12 hypersonic shock tunnel, a research project on aerodynamic force measurements was carried out with in-house built six-component strain-gauge balances (SGBs). Some of the experimental data collected at a series of angles of attack demonstrated two important issues. One issue is that voltage signals of the pitching moment indicate that low-frequency oscillations are observable at three periodical cycles within the 100-ms test duration. This means that high-accuracy experimental data could be obtained with our impulse-type balances. The other issue is that the experimental data of the JF-12 hypersonic shock tunnel is in good agreement with blowdown hypersonic wind tunnels when the angle of attack is small, however, significant differences are observed in test cases with high angles of attack for the same Mach number. It is believed that the conical shock of the test model becomes strong enough to excite gas molecular vibrations at high angles of attack, and it is one kind of real-gas effects that were demonstrated from the flight of the Space Shuttle Orbiter.

8.2.1 Development of Force Balances Used in a Shock Tunnel

In a hypersonic shock tunnel, the mechanical vibration of the model-balance-support (MBS) system occurs and cannot be damped during a shock tunnel run due to the instantaneous flow field and short test time (generally 500 μs–20 ms) (Arrington et al. 1954, Bernstein 1975, Naumann & Ende 1990, Naumann et al. 1993). For the MBS system, the lowest natural frequency of 1 kHz is sometimes required for a test time of typically 5 ms to obtain improved measurement results. The higher the natural frequencies, the better the justification for the neglected acceleration compensation. For such test conditions, many researchers proposed several special balances to measure the aerodynamic forces in impulse facilities, that is, accelerometer balance (Joarder & Jagadeesh 2003, Sahoo et al. 2003, Saravanan et al. 2009), stress-wave force balance (Sanderson et al. 1991, Mee et al. 1996, Robinson et al. 2011), free-flight measurement technique (Seiler et al. 2005, Laurence & Karl 2010, Wey et al. 2012, Tanno et al. 2013, 2014, Martinez et al. 2014), and compensated balance (Marineau et al. 2012). Owing to the very short test time, however, a mature technology has not been develop for force measurements in shock tunnels.

8.2.1.1 S-Series Impulse-Type Strain-Gauge Balances

The traditional internal SGB, which has mature technology and low cost of the strain-gauge, was considered for use in force measurements in the JF-12 long-test-duration impulse facility due to its stiff construction balance and the test duration of more than 100 ms. However, when the force test is conducted in a shock tunnel, the inertial forces lead to low-frequency vibrations of the model and its motion cannot be addressed through digital filtering because a sufficient number of cycles cannot be found during a shock tunnel run. This condition implies that there is a restriction on the model size and mass as its natural frequencies are inversely proportional to the length scale of the model. Based on these technical difficulties, S-series (sting-series) impulse-type SGBs were proposed and the measuring element structure of SGB was optimized by the finite element method (FEM). The maximum loads (i.e., normal force) were from 500 N to 12,000 N for the test models with different scales. Finite element computations were performed to analyze the vibrational characteristics of the MBS system to ensure there were enough cycles of balance signal and high measuring sensitivity, especially axial element structure, during the 100-ms test.

In the design process of the balance, we used the strain-gauge sensor to measure the aerodynamic loads in the JF-12 shock tunnel. The strain-gauge sensor has enough high-frequency response for the force test during a test period of more than 100 ms. Our experimental and computational results show that the SGB, with the optimized structures, can be used in this long-test-duration shock tunnel. Therefore, S-series impulse-type SGBs were designed and fabricated for the force tests of JF-12. The technical data of balances JF-12-ISG3-D053-S01 (hereafter referred to as S01) and JF-12-ISG6-D106-S03-II (S03-II), which are the typical balances used in the JF-12 shock tunnel, are shown in Table 8.2. S01 and S03-II are described in detail as examples of S-series impulse-type SGB. The difference between S01 and S03-II is the element structure of the axial load. The balance S01 was further optimized in the aspect of the measuring element of axial load based on the axial element of S03-II. The performances of S01 and S03-II were examined to determine which type is better for force measurement in the JF-12 shock tunnel.

Figure 8.2 shows a schematic of the two types of SGB and their measuring element details (Wang et al. 2016b). As shown in Table 8.2 and Figure 8.2, the two balances

Table 8.2 Technical data on S-series SGBs

Serial No.	Type	Components	Diameter (mm)	Length (mm)
S01	Sting	Three	53	202.5
S03-II	Sting	Six	106	315

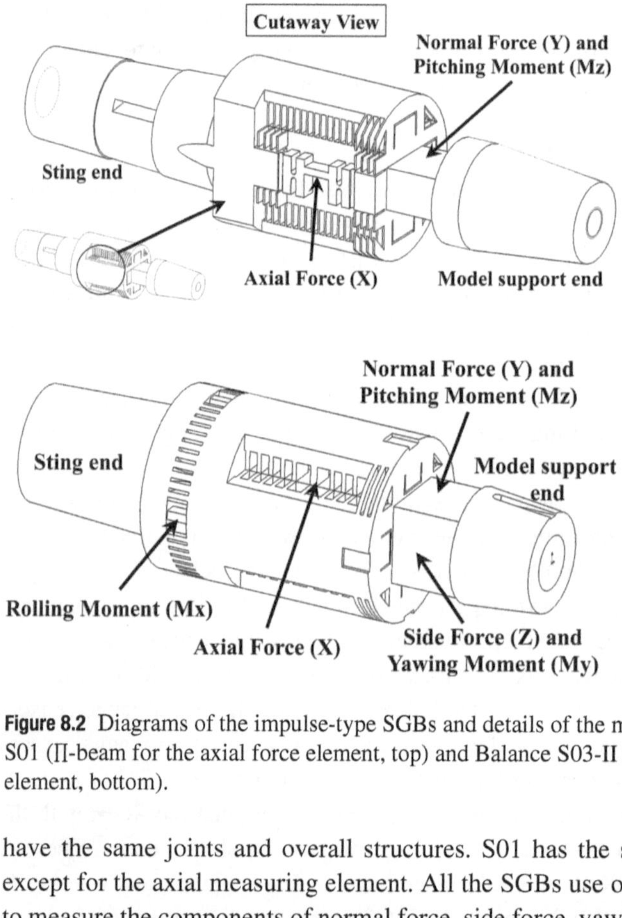

Figure 8.2 Diagrams of the impulse-type SGBs and details of the measuring element. Balance S01 (Π-beam for the axial force element, top) and Balance S03-II (I-beam for the axial force element, bottom).

have the same joints and overall structures. S01 has the same structure as S03-II, except for the axial measuring element. All the SGBs use only one rectangular beam to measure the components of normal force, side force, yawing moment, and pitching moment. The moment center is located at the center of the rectangular beam. The first one marked S01 is a three-component (i.e., axial force, normal force, and pitching moment) sting balance and the S03-II is the six-component one. Based on the structures of S-series SGB, therefore, four strain-gauges are used for the axial load element in the case of S01 and eight strain-gauges for the S03-II. These strain-gauge sensors are arranged in a Wheatstone bridge to measure the strain produced by the loads. The output voltage of a balance bridge changes as a function of the strain at the bridge location produced by the applied loads. Table 8.3 provides the details of the load range for the two balances used in the present study. The S-series balances with different limited loads are used for test models with different scales.

Table 8.3 Simultaneous component load ranges (N, N-m)

Serial No.	*X*	*Y*	*Z*	*Mx*	*My*	*Mz*
S01	1,000	2,000	–	–	–	100
S03-II	4,000	12,000	12,000	100	400	400

8.2.1.2 Design and Optimization of S-Series SGBs

To design and construct a high-stiffness SGB that can meet certain demands, all aspects of the balance technology must be investigated. However, in this study, only the structure of SGB is considered. The high demands on the balance from the point of view of the structure can be expressed as (1) low interference between elements; (2) high stiffness; (3) low stress level at the strain-gauge positions and related parts; and (4) capability to tolerate errors from temperature gradients. Among these demands, the important properties of this transport-type balance, such as stiffness, sensitivity, and interference, were investigated in the present study. The computational and experimental results show that the SGB still has plenty of scope for structural improvement to achieve higher measurement accuracy, especially that of the axial force component. As the same rectangular beam is used for measuring *Y*, *Z*, *My*, and *Mz*, the present study focuses on the design and optimization of the axial element structure.

A *Structural Design and Optimization*

The axial force element is extremely difficult to design compared to other measuring elements due to its complicated structure. For impulse facilities, several researchers have measured the axial force using special balances. The main reason is that the test time is very short, so sufficient cycles cannot be obtained during a shock tunnel run. In this study, we design different structures of the axial force element and perform a series of finite element computations so that the SGB with an optimized structure can be used for the force test in the JF-12 shock tunnel with 100 ms of test time.

In the case of S03-II, an I-beam is used to measure the axial force. In fact, based on the S03-II, the axial force element (I-beam) was further optimized and the Π-beam was designed for the case of S01 (see Figure 8.2 top).

In comparing the two types of beams through FEM computations, we find that the Π-beam shows the smallest interference when a load, the normal force or pitching moment, is acting on the moment center. At the same time, the Π-beam has the largest strain output of 336 με, while the averaged strain is only 84 με in the case of the I-beam. This result also indicates that the Π-beam has a higher sensitivity in measuring the axial load.

B *Structural Analysis by FEM*

To examine the sensitivity performance of the measuring element, the strain computations were conducted for the cases of S01 and S03-II (Wang et al. 2016b). The findings are shown in Figures 8.3 and 8.4. The first case is the axial force of 1,000 N (limited load) acting on the moment center of S01. In this case, the strain output of the

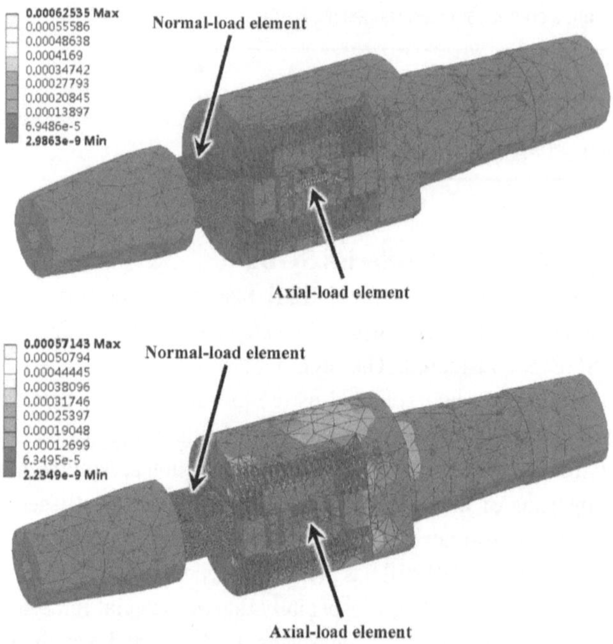

Figure 8.3 Strain contours of S01 by FEM computation. Axial force of 1,000 N (Cutaway view, top) and normal force of 2,000 N (Cutaway view, bottom).

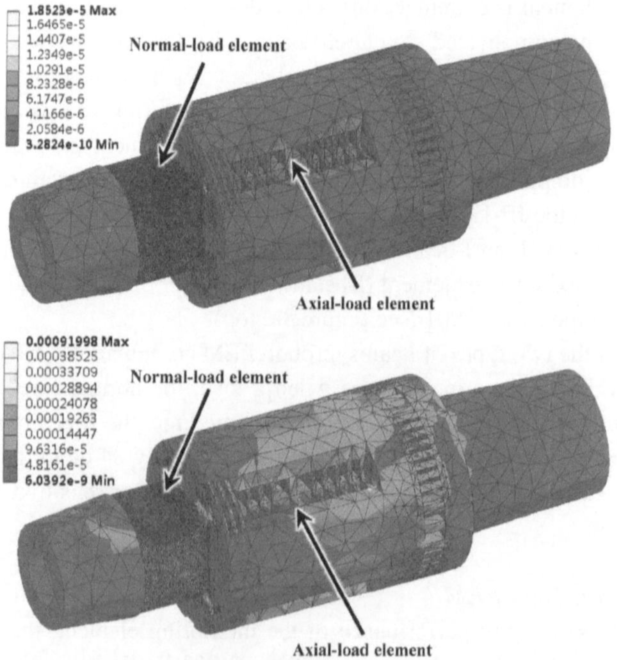

Figure 8.4 Strain contours of S03-II by FEM computation. Axial force of 4,000 N (Cutaway view, top) and normal force of 12,000 N (Cutaway view, bottom).

axial force element is 336 με, while the output of the normal force element is only 15 με. Minimal strain is generated on the rectangular beam when an axial force of 1,000 N is applied at the moment center of S01 (Figure 8.3 (top)), and vice versa (Figure 8.3 (bottom)). The effect of the axial force on the rectangular beam is very small because of the optimized axial force element. In the S03-II case, however, when the axial force of 4,000 N (limited load) is applied at the moment center, the strain of the rectangular element becomes 17 με and that of the axial load element is only 40 με. This result means that the interference of the axial force is larger in the case of S03-II than that of S01. Additionally, the measuring sensitivity of the axial force element of S03-II is also significantly lower than that of S01.

In the same manner, we calculated the cases of the limited normal load acting on S01 and S03-II (see Figure 8.4). The case of S01 shows a smaller interference from the normal force where it is less than 5%, while it is more than 70% for the case of S03-II.

C *Balance Calibration*

The S-series SGBs were calibrated prior to testing in JF-12. The calibration of an internal balance involves applying known weights to the balance and recording strain-gauge readings for each force and moment combination. In the case of sting balance S01, the calibration loads are 1,920 N, 96 N-m, and 960 N for the normal force, the pitching moment, and the axial force, respectively. The multicomponent loading method was employed in this study to calibrate the present impulse-type SGB. In the multicomponent loading calibration process, all the components are simultaneously loaded under different load combinations. An equation that describes the relation between the loading and the balance signal can be constructed for every loading, with the unknown coefficients of the calibration formula. The calibration formula of the balance is obtained by solving the set of calibration equations (AIAA 2003). For the S01 with three components, at each of the 76 loadings (total loading number), the bridge outputs are read several times and the mean values and standard deviations are computed. In the cases of S03-II and the other S-series SGBs used in JF-12, the same definition and method were employed to calibrate the balances. The structures of the present balance show good accuracy and precision in the static calibration. The calibration results show that the axial force of S01 has the highest accuracy and precision because of the optimized measuring element. Almost the same normal force errors were obtained for the two balances as the same structure was used in the measuring element, that is, the rectangular beam.

The SGBs used in the current study are calibrated in the fully automatic balance calibration system (see Figure 8.5) of the Institute of Mechanics. In order to develop and improve the force measurement capability in JF-12, a high-accuracy fully automatic calibration system for the large-load wind tunnel balances was designed and built, based on the characteristics of impulse-type balance. SGBs with a maximum load of 15 kN can be calibrated automatically with high-accuracy in this balance calibration system.

Figure 8.5 Calibration system used in the shock tunnel at the Institute of Mechanics of CAS.

8.2.2 Experimental Description under Duplicated Hypersonic Flight Conditions

The experimental work for aerodynamic force and moment measurements was carried out with the six-component SGBs that were specially designed for the JF-12 hypersonic shock tunnel to ensure that the output signals from low-frequency vibration of the balance have enough full periods within the 100-ms test duration. The test models for the experiment were selected to be a sharp cone and the AGARD Standard Model II (HB-2) (Gray 1964) because of their wide acceptance for hypersonic wind tunnel calibrations. In the tests, the average stagnation pressure and temperature were 2.5 MPa and 2,200–3,500 K, respectively. These conditions resulted in an average freestream Mach number of 7 and an average unit-Reynolds-number of approximately $0.8 \times 10^6\ m^{-1}$. In addition, the model was supported by a tail sting mounted on a support mechanism in the test section.

8.2.3 Force Measurements in the JF-12 Shock Tunnel

Experimental data have been collected from SGBs S01 and S03-II. Three typical balance signals are presented in Figure 8.6. The voltage signals indicate that the lowest-frequency oscillations observable from the pitching moment present at least three full periodical cycles within the 100-ms test duration. It means that high-accuracy

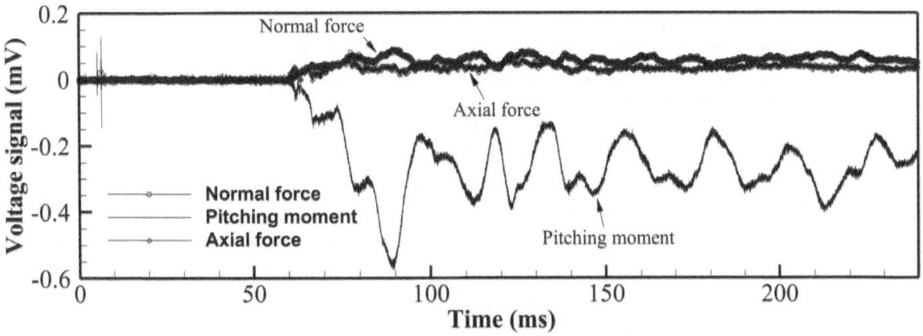

Figure 8.6 Balance signals from aerodynamic force and moment tests for Mach number 7.

pitching moments could be attained with the impulse-type balances by applying simple data processing techniques. As to axial and normal forces, their low-frequency oscillations have more than enough periodical cycles to achieve high-accuracy data. The periodicity of the force measurement data from the JF-12 shock tunnel indicates that its accuracy could be as high as that from blowdown hypersonic wind tunnels. However, the data accuracy will be ruined if the wind tunnel test time is shorter than 30 ms for this test or if the low-frequency natural vibration of the force measurement system is further lowered. It could be very difficult to discern the forces and moments on the test article if the balance signals from forced and inertial vibrations of the force measurement system are superimposed.

8.2.3.1 Standard Model I: Cones

In the experiments, two cones with 10° semi-vertex angle were used; these were made of aluminum alloy and were 0.75 m and 1.5 m long, respectively. The balances S01 and S03-II were used for the smaller and larger cones, respectively, in the force tests (Figure 8.7). The cone is a standard model and data is available in the literature. The reference area and dimensions of the models are presented in Table 8.4, including the models of the China Academy of Aerospace Aerodynamics (Zhang 2004) and the National Aeronautics and Space Administration (NASA) (Penland 1961, Ladson & Blackstock 1962).

Figure 8.8 presents a comparison of the test results for S01 and S03-II (Wang et al. 2016b). Some data were obtained from other conventional hypersonic wind tunnels (FD-03 and NASA Langley 11-in.). The theoretical result was obtained using Cheng's cone theory (Cheng 1959). The results have good agreement with the data from Langley 11-in. ($T_0 = 630$ K) (Penland 1961, Ladson & Blackstock 1962) and FD-03 ($T_0 = 600$ K) (Zhang 2004) wind tunnels. In the case of the NASA facility, the Mach number is 6.8 and the Reynolds number (based on model length) is 0.81×10^6. A comparison of the results shows that the normal and axial force coefficients obtained using S01 differs from NASA's data by 2.61% and 4.69%, respectively. For the hypersonic shock tunnel and large test model, based on the mentioned flow conditions, the test results are acceptable. Furthermore, good repeatability was observed during the shock tunnel testing

Table 8.4 Dimensions of the cones used in different wind tunnels

Wind tunnels	Length (m)	Reference area (m^2)	Weight (kg)
JF-12	0.75	0.0549	6.9
JF-12	1.5	0.2198	29.5
FD-03	0.14178	0.00196	
Langley 11-in.	0.0855	0.0007	

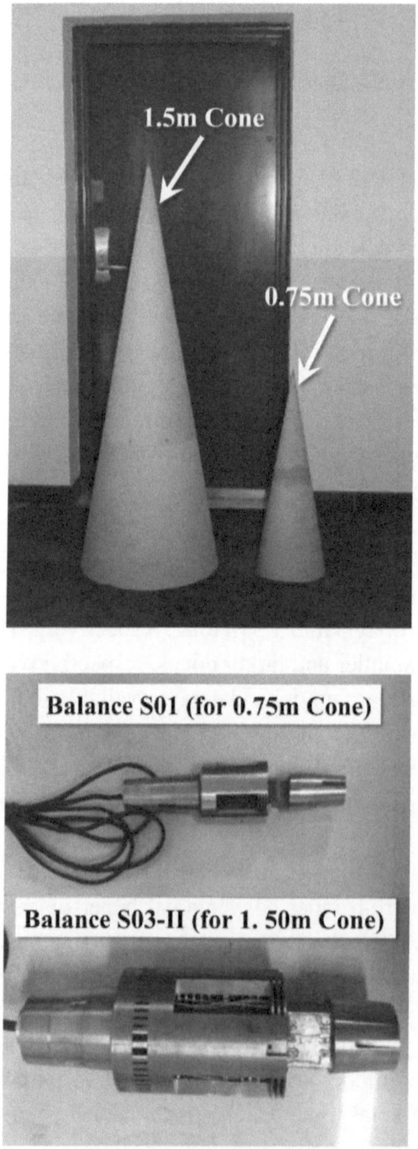

Figure 8.7 Two large-scale cones and S-series balances used in the present force tests.

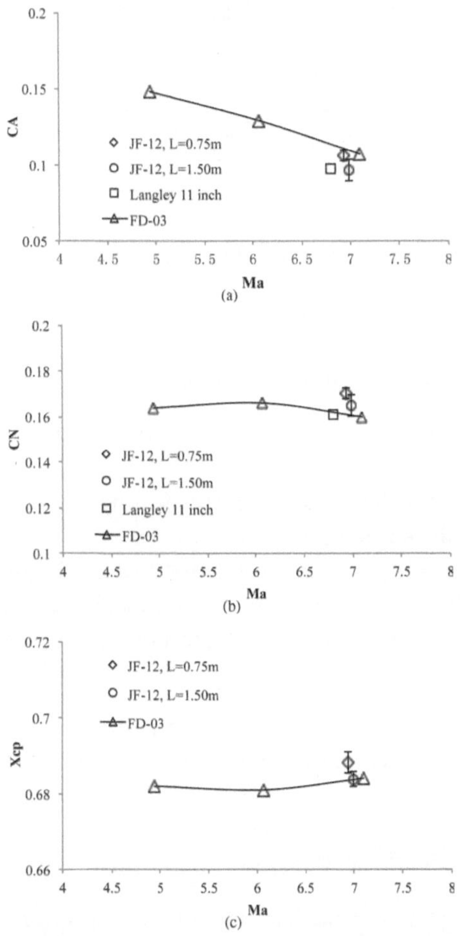

Figure 8.8 Aerodynamic coefficient comparison and center of pressure for S01 (for 0.75-m cone) and S03-II (for 1.5-m cone). (a) Axial force, (b) normal Force, and (c) center of pressure.

(Wang et al. 2016b). As shown in Table 8.5, the balance S01 shows good performance in terms of test precision. The following equation is used to calculate the standard deviation σ,

$$\sigma_R = \sqrt{\sum_{i=1}^{n} \frac{\left(R_i - \overline{R}\right)^2}{n-1}} \tag{8.1}$$

where $\overline{R}$ is the averaged value of some aerodynamic coefficient of n tests; in this study, $n \geq 5$.

Measurement uncertainty is a complex subject that involves both statistical techniques and engineering judgment. An aerodynamic force measurement is influenced by many factors, particularly in a shock tunnel. The short test time requires a fast response of instrumentation and measuring system (MBS system). These issues have implications for the uncertainty analysis. Based on a preliminary analysis, the effects (error sources) that affect the results of the force test can be summarized as follows: (1) the shock tunnel, (2) the calibration of SGB, (3) the test model, and (4) the measurements.

Table 8.5 Uncertainty results of force coefficients obtained using S01 and S03-II

S-series SGB	U_{C_A}	U_{C_N}	U_{C_M}
S01	±9.880% C_A	±8.566% C_N	±8.647% C_M
S03-II	±13.184% C_A	±9.412% C_N	±10.140% C_M

The axial force coefficient, C_A, can be calculated by the following equation:

$$C_A = \frac{X}{qS} \tag{8.2}$$

The methodology of the measurement uncertainty is as follows (Kammeyer 1999):

1) The measurement system is defined and the data reduction equations are determined.
2) The sources of uncertainty for each measurement are identified.
3) The relative significance of the uncertainty sources is assessed.
4) The precision, P, and bias, B, for each significant error source are estimated. The total bias and precision are then obtained as the root-sum-square of each contributor.
5) The precision and bias limits are propagated into the test results and combined into total uncertainties.

To take C_A as an example, therefore, the relative uncertainty in C_A can be written as (Fluid Dynamics Panel Working Group 15 1994, AIAA-Standard 1999, AIAA-Guide 2003)

$$U_{C_A} = \pm\sqrt{{P_{C_A}}^2 + {B_{C_A}}^2} \tag{8.3}$$

where P_{C_A} and B_{C_A} are the precision limit and bias limit of C_A, respectively.

Table 8.5 shows the uncertainty of test results obtained using S01 and S03-II, respectively. As shown in Table 8.5, the results for S01 have a higher accuracy in the axial component because of the improved measuring element. The test results suggest that the accuracy of S-series SGBs should be high enough for the force measurement in the JF-12 shock tunnel. The data uncertainties indicate that the test results can be considered acceptable because the force test was conducted in a hypersonic shock tunnel, where the aerodynamic force is unsteady during the test run. The present accuracy and precision of the test data are substantially increased because the test duration is more than 100 ms.

8.2.3.2 Standard Model II: HB-2 of AGARD

In the force tests, a standard model which has a blunt forebody, HB-2 (Gray 1964), was employed for measuring the drag using the current balance system in the JF-12 shock tunnel. At the same time, the drag data from the HB-2 test can also be used to further analyze the viscous interaction.

Figure 8.9 shows the HB-2 test model for the drag test using the present SGB in the JF-12 shock tunnel. HB-2 is a standard hypersonic ballistic correlation model, which has been examined to establish reference characteristics for use in the correlation of

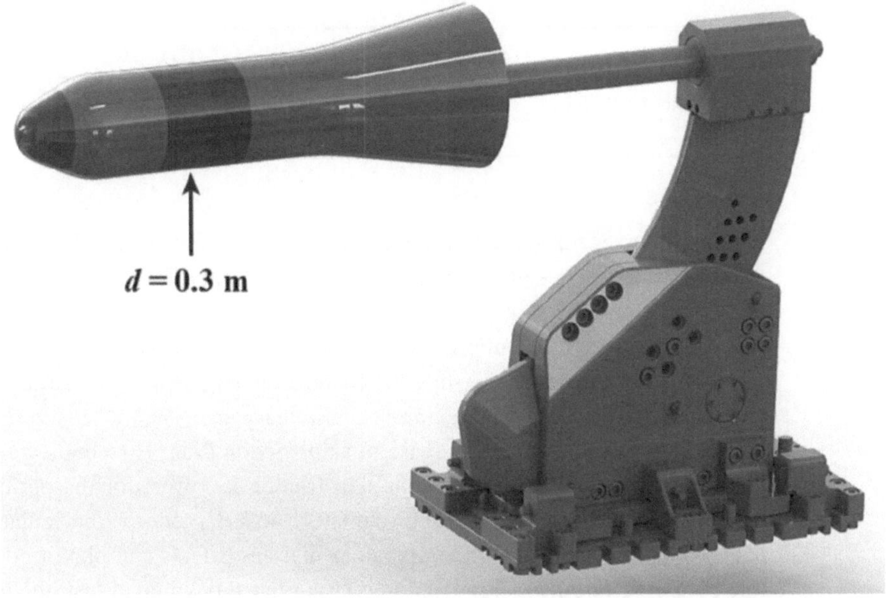

Figure 8.9 Standard force model HB-2.

experimental results among various aerodynamic facilities. The interest in a standardized ballistic correlation model was first voiced at the joint meeting of AGARD (The Advisory Group for Aeronautical Research and Development) and STA (Supersonic Tunnel Association) held in France in September 1959 (Gray 1964). The current model used in this study has a forebody diameter d of 0.3 m and a length that is 4.9 times the diameter d.

In the force measurement of the HB-2 model, the standard deviation of the drag coefficient is approximately 1.0%, and $U_{C_D} \leq \pm 5.0\%\ C_D$ (95% confidence interval) due to the long test duration of the JF-12 shock tunnel, which is greater than 100 ms during a shock tunnel run. Therefore, the accuracy of the current balance system has been further verified. Furthermore, compared with the data from VKF (tunnel E), the viscous parameter $M_\infty^3 / \sqrt{Re_L}$ is larger in JF-12's test due to a higher total temperature, which is more than 3,400 K (see Table 8.6). A higher-enthalpy condition leads to a lower Reynolds number in JF-12 so that the viscous interaction becomes larger. Here, the viscous interaction is analyzed using the parameter $M_\infty^3 / \sqrt{Re_L}$, which can be used for the evaluation of the viscous interaction on aerodynamic characteristics. If the value of $M_\infty^3 / \sqrt{Re_L}$ is low, the actual pressure distribution is mainly governed by inviscid flow effects. In contrast, a high value indicates that viscous interactions should be important (MacWherter et al. 1986).

The hypersonic viscous interaction is analyzed in the works of Anderson (2006) and Hirschel (2005). Anderson (2006) shows that shear stress is a viscous flow phenomenon and the skin friction drag can be obtained by calculating the shear stress

Table 8.6 Comparison of the drag data in the VKF (the von Karman gas dynamics facility) 12 in. × 12 in. hypersonic tunnel E and the JF-12 hypersonic shock tunnel

Wind tunnels	d (m)	M_∞	Re_d	T_0 (K)	C_D
VKF (Buttsworth 2001)	0.0635	7.07	6.0×10^5		0.5320
JF-12	0.3	6.98	0.9×10^5	3,400	0.5398

distribution over the surface of hypersonic aerodynamic shapes. In his work, a conclusion is that viscous interactions can have important effects on the surface pressure distribution, and the skin friction and heat transfer are increased by viscous interaction. The pressure distributions are always somewhat insensitive to chemically reacting effects, as noted earlier, while the chemical reactions have the strongest effect on temperature. Therefore, in the current preliminary analysis, the reason for the higher drag is considered to be an increase in skin friction drag. Hirschel's work (Hirschel 2005) also shows that the turbulent skin friction is lowest for the perfect gas, adiabatic wall case, and largest (approximately 45% larger) for the radiation cooling case, which means that high-temperature real-gas effects can play a non-negligible role regarding turbulent skin friction. During the shock tunnel test, the test flow did not reach equilibrium in the boundary layer, where the temperature continued to rise due to the strong effect of chemical reactions. At the same time, the skin friction was also increased due to high-temperature real-gas effects. Obviously, this phenomenon becomes inconspicuous in the results of conventional hypersonic wind tunnel tests because of the very low total temperature.

8.2.4 Discussions of High-Temperature Effects on Aerodynamic Force and Moment

A series of experiments were carried out with various changes in the angle of attack at Mach 7. The reason for such an experimental design is to examine the effect induced by different post-shock temperatures at different incidence angles. It is believed that the temperature at a small angle is below the critical temperature at which air molecular vibrations are excited, but it will be raised higher than the critical temperature when the incidence angle is high enough to induce a strong conical shock wave at the windward side of the test model. It is one kind of real-gas effects that the Space Shuttle Orbiter experienced.

The experimental data of the normal force coefficient are plotted in Figure 8.10 with the corresponding results from a conventional hypersonic wind tunnel (Chen 2000). By examining the discrepancy between the data from the two wind tunnel tests, it could be concluded that the normal force coefficient obtained from the JF-12 hypersonic shock tunnel is smaller than that from the FD-07 hypersonic wind tunnel. The discrepancy is increased as the incidence angle is set to be larger and reaches about 10% at an incidence angle of 15°. The axial force coefficient is plotted in Figure 8.11 and the curve from the JF-12 shock tunnel is located below the one from FD-07.

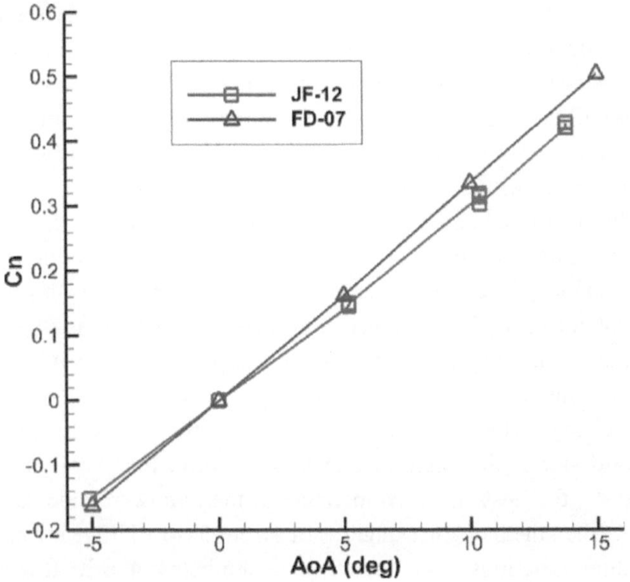

Figure 8.10 Experimental data of the normal force coefficient for Mach number 7.

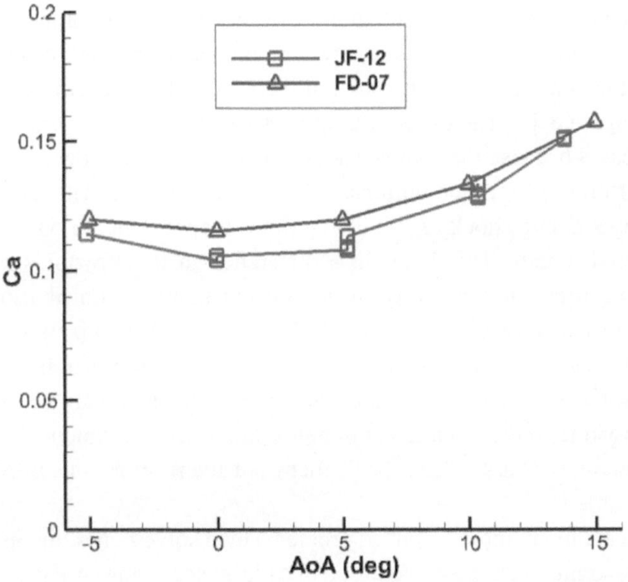

Figure 8.11 Experimental data of the axial force coefficient for Mach number 7.

By considering the differences in wind tunnel test conditions, several main physical issues are discussed below for further investigation.

The key difference in wind tunnel test conditions is the total temperature. It is 2,200 K for the JF-12 hypersonic shock tunnel and about 800 K for the FD-07 conventional hypersonic wind tunnel. This temperature difference can lead to two main physical issues that will affect more or less the aerodynamic forces acting on the sharp-cone model. The first issue is the high-temperature boundary layer development. The boundary layer will get thicker when the temperature is higher. The thicker boundary layer will reduce the drag force and this effect is demonstrated in Figure 8.11. The other is the pressure distribution along the cone model surface, and it will cause differences in both the normal and axial forces. There are two important things that are related to the pressure variations. One of them is the temperature-induced pressure change in the boundary layer. However, according to the boundary-layer theory, the pressure gradient normal to the solid surface is small and could be assumed to be zero. Furthermore, there is no stagnation area for the sharp cone model and the stagnation pressure change related to chemical reactions can be ignored. Another is the post-shock pressure change. The post-shock temperature at the windward side is different from the leeward side if the angle of attack is not zero. For the JF-12 hypersonic shock tunnel tests, the post-shock temperature at the windward side can vary from 350 K to 700 K when the incidence angle varies from 5° to 15°. Meanwhile, the post-shock temperature variations at the leeward side are below 400 K. It is well known that gas molecular vibrations are excited significantly when the static temperature is raised to above 500 K and this will result in pressure-specific heat change and energy transfer from mechanical to internal energy. For the JF-12 hypersonic shock tunnel tests, gas molecular vibrations exist at the windward side but not at the leeward side. This physical phenomenon will lead to pressure imbalance behind the conical shock wave around the cone model, and results in the normal force coefficient difference, as shown in Figure 8.10. The difference appears to be larger and larger as the angle of attack increases because the conical shock wave is getting stronger and stronger.

Another difference in test conditions is the test model scale. The test model used in the JF-12 hypersonic shock tunnel is five times bigger than the one in the FD-07 hypersonic wind tunnel. This leads to a difference in the Reynolds number and affects the drag force in some way. It is believed that the contribution from the Reynolds number difference is included in Figure 8.10. It is puzzling that predictions from numerical simulations agree well with the experimental data from the FD-07 hypersonic wind tunnel but are underestimated in the JF-12 experimental data. It is interesting to recall Park's comments on unsolved problems in hypersonic aerothermodynamics (Park 2013). Still, there are many more unsolved problems than solved problems.

While there are many more unsolved problems than solved ones, the first unsolved problem is not so much an unsolved problem as it is, according to Park (2013, p. 3), the "cavalier attitude of the scientists". This problem concerns the definition of reaction rates. Many researchers believe that the rate of reaction is the sum of all state-to-state transitions from an initial state to a final state. However, this is not true. A reaction rate is the sum of only those components of the state-to-state transition rates that are proportional to the initial state density under a quasi-steady-state condition (Park 2013).

Due to a series of force tests carried out under duplicated hypersonic flight conditions in the JF-12 hypersonic shock tunnel, some remarks can be summarized as follows:

1) The experimental data at duplicated hypersonic flight conditions show obvious discrepancies from those of cold hypersonic wind tunnels. The main physical issue is considered to be due to gas molecular vibrations, one of the key phenomena belonging to the so-called real-gas effects.
2) The cold wind tunnel tests can be numerically simulated by computational fluid dynamics (CFD) techniques with acceptable accuracy, but modeling thermochemistry in hypersonic flows is still a challenge. The thermochemical reactions should be taken at the correct temperature at a proper rate and a proper location.
3) Gas molecular vibrations, oxygen dissociation, nitrogen dissociation, and ionizations are dominated respectively at different temperature regions. It is necessary to classify their effects and model them separately to understand real-gas effects in hypervelocity flows.

8.3 Aerothermal Heating on Hypersonic Vehicles

Aerodynamic heating is one of the main factors that must be considered in the design of hypersonic vehicles. For example, the heat flux distribution over the vehicle surface can provide an important basis for determining the thermal protection system (TPS). From the point of view of aerodynamics, the hypersonic flow associated with a hypersonic vehicle is usually characterized by the presence of strong shock waves, thin boundary layers, gas thermochemistry, and their interactions. High-temperature gas effects strongly influence the forces acting on the surface and the heat flux across the surface of a hypersonic vehicle. Accurate prediction of these effects is critically needed for the design of any hypersonic vehicle.

To improve the reliability of hypersonic vehicles, technological advancements should be applied during the design and verification phases. The essentially required technologies in the field of aerodynamics during the whole R&D process of a hypersonic vehicle include CFD simulation, wind tunnel test, and flight test. Looking through the history of the Space Shuttle Orbiters, one can find that the TPS was designed mainly based on wind-tunnel data. Despite the remarkable progress in CFD technology and its irreplaceable application in the development of hypersonic vehicles by far, wind-tunnel tests are still required. Two of the main purposes of conducting ground-based tests are as follows. One is to define the aerothermodynamic environment of a hypersonic vehicle in the design phase. The other is to obtain flow field data for the validation and verification of newly developed numerical methods. In the 1960s, two research groups were assigned to develop analysis-experiment integrated techniques for the parameter correlation of the heat-transfer rates at the stagnation point of a model under atmospheric reentry conditions. Their results were dramatically different from each other. On the other hand, each group proved that the experimental data were consistent with its analytical data. Bertin and Cummings quoted the aforementioned example to stress that the models used to represent physical processes in hypersonic CFD codes must be validated carefully with reliable experimental data (Bertin & Cummings 2006).

Ground-based test facilities can simulate one or some of the following flow parameters: (1) the freestream Mach number, (2) the freestream Reynolds number, (3) the freestream velocity, (4) the static pressure, (5) the total enthalpy of the flow, (6) the test gas composition, (7) the wall-to-total temperature ratio, and (8) the thermochemistry. Unfortunately, none of the ground-based facilities existing in the world is capable of duplicating the hypersonic flight conditions by simulating all the parameters simultaneously (Bertin & Cummings 2006). It is difficult to match the thermochemistry of high-temperature gases associated with hypersonic flight in ground-based facilities (Bertin & Cummings 2003). This is an important consideration.

Shock tunnels are the available facilities that can generate high-enthalpy flows to simulate the aerothermodynamic environment for hypersonic tests (Schultz & Jones 1973). These tunnels can use the following driving techniques, that is, detonation driver (Yu et al. 1992, Jiang 2014, Jiang et al. 2012, 2015, Jiang & Yu 2017), free-piston driver (Stalker 1967, Hornung 2010), and light-gas driver (Holden et al. 2007). As the test duration of these kinds of facilities is generally in the range of a few to tens of milliseconds, the facilities and the test models can withstand the thermal loads. The following equation gives the approximate heat transfer rate at the stagnation point of a model nose (Bertin & Cummings 2003),

$$\dot{q}_s = C\rho_\infty^2 u_\infty^3 / R_N^{0.5} \, . \tag{8.4}$$

From the equation, one can find that aerodynamic heating is a severe problem if the velocity or total enthalpy of a hypersonic flight is simulated in the test facility. Although the test duration of a shock tunnel is inversely proportional to u_∞ generally, the heat transfer rate may lead to thermal destruction to test models and facility structures. Nevertheless, high-enthalpy shock tunnels are the only choice for high-enthalpy flow tests and have provided us with massive experiment data by far.

Since the effective test duration of high-enthalpy shock tunnels is usually on the order of milliseconds, it is a great challenge to develop heat flux sensors with high sensitivity, rapid response, and robustness. Thermocouples are the qualified sensors that can be used in the harsh environment of hypersonic test facilities. According to the Seebeck effect, a thermocouple measures the surface temperature's change with time from which the corresponding heat transfer rate can be calculated. The Seebeck effect refers to an electromotive force whenever there is a temperature gradient in a conductive material. Under open-circuit conditions without internal current flow, the gradient of voltage (∇V) is directly proportional to the gradient in temperature (∇V):

$$\nabla V = -S(T)\nabla T \tag{8.5}$$

where $S(T)$ is a temperature-dependent material property known as the Seebeck coefficient.

The temperature dependency indicated by $S(T)$ of the thermocouple causes the temperature dependence and the nonlinearity of the heat flux sensor. However, a well-designed sensor may have a lower temperature dependence and better linearity than expected. As shown in Figure 8.12, nickel-alloy-based thermocouples

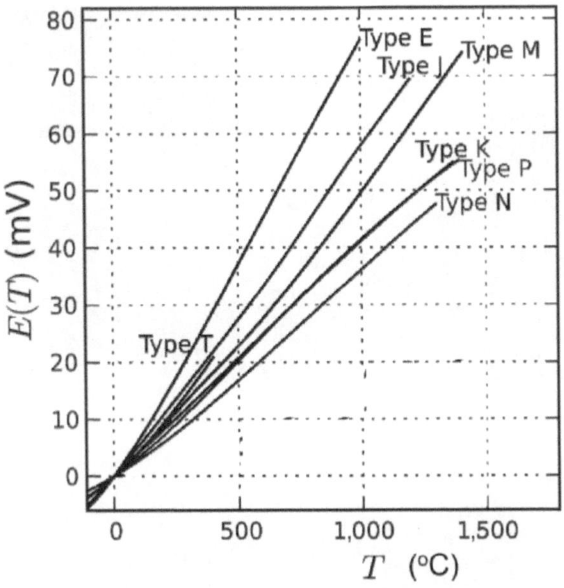

Figure 8.12 Characteristic functions of thermocouples (types E, J, K, M, N, T).

have excellent performance, which includes type E (chromel–constantan), type J (iron–constantan), etc.

The thin-film resistance thermometer and coaxial thermocouple are the two most commonly used surface thermometers that operate on the theoretical basis of the thermocouple. Figure 8.13 depicts coaxial thermocouples with different external diameters, as well as a sketch of the structure. The thin-film resistance thermometer has a higher sensitivity but poorer robustness than the coaxial thermocouple. Although a variety of sensors are available for heat-transfer measurement in high-enthalpy flow, there are many experimental technical problems, such as large dispersion and low precision of the acquired data (Buttsworth 2001). The measurement error in the region with complicated flow structures such as shock-wave/boundary-layer interaction is still a challenging problem at present. Moreover, due to high-temperature oxidation and scouring effects by the high-speed flow, the repeatability of the sensor is poor. Therefore, further exploration is needed to improve precision, robustness, and repeatability. The state-of-the-art research to improve the thermocouple performance for transient measurements can be found in the literature (Sanderson & Sturtevant 2002, Tashiro et al. 2005, Li et al. 2017).

The output of a thermocouple is not the heat transfer rate but the temperature signal $T(t)$ varying with time t. The determination of the heat flux from the time-dependent temperature signal is based on the theory of one-dimensional heat conduction into a semi-infinite body. The one-dimensional heat conduction theory yields the following relation between the surface heat flux $\dot{q}_s$ and the surface temperature signal $T(t)$

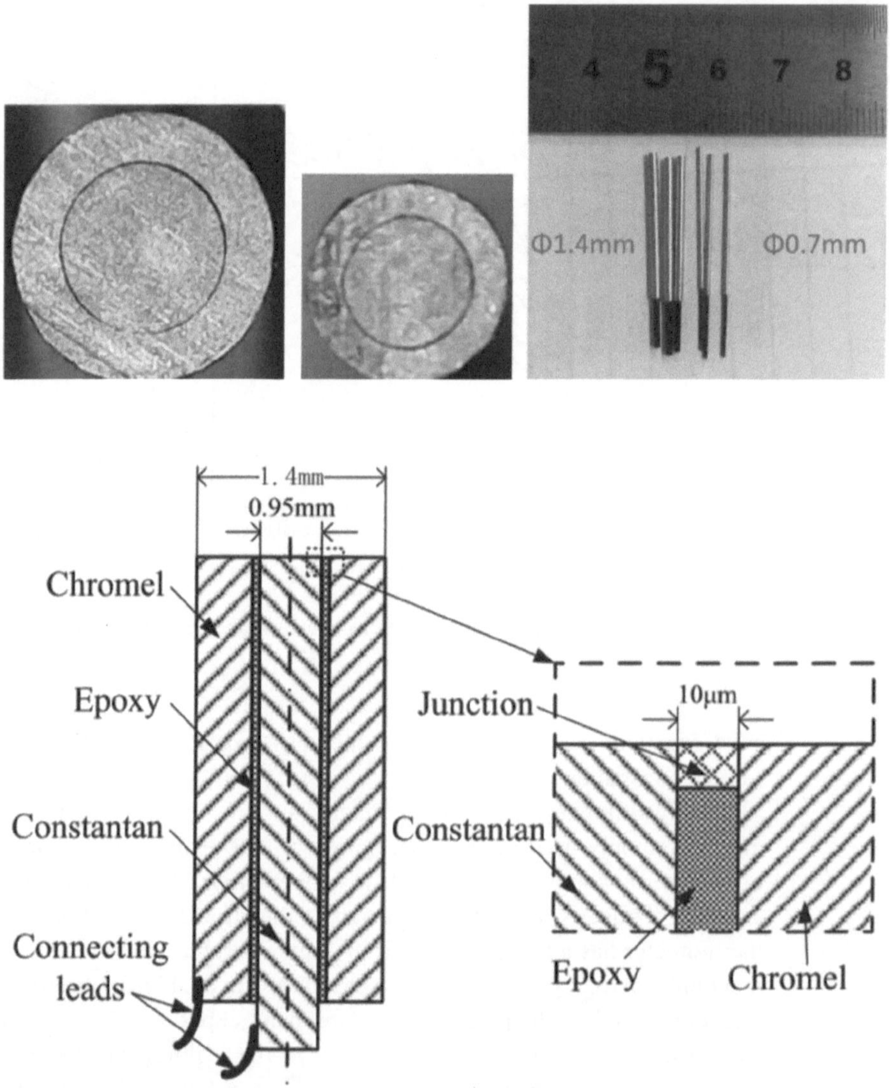

Figure 8.13 Coaxial thermocouple and a sketch of its structure.

$$\dot{q}_s = \frac{\sqrt{\rho c k}}{\sqrt{\pi}}\left[\frac{T(t)}{\sqrt{t}} + \frac{1}{2}\int_0^t \frac{T(t)-T(\tau)}{(t-\tau)^{3/2}}\mathrm{d}\tau\right]. \tag{8.6}$$

The formula given in Equation (8.6) holds for the coaxial thermocouple as well as for the thin-film thermometer as long as the principle of a semi-infinite body is valid. Here, ρ, c, k respectively represent the density, the specific heat capacity, and the thermal conductivity of the junction material. For discrete data generally acquired in wind tunnel experiments Equation (8.6) can be transformed to

$$\dot{q}_s(t_n) = \frac{2\sqrt{\rho c k}}{\sqrt{\pi}} \sum_{i=1}^{n} \frac{T(t_i) - T(t_{i-1})}{(t_n - t_i)^{1/2} - (t_n - t_{i-1})^{1/2}}. \tag{8.7}$$

Note that the response time of the coaxial thermocouple is proportional to the square of the junction scale as written in Equation (8.8). Therefore, the isolation material between the two metals should be as thin as possible for the transient measurement application (Li et al. 2017).

$$\tau_{res} = \frac{\rho c d^2}{k}. \tag{8.8}$$

Jiang and Yu et al. developed a large-scale detonation-driven shock tunnel with a laboratory series number of JF-12. The exit diameter of the JF-12 nozzle is 2.5 m while the test section diameter is 3.5 m which can accommodate large-scale test models. JF-12 can duplicate flight conditions at altitudes of 25–50 km, Mach numbers 5–9, with a test time longer than 100 ms (Jiang et al. 2012, Jiang 2014, Jiang & Yu 2017). The technical breakthrough in the key detonation driver techniques achieved during the development of JF-12 stems from decades of research on detonation-driven shock tunnels (Yu et al. 1992).

The facility of JF-12 has a series of advantages, such as the capability to generate high-enthalpy test flows, long test time, low operation cost, and good scalability, thus making it a prospective high-enthalpy facility. On the other hand, challenges still exist in the operation, such as the limitations of total temperature and pressure, surface erosion caused by the high-temperature detonation products and test gases, hydrogen brittleness, and cleaning post-operation (Jiang et al. 2020) Such a harsh test environment gives rise to a great challenge to the survivability of surface temperature sensors such as thin-film gauges and thermocouples. In general, thin-film gauges (Rose 1958, Willeke & Bershader 1973) are capable of measuring surface temperature changes on the order of one-tenth of a °C and are much more sensitive than thermocouples. However, the thin metallic film of thin-film gauges may be destroyed by the impact of small particles carried by the flow or by excessive thermal stresses due to high heat fluxes. On the other hand, thermocouples are one order of magnitude less sensitive than thin-film gauges, but are much more structurally robust, easier to install and much better in survivability in harsh environmental conditions and extremely high heat fluxes. Therefore, for measurements of high heat fluxes in harsh conditions, coaxial thermocouples represent a better choice.

Recent research indicates that reducing the isolation layer between the two electrodes can improve the response property of a coaxial thermocouple of type E, as shown in Figure 8.13, for the measurement of transient temperature (Li et al. 2017). In addition, an integral design of thermocouples in which a chromel wire is embedded in a well-shaped substrate of constantan is reported to improve the spatial and temporal resolutions of the measurement (Wu et al. 2014). The aforementioned thermocouples have been applied as heat flux sensors in the aerothermodynamic heating tests in the JF-12 hypersonic shock tunnel. Both of them are found to perform quite well in the harsh test conditions. As an example, the state-of-the-art high-resolution heat flux sensors are used to measure the stagnation-point heat flux in JF-12 with a series of spherical models as given in Figure 8.14.

The spherical test models with different radii are assembled on a rake, as shown in Figure 8.15. A schlieren image is also given in the figure which shows the bow shock waves over the models. Several shock waves can be seen around the third (counted from the top) model and this is because there are a total of 21 models assembled along the horizontal span of the rake. The detailed test flow conditions under a nominal freestream Mach number of 7 are listed in Table 8.7. Some of the temperature signals are selected and depicted in Figure 8.16.

Coaxial thermocouples with specific diameters of 1.4 and 0.7 mm, and integral sensors are applied in the tests. The time-dependent temperature data as given in Figure 8.16 are acquired using a sampling frequency of 200 kHz. On the whole, the heat flux sensors perform well with an acceptable range of signal-to-noise ratio and rapid response property. The "spiky" structures with very high impulse frequency along the curves may be caused by nonideal conditions associated with the junction structure of the sensor and/or the test flow conditions. One of these conditions is probably the impact of small particles carried by the test flow. Nevertheless, these high-frequency noises contribute a negligible part to the calculated heat flux by the integration using Equation (8.6).

Table 8.7 Typical test conditions of the JF-12 shock tunnel

P_0 (MPa)	T_0 (K)	H_0 (MJ/kg)	M_∞	T_∞ (K)	P_∞ (Pa)	ρ_∞ (kg/m^3)	U_∞ (m/s)
2.45	2,081	2.35	7	217	447	0.0072	2,067

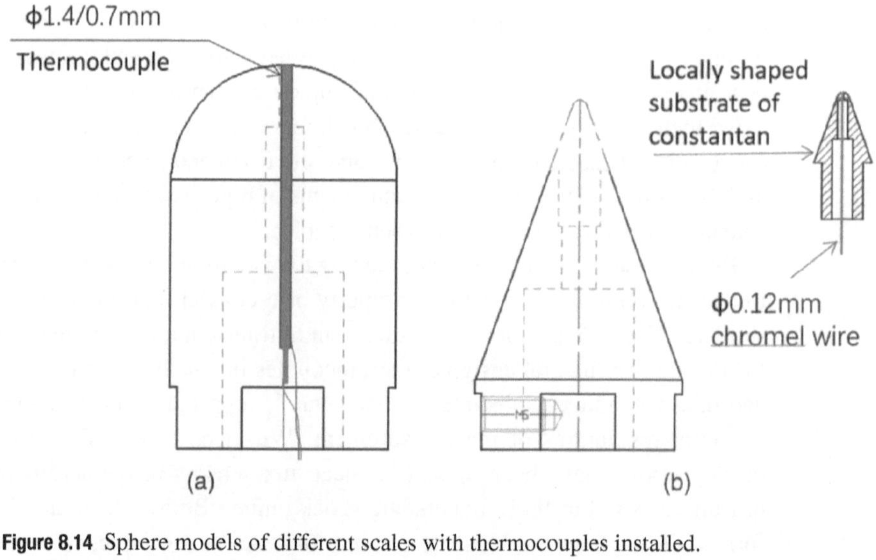

Figure 8.14 Sphere models of different scales with thermocouples installed.

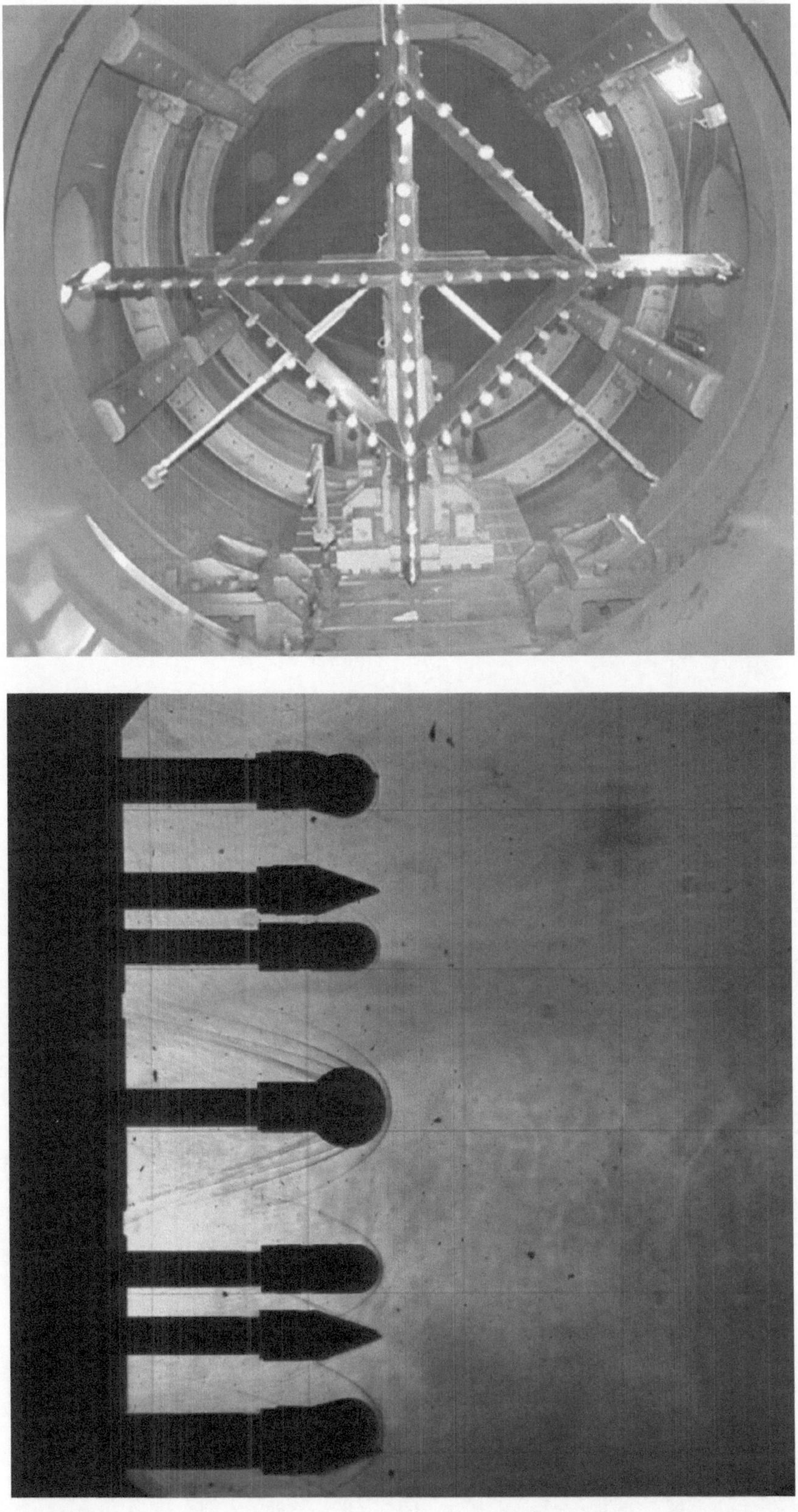

Figure 8.15 Test models in the detonation-driven shock tunnel JF-12 and schlieren.

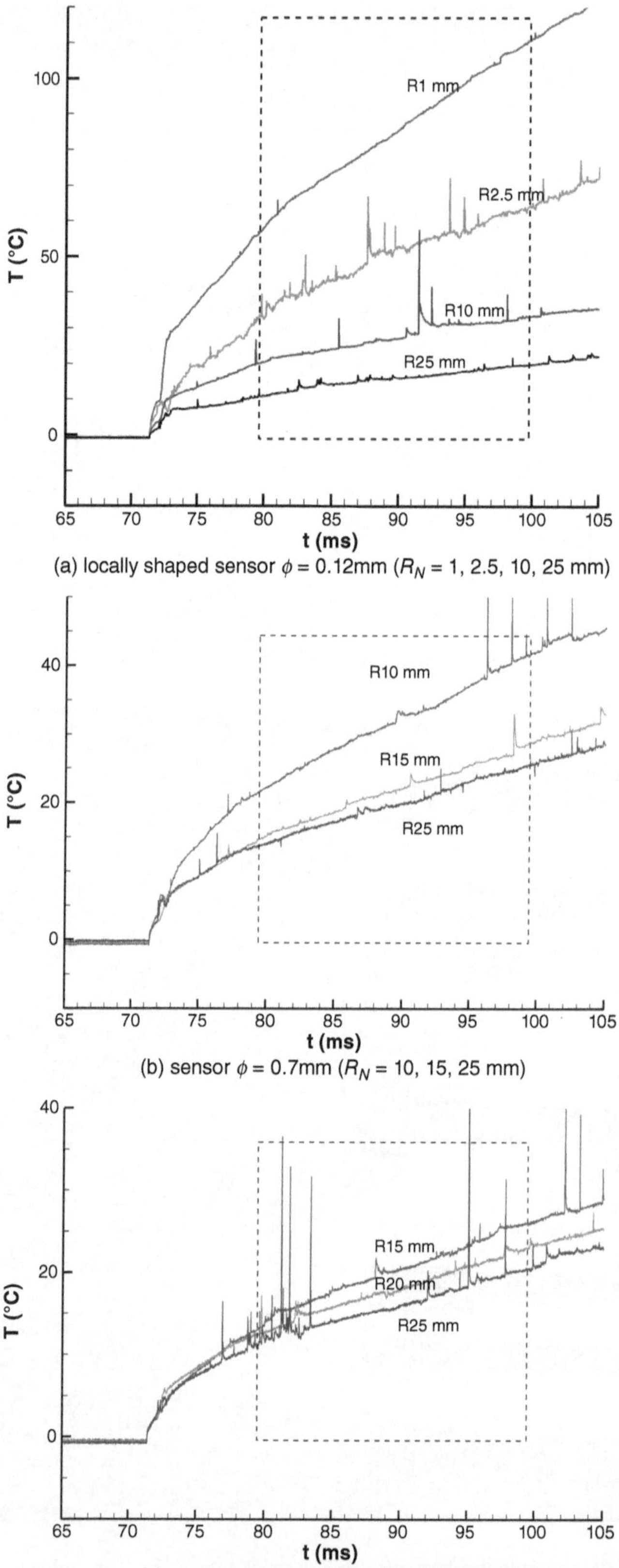

(a) locally shaped sensor ϕ = 0.12mm (R_N = 1, 2.5, 10, 25 mm)

(b) sensor ϕ = 0.7mm (R_N = 10, 15, 25 mm)

(c) sensor ϕ = 1.4mm (R_N = 15, 20, 25 mm)

Figure 8.16 Stagnation-point temperature signals with thermocouples of different specifications.

Equation (8.4) indicates that the stagnation-point heat flux is inversely proportional to the square root of the nose radius, R_N. However, looking carefully into the curves as depicted in Figures 8.16(b) and (c), one can see that the correlation is not being reflected in the figure during the first 8 ms after the jump in the temperature curves, especially for models with larger radii. This duration is considered to be an uncertain interval which is probably associated with the uncertainties originating from the thermodynamic property of the thermocouple junction and/or the buildup process of the flow field around the test model. The data during the next 20 ms as framed by the dashed box in each curve are employed to calculate the heat flux to reduce the effect of the uncertain interval. The uncertain interval for the locally shaped sensor is less than approximately 3 ms as can be seen in Figure 8.16(a). The test time of the JF-12 shock tunnel is longer than 100 ms which is long enough for the thermocouples to overcome the effects of the uncertain interval.

The stagnation-point heat fluxes of the spherical test models are depicted in Figure 8.17 in which the theoretical values obtained using the Fay–Riddel correlation (Fay & Riddel 1958) are also included for direct comparison. The experimental data coincide with the theory with acceptable deviation. In the aforementioned tests, as the total temperature T_0 = 2,081 K is not so high to trigger the dissociation of air molecules, the flow state within the stagnation region does not appear to have significant chemical reactions other than vibrational excitation.

When an orbital vehicle enters an atmosphere at hypervelocity, the flow field around it features thermochemical nonequilibrium induced by the extremely high temperature behind the strong shock wave. The aerothermodynamic performance depends on the degree of flow nonequilibrium. One of the challenges associated with hypervelocity entry physics lies in the simulation of such test flow conditions in ground-based facilities. The primary difficulty is due to the extremely high total enthalpy or total temperature required to generate the hypervelocity test flow. Even though such a high-enthalpy flow can be realized in some specially designed shock

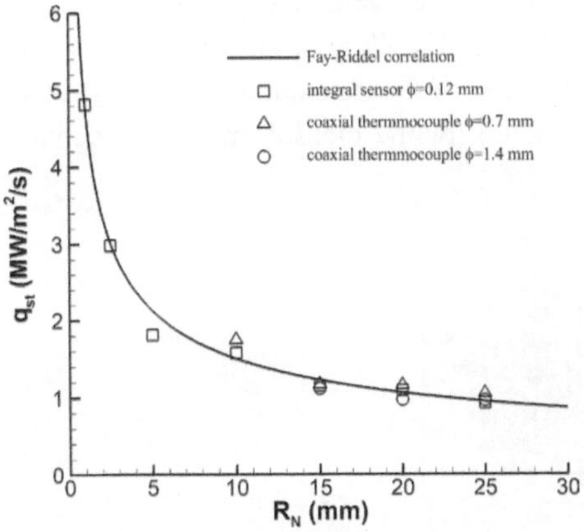

Figure 8.17 Stagnation-point heat flux obtained in JF-12 for spherical test models (test flow conditions: $M_\infty = 7$, $H_0 = 2.35$ MJ/kg).

tunnels, for example, a shock-expansion tube/tunnel, the pulse-operation mode with an extremely short test time leads to a lot of challenges in flow diagnostics and measurement.

JF-16 is a detonation-driven shock-expansion tunnel that was built based on a forward detonation-driven shock tube (Gao 2008, Wu 2012, Jiang et al. 2015, Zhou et al. 2016a, 2016b). A detonation driver is capable of generating high-enthalpy flows due to the high speed of sound and pressure of the driver gas. The original JF-16 facility is operated in a shock-expansion tube mode (Gao 2008, Wu 2012, Zhou et al. 2016a). An upgrade has been completed by appending a conic nozzle to the original shock-expansion tube (Zhou et al. 2016b). Through the steady expansion in the nozzle, the test flow temperature is reduced while the Mach number is increased. The uniform test flow size reaches a diameter of 170 mm after the aforementioned upgrade of the JF-16 shock-expansion tunnel (Hu et al. 2017). A blunted half-sphere model with a diameter of 50 mm, as shown in Figure 8.18, and a blunted half-cylinder model with the same diameter are used for the flow visualization and stagnation-point heat flux measurements. Flow structures are captured directly from the self-illumination of radicals at high-energy states downstream of the strong shock wave. The resolution of the camera (FASTCAM SA4) is 1024×1024 with a frame rate of 500 kfps.

A series of computations are conducted and shown in Figure 8.19 for comparison with the experiments. The agreement is acceptable which implies the test flow is well predicted. Further quantitative comparison about the standoff distances of the bow shock wave indicates obvious discrepancies which may be caused by uncertainties in the flow diagnostics. The test time of the shock-expansion tunnel is extremely short, only under the order of 100 μs (Gao 2008, Wu 2012, Jiang et al. 2015, Zhou et al. 2016a, 2016b, Hu et al. 2017). This leads to difficulties in flow measurement since the response time of a conventional transducer may exceed such a short duration. A specially designed spherical model has been applied for stagnation heat measurement. As shown in Figure 8.18, the spherical model of 50 mm in diameter, along with a coaxial thermocouple mounted at the stagnation point, is coated with a film of copper (for full catalytic surface chemistry kinetics model) with an averaged thickness of 300 nm (Hu et al. 2017, Zhou et al. 2020). Such a technique can significantly reduce the response time of the transducer. Several runs with different total enthalpies have been conducted for the measurement of stagnation heat flux.

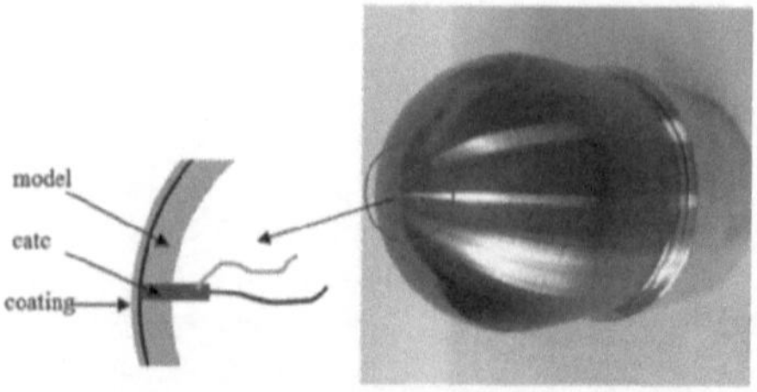

Figure 8.18 Specially designed model for stagnation-point heat flux measurement in hypervelocity test flows (catc: co-axial thermal couple).

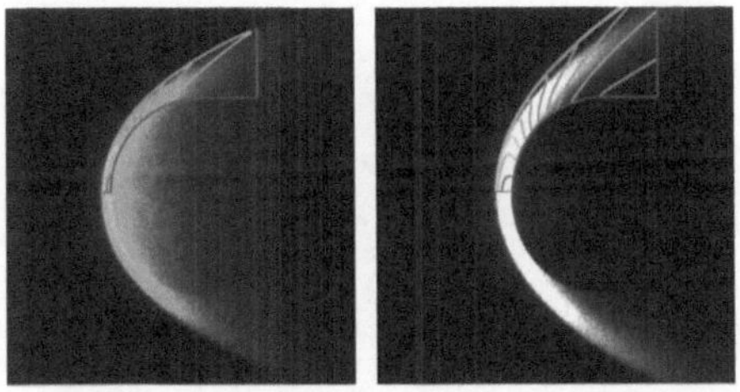

Figure 8.19 Comparison between experiment and computation: (left) Blunted-sphere model; (right) Blunted-cylinder model ($U_\infty = 7.9$ km/s).

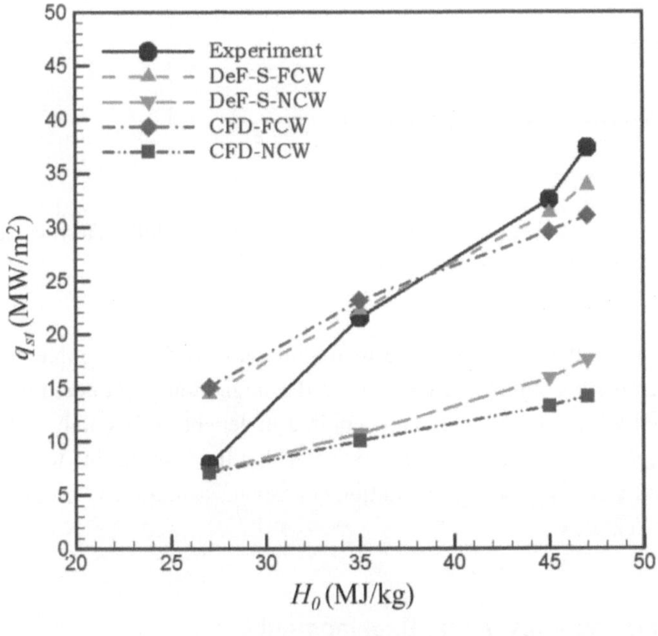

Figure 8.20 Comparison between experiment and computation: (left) Blunted-sphere model; (right) Blunted-cylinder model ($U_\infty = 7.9$ km/s, FCW: full-catalytic wall, NCW: non-catalytic wall).

The comparison among experiments, computations, and De Filippis–Serpico correlations (De Filippis et al. 2005) is given in Figure 8.20. The velocities of the test flows are 7.1, 7.9, 8.8, and 9.0 km/s, corresponding to a series of total enthalpies of 27, 35, 45, and 47 MJ/kg, respectively.

For the test condition with a total enthalpy of 27 MJ/kg, the measured data agree with the predicted data using a non-catalytic wall condition. When the total enthalpy exceeds 35 MJ/kg, the experiment and prediction agree with each other under the full-catalytic wall condition. Such a finding implies that a strong chemical nonequilibrium is present in the stagnation region and recombination of atoms occurs at the cool model surface due to the catalytic effect of the Cu film.

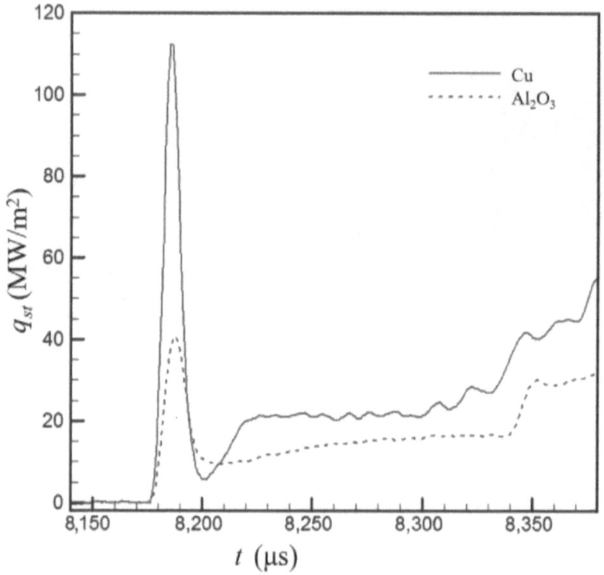

Figure 8.21 Comparison between stagnation-point heat fluxes for fully catalytic (Cu) and non-catalytic (Al_2O_3) surface conditions ($U_\infty = 7.9$ km/s).

If the coating material is replaced with aluminum oxide to simulate non-catalytic surface chemistry, a significant decrease in the stagnation-point heat flux is captured as compared with experimental data acquired under the fully catalytic surface condition in Figure 8.21. Such a finding indicates that the heating due to recombination under the fully catalytic surface condition contributes a significant portion to the total convective heat flux.

8.4 Hypersonic Boundary-Layer Development

The flow of gas at hypersonic speed near a solid surface leads to the formation of a boundary layer whose thickness greatly exceeds the thickness of the boundary layer formed at low speeds. The concept of the boundary layer implies that a flow at a high Reynolds number can be divided up into two regions. In the bulk of the flow region away from the solid surface, the viscosity can be ignored, and this is called the inviscid external flow. The second region is the boundary layer at the wall where the viscosity must be considered (Schlichting & Gersten 2017). Basically, hypersonic speeds do not lead to fundamentally new results for boundary-layer flows in themselves. What is significant is that the magnitude of certain variables becomes much larger when the Mach number of the external flow becomes large (Rasmussen 1994). Hypersonic speeds mean high temperature levels and the consequent high coefficients of viscosity in the boundary layer. As a result, the effect of viscosity at hypersonic speeds is not confined to the interior of the boundary layer and leads to substantial changes in the entire flow field. It is worth noting that the boundary-layer thickness becomes much thicker for hypersonic speeds, which changes the "effective" shape of the body (Mikhailov et al. 1971). When this happens, the boundary layer interacts with the inviscid external flow and produces changes in the pressure which in turn are impressed back again on the boundary layer. Hypersonic boundary layers are usually studied based on the theory of interaction of the boundary layer with the outer stream.

8.4.1 Hypersonic Boundary Layer

Hypersonic boundary layers are important in considering flows over spacecraft, ballistic missiles, and hypersonic aircraft. The following features are typical for hypersonic boundary layers (Anderson 2006): (1) Thin shock layers. For a given flow deflection angle, the density increase across the shock wave becomes progressively larger as the Mach number is increased. For the same mass flow, this means that the distance between the body and the shock wave can be small. (2) Entropy layer. The entropy of the flow increases across a shock wave, and the stronger the shock, the larger the entropy increase. (3) Viscous interaction. The properties of the hypersonic boundary layer are mainly affected by the increase in temperature. The increase in temperature leads to an increase in the viscosity coefficient, which increases the thickness of the boundary layer. In addition, an increase in temperature will result in a decrease in density, which means that the thickness of the boundary layer must increase for mass flow through the boundary layer. The coupling of the two leads to a sharp increase in the thickness of the boundary layer, which causes a given body shape to appear much thicker than it really is, which greatly interferes with the external inviscid flow. At the same time, the external inviscid flow will in turn affect the development of the boundary layer. (4) High-temperature flows. The extreme viscous dissipation in the hypersonic boundary layer can create very high temperatures. It is sufficient to excite the vibrational energy internally within molecules, leading to the dissociation and even ionization of the gas. (5) Low-density flow.

The simplest example of the boundary layer is the flow along a flat plate. As shown in Figure 8.22, two regions can be distinguished, between which there is actually no sharp division:

1) A thin layer right beside the body surface where the velocity gradient normal to the wall $\partial u/\partial y$ is very large (boundary layer). In the thin boundary layer at the wall, the viscous shear force $\tau = \mu\, \partial u/\partial y$ can reach an order of magnitude equivalent to the inertial force.
2) The inviscid external flow region outside this thin boundary layer. In this region, the velocity has no gradient, the viscosity can be ignored, and the flow is frictionless and potential.

After a dimensional analysis of the two-dimensional compressible fluid Navier–Stokes governing equations, the high-order terms are small and are ignored. By making the combined assumptions of $\delta < L$ and $Re \geq 1/\delta^2$, the complete Navier–Stokes

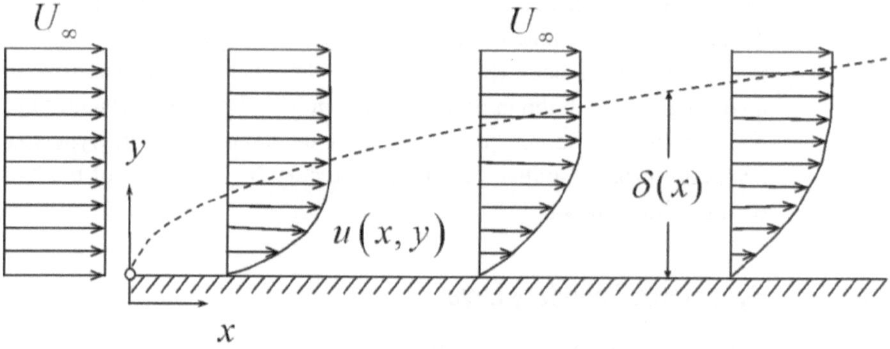

Figure 8.22 Boundary layer on a flat plate at zero incidence (schematic).

equations can be reduced to a simpler form that is applied to the boundary layer (Lu & Marren 2002). Further introducing the Illingworth transformation, the boundary-layer equations of the plate are transformed into:

$$\begin{cases} (Cf'')' + ff'' = 0 \\ \left(\frac{C}{\Pr} g'\right)' + fg' + C\frac{u_e^2}{h_e}(f'')^2 = 0. \end{cases} \tag{8.9}$$

Here, $C = \rho\mu / \rho_e\mu_e$ is the Chapman–Rubesin parameter, and ρ_e, u_e, μ_e, h_e are the density, velocity, viscosity coefficients, and enthalpy, respectively. The subscript "*e*" refers to the boundary-layer edge condition. The full flow field parameters can be solved by the shooting method.

For a self-similar boundary flow of a perfect gas, the dimensionless heat transfer *St*, is given by:

$$St = \frac{q_w}{\rho_e u_e (h_r - h_w)} \tag{8.10}$$

Here, q_w is the wall heat transfer, $h_r = h_0 + 0.5\left(\Pr^{1/2} - 1\right)u_e^2$ is the recovery enthalpy, $h_w = c_p T_w$ is the enthalpy at the wall and c_p is the specific heat at constant pressure.

High-speed aircraft are generally composed of an axisymmetric slender body and several groups of small control surfaces. Its typical main body is a cone, as shown in Figure 8.23, or a body of revolution, such as rockets and missiles. For a cone with a 0° angle of attack, when flying at a supersonic speed, the physical quantities of the flow field on each generatrix, such as p_e and u_e, have the same value.

Because the differences from a plane boundary layer are so little, the axisymmetric boundary-layer equations can be simplified to the plane boundary-layer equation through the Mangler transformation. It basically allows the calculation of the boundary layer on a body of revolution to be reduced to the calculation of the boundary layer on a suitable plane body. According to the Mangler transformation (Rasmussen 1994):

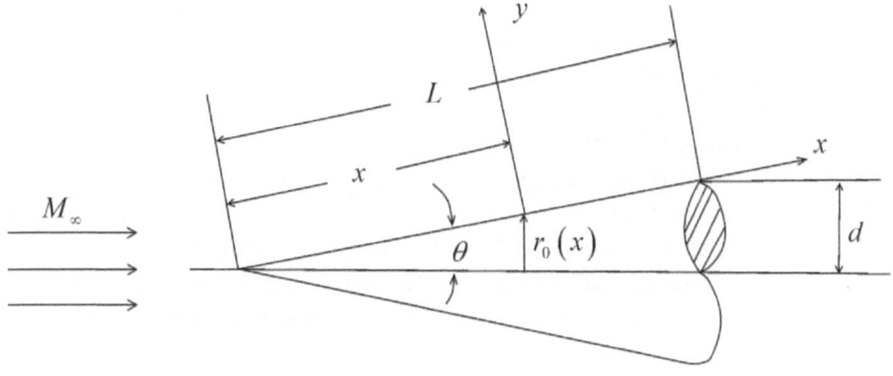

Figure 8.23 Boundary layer on a cone at zero incidence (schematic).

$$\left.\begin{aligned} X &= \int_0^x \left(\frac{r_0}{d}\right)^2 dx = \frac{L}{3}\left(\frac{x}{L}\right)^3 \\ Y &= \frac{r_0}{d} y \\ U(X,Y) &= u(x,y) \\ \overline{T}(X,Y) &= T(x,\ ) \end{aligned}\right\} \tag{8.11}$$

Under this transformation, the wall heat flow on the cone is:

$$q_w = \left(k\frac{\partial T}{\partial y}\right)_w = \frac{r_0}{d}\left(k\frac{\partial \overline{T}}{\partial y}\right)_w, \tag{8.12}$$

the corresponding heat transfer is:

$$St(x) = \frac{q_w}{\rho_e u_e (h_r - h_w)} = \frac{r_0}{d}\overline{St}(X) = \sqrt{3}\overline{St}(x) \tag{8.13}$$

Among them, $\overline{St}(x)$ is the local heat flow coefficient of the flat plate. That is the coefficient obtained by the flat plate formula using the outflow condition of the cone. It can be seen that the heat transfer rate of the cone is $\sqrt{3}$ times that of the flat plate.

8.4.2 Boundary-Layer Transition

Hypersonic laminar–turbulent boundary-layer transition is one of the most important issues in the design of hypersonic aircraft. The boundary-layer transition affects the heating, skin friction, drag, and moments that a vehicle may experience in flight. The heating rates of the transitional or turbulent boundary layer may be three times or more than those under laminar conditions. The early boundary-layer transition can cause the ablation rate of the heat shield to exceed expectations and cause vehicle failure. If the designer adopts the worst-case scenario, the heat shield may be excessively heavy, thereby reducing the vehicle's mobility and cruising range. The boundary-layer transition may not occur symmetrically, and the vehicle may be subjected to asymmetric loads. If the control system is not strong enough, the vehicle may be lost.

The problem of the transition from laminar to turbulent flow in the boundary layer has attracted the interest of researchers for countless years. Despite a lot of research effort devoted to it, only partial knowledge of hypersonic boundary-layer transition has been obtained so far due to the complexity of the physical mechanism in hypersonic boundary-layer flow, and our complete understanding of the problem is still far away. The insufficient understanding of the hypersonic boundary-layer transition mechanism makes it difficult to accurately predict the transition location, which further aggravates the difficulty of the deployment of the thermal protection system and the trajectory design of hypersonic aircraft (Tani 1969). Most of the difficulties in understanding lie in the many factors that affect transition, but these factors are not independent of each other. And how these factors interact with physical mechanisms is still not well understood.

The boundary-layer transition of an aircraft in a high-altitude low-background disturbance environment is generally natural. The transition caused by small disturbance instability is called natural transition. If there is a large disturbance in the boundary layer from the beginning, so that there is no slow evolution process from the disturbance to the transition, this transition is called a bypass transition. According to previous studies, the basic process of the natural transition of the boundary layer is shown in Figure 8.24. The flow after the leading edge of the boundary layer is laminar, but there are some small disturbances. In the process of propagating downstream, some disturbances will gradually increase, and when they increase to a certain level, they will trigger a transition, turning the laminar flow into a turbulent flow. The entire natural transition process can be divided into the following three stages:

1) Laminar stage: the process (I) of disturbance receptivity, the process (II) of disturbance linear evolution, and the process (III) of disturbance weak nonlinear evolution.

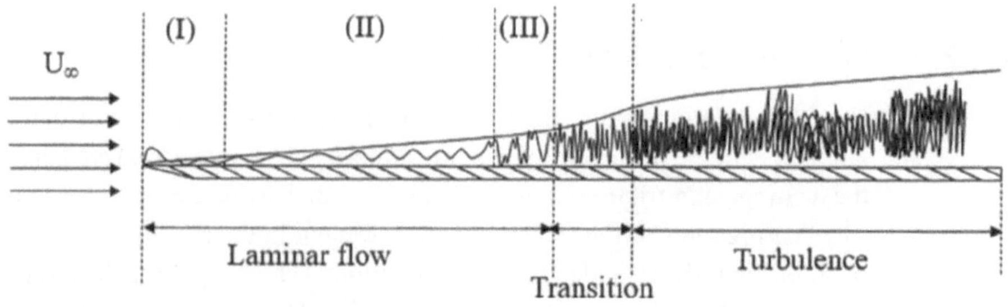

Figure 8.24 Sketch of the transition process of the boundary layer.

2) Transition stage: the process of strong nonlinear interaction between disturbances. It is manifested by strong pulsation in the flow field, which can even exceed the turbulent stage.
3) Turbulent stage: the disturbance wave breaks and the flow develops into full turbulence.

The entire transition process actually involves several issues: (1) Where does the disturbance in the boundary layer come from? (2) How does the disturbance in the boundary layer evolve? (3) To what extent the disturbance evolves to trigger the transition? (4) What is the mechanism of transition? The basic problems that need to be solved in the transition research are summarized as follows:

1) Receptivity process:
It mainly studies the mechanism of how external disturbances induce the internal disturbances of the boundary layer. The dispersion relationship between the pulsation in the free flow and the wave mode in the boundary layer is different. External disturbances cannot directly induce disturbance waves inside the boundary layer, as shown in Figure 8.24, but must be stimulated by a certain mechanism. Receptivity denotes the means by which a particular forced disturbance enters the boundary layer and the nature of its signature in the disturbance flow. If the initial disturbances are sufficiently large, they can grow by forcing mechanisms to nonlinear levels and lead directly to turbulent flow. If they are small, they will tend to excite free disturbances in the boundary layer.

2) Disturbance evolution:
The evolution of disturbances is divided into linear and weakly nonlinear stages. The growth of small disturbances in the boundary layer can be predicted according to the linear theory of flow stability; but when the disturbances increase to a certain degree, nonlinear effects appear, which need to be described by nonlinear theory.

3) Transition process:
In the transition process, when the disturbance increases to a certain extent, the disturbance wave will be broken under the action of nonlinearity, and the laminar flow will eventually develop into a turbulent flow on different paths (Morkovin et al. 1994). Therefore, judging from where the transition occurs and determining the extent to which the disturbance evolves to trigger transition is a problem that needs to be solved in transition prediction research. Since the transition depends not only on the initial amplitude of the disturbance, but also on other characteristics such as the frequency and shape of the disturbance, this increases the difficulty of transition prediction. At present, no mature research has resulted in this area, which means one has to mainly rely on empirical or semiempirical methods.

Compared with incompressible flow, hypersonic transition processes are less understood due to additional complexity, as listed below, such as the second and higher instability modes, the applicability of similarity parameters, and the nonlinear coupling of different processes. This problem poses a huge challenge to theory, experiment, and calculation.

1) Higher instability modes: One such disturbance that may cause transition on axisymmetric geometries near a 0° angle of attack for hypersonic aircraft is the second-mode instability. This instability is similar to an acoustic wave captured between a surface and the sonic line within the boundary layer. The amplified frequency range largely depends on the thickness of the local boundary layer. During the transition from laminar to turbulent flow in the compressible boundary layer, the second-mode instability appears at a very high frequency range.
2) Nonlinear: As an instability packet convects in the downstream direction, linear and non-linear growth occurs until the pressure fluctuations reach a sufficient magnitude to cause breakdown in the laminar boundary layer and transition begins. Unlike the nonlinear behavior in subsonic flow, the nonlinear process in the hypersonic regime may last a long distance and have a significant impact on the aerodynamic performance of the aircraft. However, due to inherent nonlinearity, the nonlinear behavior of hypersonic instability waves cannot be predicted by classic linear stability theory, resulting in few studies on the nonlinear interaction of hypersonic instability waves.
3) Similarity parameters: The transition Reynolds number obtained from the wind tunnel experiment will systematically change with the increase of the stagnation pressure or the unit Reynolds number. This is known as the "unit Reynolds number effect." Although most of the unit Reynolds number effect seems to have been caused by the variable acoustic in the boundary layer of the tunnel wall, there are likely to be different unit Reynolds number effects corresponding to different transition-sensitive characteristic lengths. In fact, similarity parameters such as Reynolds number or Mach number differ from each other in different facilities due to differences in their capabilities, which makes general analysis, extrapolation, comparison, or full use of experimental data very difficult (Wang et al. 2016a). Therefore, in order to establish effective comparisons between different ground facilities and promote successful inferences from ground to flight, there is still a great need for databases and related sets of principles obtained from the original model.

In order to fully understand hypersonic laminar–turbulent boundary-layer transition, currently, researchers use theoretical analysis (Alexander 2011, Casper et al. 2014, Kuehl 2018), numerical simulation (Zhong 2009, Li et al. 2018) and experimental methods (Zhu et al. 2018, Liu et al. 2019, Xiong et al. 2020) to study hypersonic boundary-layer transition. Due to many factors and complex mechanisms, theoretical analysis and numerical simulation are difficult to consider all factors. Experimental data with corresponding numerical analysis can give a better understanding of what we studied. Flight testing has problems such as high cost, lack of measurement technology, and high risk. Therefore, wind tunnel experiments will still be one of the most important approaches for hypersonic boundary-layer transition mechanism study in the foreseeable future.

8.4.3 Experimental Methods

High-fidelity measurements are required to make better flight predictions based on wind tunnel measurements and verify boundary-layer stability calculations. The shock tunnel is the main ground simulation equipment for hypersonic aerodynamics due to its large simulation range, strong simulation ability, large model scale, and multiple measurement methods. The most common research carried out in wind tunnels involves measuring aerodynamic forces and aerodynamic heating on models. Shock tunnels need to develop suitable measurement techniques based on their characteristics (high air velocity, low gas density, millisecond test time) to study hypersonic boundary-layer transition. At present, there are many technologies used to measure the transition of the hypersonic boundary layer in shock tunnels. These technologies are basically based on the different physical characteristics of the laminar–turbulent boundary layer (such as heat flow, pressure, density, and velocity) to observe and assess the results. The hypersonic boundary-layer transition test mainly uses thermocouple sensors, platinum film thermal resistance sensors, hot wire technology, pulsating pressure sensors, ALTP, etc. for local point measurement; at the same time, it uses phosphorescent heat map technology, PLIF, high-speed schlieren, and infrared heat. Graphic technology, liquid crystal, temperature-sensitive paint, pressure-sensitive paint, etc. are used to perform global surface measurements.

8.4.3.1 Point Measurement Technology

Point measurement technology is mainly based on the spatial distribution characteristics of physical parameters along the flow direction to judge the transition, and uses the time-varying characteristics of the single-point sensor curve to obtain stability evolution information. After the boundary-layer transitions from laminar flow to turbulent flow, the physical parameters will change, so the transition position can be judged based on the physical quantity distribution measured by multi-point sensors. Single-point experimental measurements, especially those with unsteady or dynamic sensors, can provide critical verification data for boundary-layer stability and transition calculations. The flow environment in a transitional hypersonic boundary layer presents many challenges for unsteady instrumentation. Intrusive instruments such as hot wire anemometers or surface-based dynamic sensors must be able to withstand the high-temperature and high-enthalpy conditions common to hypersonic flows. For scale model testing conducted in ground-based facilities, where the boundary-layer thickness on the model is small (on the order of millimeters), unsteady sensors must have high spatial resolution, high bandwidth (1 MHz), and high sensitivity to resolve the instability waves in the transition process. At present, our unsteady measurement techniques rarely meet these requirements, and each available measurement technology has both strengths and weaknesses.

The hot wire anemometer is a mature technology for testing and measuring low-speed flow. It is divided into two types based on different working principles: constant temperature and constant current. A metal wire is placed in the flow field, and the wire is heated by an electric current to make its temperature higher than the temperature of the fluid (called a hot wire). When the fluid flows perpendicular to the metal wire, some heat is taken away, and the temperature of the metal wire drops. The amount of heat dissipation is related to the flow rate. The amount of heat dissipation causes the temperature of the hot wire to change which changes the electrical resistance, and the flow rate signal becomes an electrical signal

(Comte-Bellot 1976). It can not only measure a stable flow, but also a pulsating flow field that changes with time and space, and it is also suitable for supersonic flows. The basic working principle of the hot wire anemometer is shown in Figure 8.25. Hot wire anemometry has been used to quantify the disturbances in supersonic and hypersonic wind tunnels. But these measurements are limited to lower hypersonic Mach numbers where the temperature or enthalpy is low enough to ensure the survival of the hot-wire sensor. In addition, the hot wire usually does not have enough bandwidth to resolve the instability wave of interest, and it is difficult to calibrate the hot wire in a hypersonic flow.

The more commonly used thermocouple sensors include coaxial thermocouples and thin-walled thermocouples (integrated). Shown in Figure 8.26(a) is the end face of the coaxial thermocouple sensor. The principle of thermal measurement is based on the Seebeck effect. As shown in Figure 8.26(b), at a certain temperature, the free electron density in two different conductor materials will be different. At their junction, due to the diffusion of free electrons, a thermoelectromotive force is generated. The magnitude of the thermoelectromotive force is related to the properties of the two conductor materials and the temperature of the contact point. The temperature change of the node is obtained by measuring the potential difference generated by the heating of the metal. A coaxial thermocouple measures the heat flow based on the Laplace transformation of the one-dimensional infinite heat conduction problem to obtain a relationship between the heat flow and the temperature, and further establishes the relationship between the electric potential and the heat flow. The coaxial thermocouple has a strong anti-erosion ability, high-frequency response, and reusability. It is an important measurement method in the aerothermal test environment in hypersonic wind tunnels. In order to improve the performance of the coaxial thermocouple and reduce the response time of the sensor, the thermocouple sensor is installed flush with the test model and then polished to form a junction (Li et al. 2018, Wang et al. 2019). Shown in Figure 8.27 is the heat flow distribution on the centerline of a large-scale flat plate with a length of 3.2 m, which is measured using more than 230 coaxial thermocouple sensors in the JF12 shock tunnel. The experiment captured the fine heat flow distribution in the process of hypersonic boundary-layer transition, and its repeatability was good.

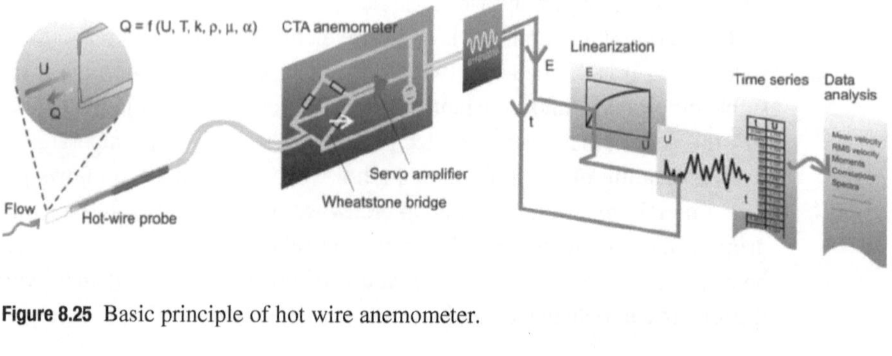

Figure 8.25 Basic principle of hot wire anemometer.

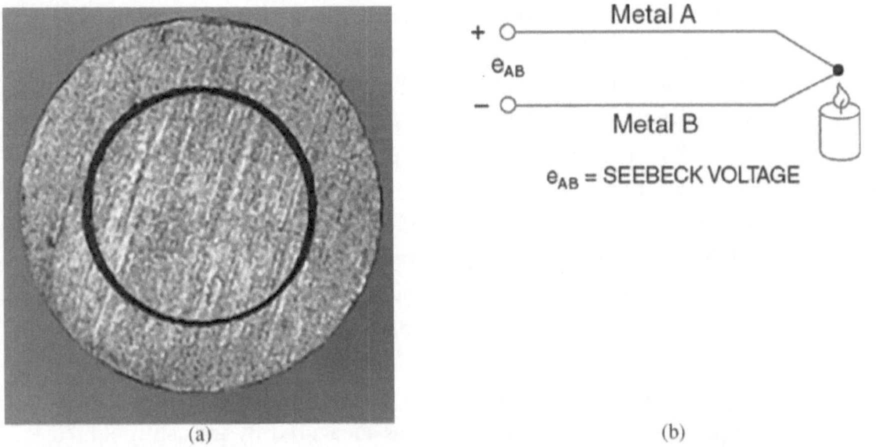

(a) (b)

Figure 8.26 Thermocouple sensor: (a) Top view of the thermocouple sensor; (b) Seebeck effect schematic.

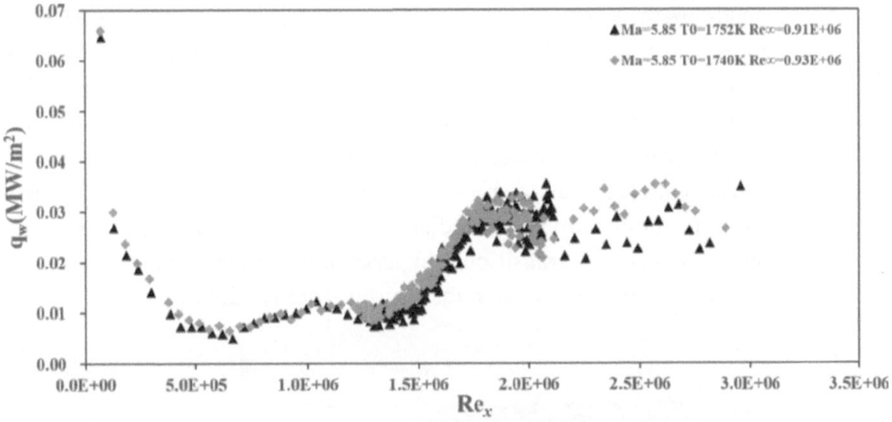

Figure 8.27 Large-scale flat plate centerline heat flow measurement using thermocouple sensors.

The principle of the thin-film sensor is to use a platinum film plated on the insulating substrate as a temperature-sensitive resistance film to measure the temperature change, and to use the thermoelectric simulation network to obtain the surface heat flux density of the measured object. It has good stability, large measurement range, fast response, high sensitivity, high precision, good repeatability, static calibration, and simple and reliable use. It is another important measurement method in the aerothermal test environment in hypersonic wind tunnels. Li et al. (2016) assembled multiple thin-film sensors into one piece and reduced the distance between each sensor to 2.5 mm to measure the transition process on a flat plate.

For the typical incoming flow condition of a hypersonic transition experiment in wind tunnels, the frequency of the disturbance wave will be around several hundred kilohertz, and its harmonics will be close to megahertz or even higher. In order to be able to measure the transition of the boundary layer in a pulse wind tunnel, the sensor needs to have the characteristics of fast response, high sensitivity, and high natural frequency. At present, the sensor that can meet the demand is the PCB-type high-frequency piezoelectric sensor, as shown in Figure 8.28. This type of sensor has a pressure resolution of 7 Pa and a response frequency of more than 1 MHz. Due to the AC coupling characteristic of the sensor, it is only suitable for measuring the pulsating component rather than the absolute value of pressure. Therefore, it can be used to measure the pulsation signal of hypersonic boundary-layer instability and transition to turbulence, and to interpret the instability mode and transition (Chynoweth et al. 2017, 2019, Marineau et al. 2019). The frequency spectrum of the sensor during the laminar flow stage on a large-scale flat plate with a length of 3.2 m, as measured by a flush-mounted PCB sensor, is shown in Figure 8.29. The test results show that the frequency spectrum can be divided into three frequency bands. The low-frequency band is concentrated at 13–93 kHz, the middle-frequency band is concentrated at 250–320 kHz, and the high-frequency band is concentrated at 640–780 kHz. The dominant frequency bands are the low- and high-frequency bands. The spectral-spatial distribution contours reflect the evolution of the power spectrum along space, as shown in Figure 8.30. The three-frequency-band distribution structure of the mode waves in the boundary layer is clearly observed. Among them, the bandwidth of the middle/low-frequency band is narrow, and the energy of the corresponding frequency band is concentrated. Fast-response piezoelectric pressure transducers have been utilized in a large number of studies for measurements of the surface pressure fluctuations induced by transitional hypersonic boundary layers. But these transducers are difficult to calibrate and may suffer from spatial averaging.

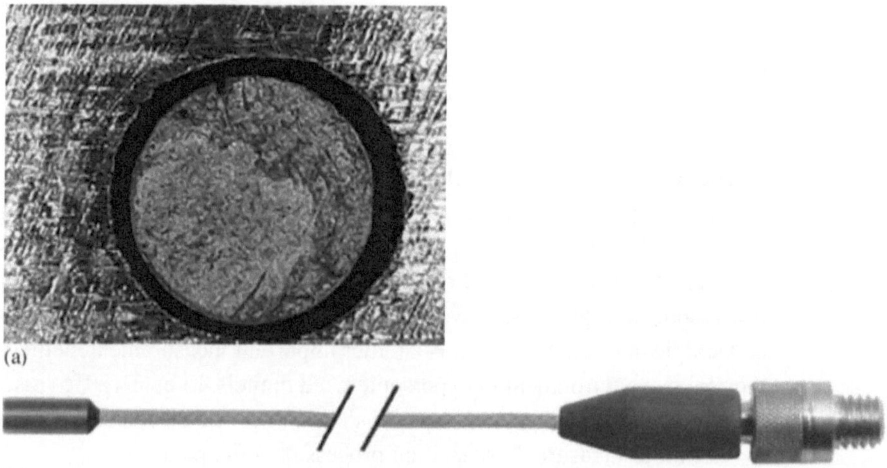

(a)

(b)

Figure 8.28 PCB sensor: (a) Top view of the PCB sensor; (b) PCB sensor panorama.

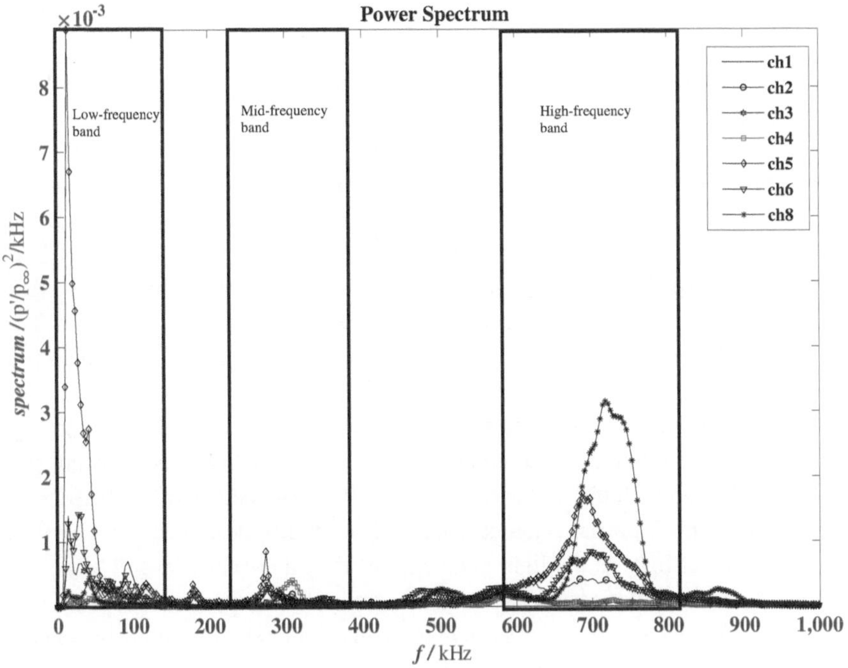

Figure 8.29 Power spectrum distribution.

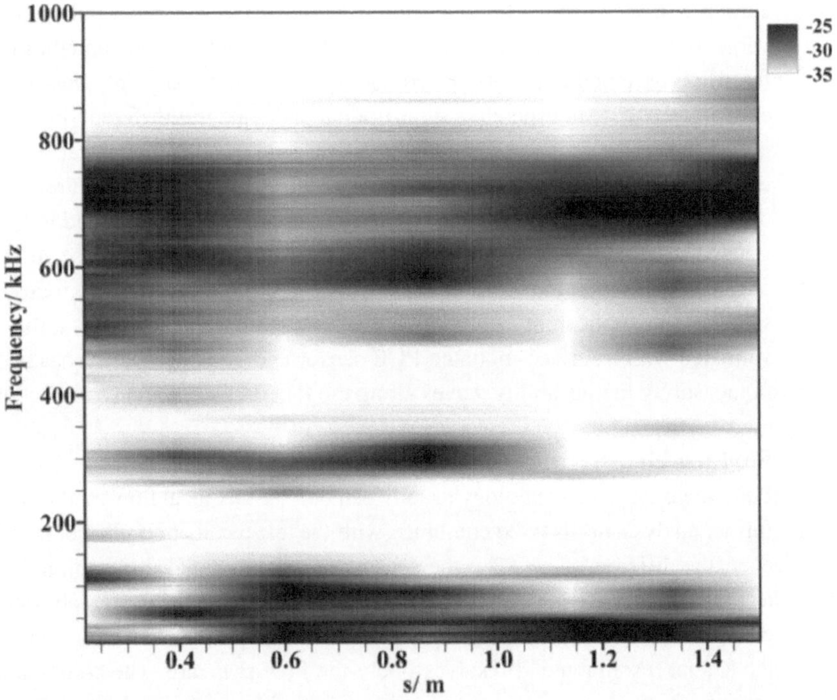

Figure 8.30 PSD spatial distribution contour.

The atomic layer thermopile (ALTP) is a heat flux sensor that can be used to measure surface heat-flux fluctuations induced by transitional hypersonic boundary layers. These sensors are reported to have a bandwidth of 1MHz, a spatial resolution of 1 mm^2, and a linear static response from milliwatts to kilowatts per square centimeter. The sensor is a highly anisotropic film with a layered structure sprayed on a base, the film thickness is about 500 nm–700 nm. Compared with the hot wire anemometer, the ALTP sensor can adapt to the pulse-type wind tunnel at high enthalpy, high speed, and with a short running time, and the response frequency is much higher. ALTP sensors can be used for the measurement of the mode waves and their harmonic components (Roediger et al. 2009).

The focused laser differential interferometer (FLDI) is an appropriate diagnostic for making high-speed wind-tunnel boundary-layer measurements because of its high-frequency response, adequate spatial resolution, and ability to repeatedly produce quantitative measurements of density fluctuation. The working principle of the FLDI can be found in Parziale et al. (2014). The details of the FLDI setup are shown in Figure 2 in that reference but will be briefly described here. First, a laser emits coherent light as the light source; after that, a concave lens diverges the laser beam through a polarizer and a Wollaston prism before it reaches the convex lens. As the divergent laser beam passes by the Wollaston prism, the single beam is separated into two orthogonally polarized beams with a small separation angle. These two separated beams will converge with the aid of the convex lens. Due to the small separation angle of the Wollaston prism, these two orthogonally polarized beams will be separated by a certain distance at the focal point. Before these two orthogonally polarized beams converge to the focal point, they share almost the same optical path, and thus no phase difference is assumed to be introduced. However, at the focal point, these two beams are converged and do not share the same optical path; and this different optical path region is defined as the probe of the FLDI system. Due to such an optical setup, the FLDI is also theoretically applicable for the spanwise disturbance measurement by tuning the focus location of the system, but special attention should be paid to the sensitivity issue. The development of an instability wave along both streamwise and wall-normal spatial resolution of the observation was measured by traversing the focus of the FLDI system. The applicability of the FLDI technique on hypersonic instability waves was validated with a surface-mounted PCB sensor and good agreement has been achieved in measuring the instability waves along the flared-cone model (Xiong et al. 2020).

8.4.3.2 Surface Measurement Technology

Point measurement technology usually can only reflect local flow field information, and further analysis needs to be combined with the test results of multi-point sensors. However, the differences between the sensors and their spatial distribution will affect the test results. Surface measurement technology can obtain large-area physical parameter changes during the process of hypersonic boundary-layer transition, which improves the spatial resolution and resolves finely the flow structure. The combination of point measurement and area measurement can finely observe the evolution of the flow structure and physical parameters in the process of the hypersonic boundary-layer transition.

Typical global surface measurement techniques in shock tunnels usually include the following: nonintrusive schlieren/shadowgraph flow visualization technology based on density changes; contact display technology for sensitive materials based on changes in wall physical parameters; optical display technology for spreading tracer particles. These surface measurement techniques have little interference with the flow field and require a high degree of optical access to the flow field. They can observe the evolution of physical parameters in the boundary-layer transition process on a large area, and can clearly observe the flow structure. Therefore, they are irreplaceable measurement techniques in hypersonic boundary-layer transition experiments.

The schlieren method is one of the most frequently applied optical visualization systems in wind tunnels since it combines a relatively simple optical arrangement with a high degree of resolution. This kind of technology is applied in the wind tunnel to visualize the hypersonic flow field and obtain the wave structure around the aircraft. As shown in Figure 8.31, the principle of the schlieren method is that when the light source passes through a flow field with a density gradient, it will deflect, and the knife edge placed on the focal plane of the lens will block the light deflected in a certain direction. Therefore, the light intensity distribution after passing through the focal plane can show the density distribution in the corresponding flow field. For instance, schlieren visualization taken in a shock tunnel shows three typical boundary layers, that is, laminar, transitional with second mode wave, and turbulent, on the surface of the cone model (Li et al. 2019).

The temperature-sensitive paint (TSP) provides global surface heat transfer to reveal detailed information about the shape of the transition front. TSP is a large-area heat flow measurement technology. It is an optical measurement technology that uses the characteristics of temperature-sensitive materials that change with temperature to measure the temperature distribution of the model surface. The operation process is to spray temperature-sensitive phosphorescent material on the surface of the model, use a high-power ultraviolet light as the excitation light source to irradiate the model, and collect the luminous intensity in a specific band on the model surface through the camera. According to the relationship between the calibrated light

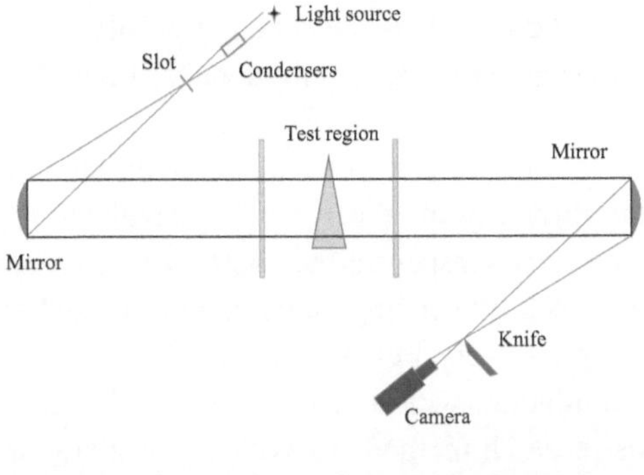

Figure 8.31 Schematic of schlieren optical setup.

intensity and temperature, the temperature distribution on the surface of the model during the test is obtained, and then the large-area heat flow distribution is obtained. When the boundary-layer transition experiment is conducted in the shock tunnel, the observation of the laminar-turbulent transition flow state as well as the momentum and energy capability of the fluid is enhanced. The difference in heat transfer between laminar and turbulent boundary layers is reflected in aerodynamics. Thermally, the transition to turbulence will cause a significant increase in surface heat flow. For instance, experimental images can be quantitatively analyzed to obtain the large-area temperature and heat flow distribution on the surface of the cone model (Li et al. 2019). According to the difference in heat transfer between the laminar and turbulent boundary layers, the transition of the boundary layer on the surface of the model is analyzed to obtain the transition position.

The nano-tracer planar laser scattering (NPLS) system (Yi et al. 2009) includes a computer, a synchronizer, a CCD camera, a Q-switched Nd:YAG laser, a nanoparticle selector, a beam articulated arm, and a cylindrical lens. According to the theory of light scattering by particles, we know that if the distance between particles is evidently larger than the particle diameter, the scattering of different particles is independent. Therefore, the scattering light intensity of nanoparticles received by the CCD is in proportion to the local concentration of nanoparticles on the plane of the laser sheet. If the scattering laser intensity falls into the linear scope of the CCD, the grayscale of the flow image is also in proportion to the concentration of nanoparticles. Due to the good flow-tracing ability of nanoparticles, if the concentration distribution of particles on the inlet of the flow field is uniform, the variation of particle concentration will reflect the density variation and mixing structure of the flow field. Thus, the NPLS method is appropriate for the study of the compressible flow field and mixing flows such as the mixing layer, jet, and transverse injection in which the density or concentration gradient are rather large. When measuring the flow field with NPLS, the nanoparticles are injected into and mixed with the inflow of the flow field. While the flow is established in the observing window, the synchronizer controls the laser pulse and CCD to ensure synchronization of the emission of scattering laser by nanoparticles and the exposure of the CCD. Many flow images show that nanoparticles can follow shock waves, expansion waves, slip lines, Mach disks, and turbulent boundary layers in supersonic flow (Zhao et al. 2009), which has proven the excellent flow-tracing ability of nanoparticles in supersonic flow. Thus, NPLS can meet the requirements of flow visualization of supersonic mixing layers.

In short, due to the complexity of the hypersonic transition problem, the diversity of influencing factors, and the difficulty of research, it is unrealistic to expect that its mechanism and laws can be fully grasped in the short term. In future transition studies, it is recommended that each influencing factor should be studied separately to avoid interference from multiple factors. Future research should also combine theoretical analysis, numerical simulation, wind tunnel testing, and flight testing, and use the advantages of various research methods to avoid their shortcomings, while complementing each other.

8.5 Supersonic Combustion and Scramjet Testing

The effect of the vitiation of the incoming flow is a critical problem for the testing of air-breathing engines in wind tunnels, especially for scramjets operating in hypersonic conditions. The pure-air test flow produced by hypersonic shock tunnels is of great importance to assess the correct performance of a scramjet. Scramjet tests were conducted in the JF-12 shock tunnel, which provides over 100 ms of test duration and pure-air hypersonic test flow.

8.5.1 Ignition Delay of Fuel for Scramjet

The velocity of the airflow in the supersonic combustor of a scramjet is high, so the ignition delay of the fuel is very important for the design of scramjet engines. The ignition delay time (also called the induction time) for typical fuels can be found in the literature. The ignition delay of hydrogen in oxygen at a temperature range of 800–2,400 K and a pressure range of 1–3 atm. can be empirically expressed in Equation (8.14). A similar expression for methane-oxygen (Spadaccini & Colket 1994) is given in Equation (8.15).

For hydrogen: $$\tau(\mu\text{s}) = 1.54\times10^{-4}[H_2]^{0.14}[O_2]^{-0.56}\exp\left(\frac{8643}{T}\right) \tag{8.14}$$

For methane: $$\tau(\mu\text{s}) = 2.21\times10^{-8}[CH_4]^{0.33}[O_2]^{-1.05}\exp\left(\frac{22659}{T}\right) \tag{8.15}$$

The square-bracketed quantities in the above formula are the mole concentrations of the reactants and the unit is mol/cm^3. According to the above equations, the ignition delay of hydrogen and methane at the condition of p = 1 atm. and equivalence ratio (ER) = 1.0 is illustrated in Figure 8.32. At the same pressure and temperature, the ignition delay of the methane fuel is far longer than that of the hydrogen fuel. For the high-Mach-number scramjet working above Ma = 7, hydrogen fuel or a mixed fuel of hydrogen is necessary for the design of scramjets.

8.5.2 The Configuration of the Supersonic Chamber

A scramjet model was designed for the test using pure air as the medium in the JF-12 wind tunnel, as shown in Figure 8.33. The scramjet model is 172 mm in width and about 2.2 m in length, and the designed Mach number of the test model is Ma = 7.0.

The inlet of the model is three dimensional with both bottom and lateral compression. The capture height is 157 mm and the width is 140 mm. The bottom has a two-wedge compression with compression angles of 8° and 16°. The inner compression ratio of the inlet is 1.3, as shown in Figure 8.34.

A numerical simulation of the inlet flow field was conducted to validate the design of the inlet. The results show that the shock wave compression of the inlet is rather good at the given inflow condition, as shown in Figure 8.35. The shock waves interact just at the upper leading edge of the inlet. In the inner channel of the model,

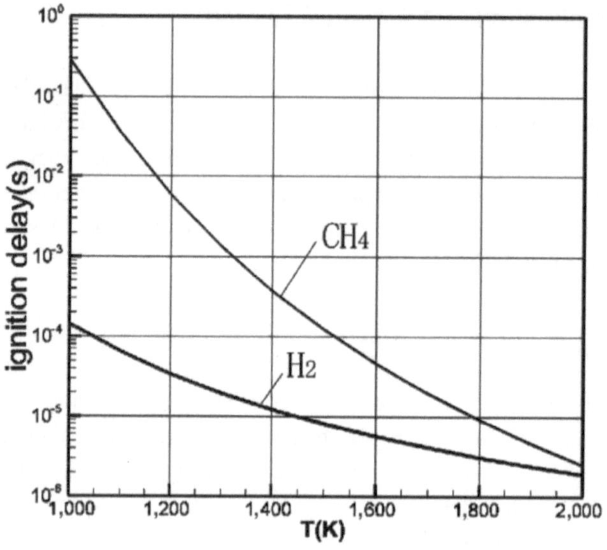

Figure 8.32 The ignition delay of hydrogen and methane (p = 1 atm., ER=1.0).

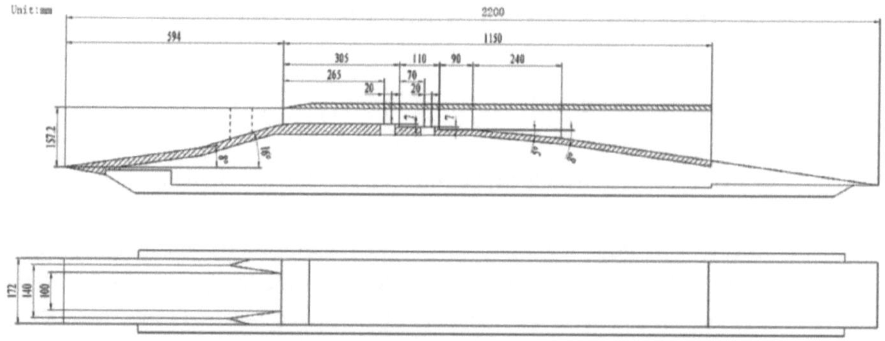

Figure 8.33 The scramjet test model in the JF-12 shock tunnel.

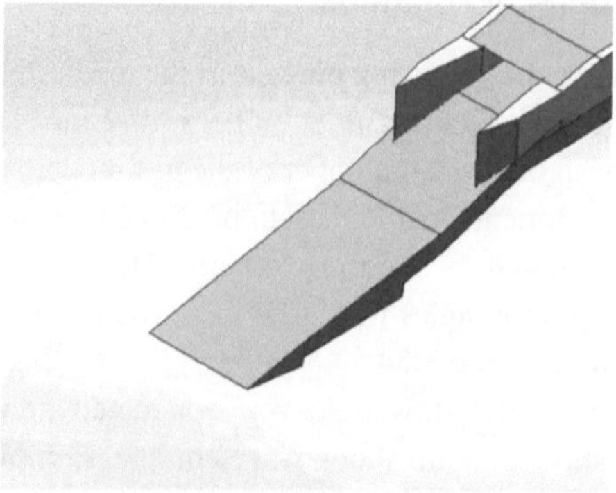

Figure 8.34 The inlet configuration of the scramjet model.

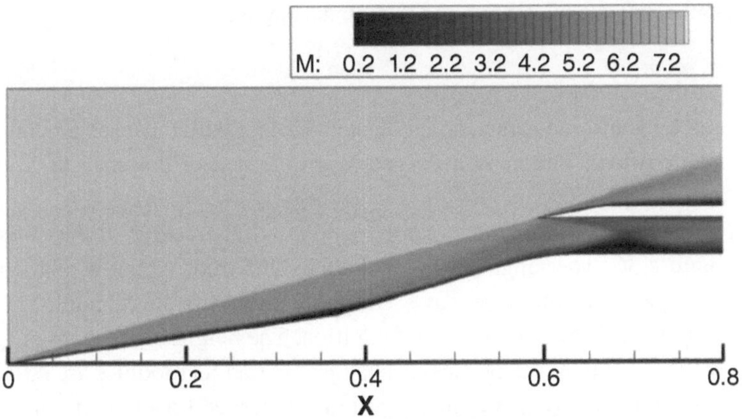

(a) Mach number contours in the inlet flow field.

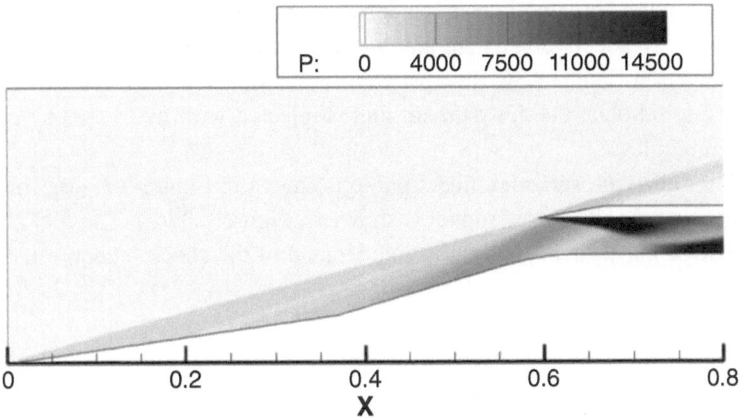

(b) Pressure contours in the inlet flow field.

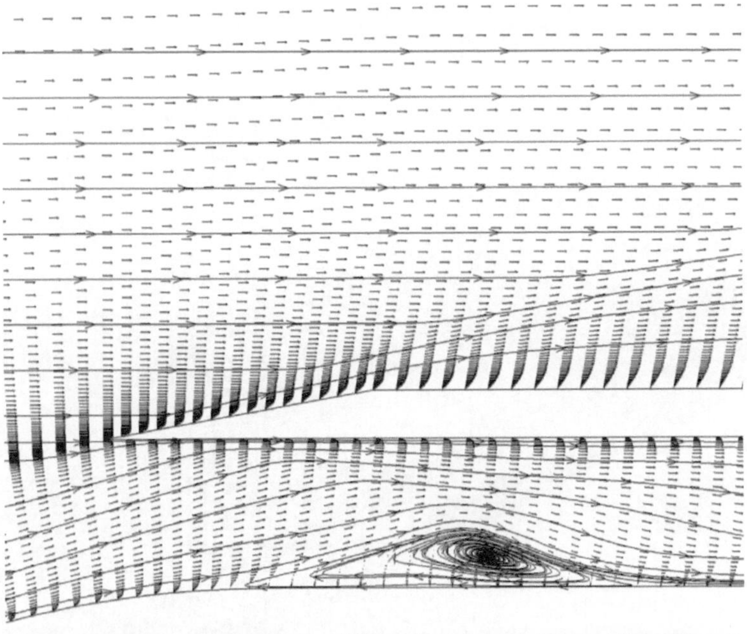

(c) Locally detached flow in the inlet induced by the three-dimensional shock-wave/boundary-layer interaction.

Figure 8.35 Numerical flow field of the inlet at $Ma = 7$ test condition.

there is a local three-dimensional detached zone, which causes a nonuniformity in the inlet outflow. The mass-averaged Mach number of the inner flow at the exit of the inlet is about 3.5.

The configuration of the isolator section and combustor of the model is shown in Figure 8.36. The length of the isolator is 265 mm, which is applied to prevent the propagation of pressure waves from the combustor to the inlet. The combustor adopts the two-backward-step configuration. The height of the steps is 7 mm. At the upstream position of the two steps, changeable fuel jet modules are designed for the hydrogen fuel supply. An igniter position is reserved for the ignition of the fuel. In this study, hydrogen fuel is used, but the igniter is not adopted. At the downstream of the combustor, two expansion sections are designed, with expansion angles of 5° and 8°, respectively. There are two fuel jet modules in the supersonic chamber. As shown in Figure 8.36, the upstream fuel is injected with five 1.5-mm holes and four 2-mm holes; the downstream fuel is injected with five 1.0-mm holes and four 1.5-mm holes.

The full-sized scramjet model photo is shown in Figure 8.37(a). The installation of the model in the wind tunnel is shown in Figure 8.37(b). The fuel system is calibrated to match the synchronization demand of the shock tunnel test. The pressure

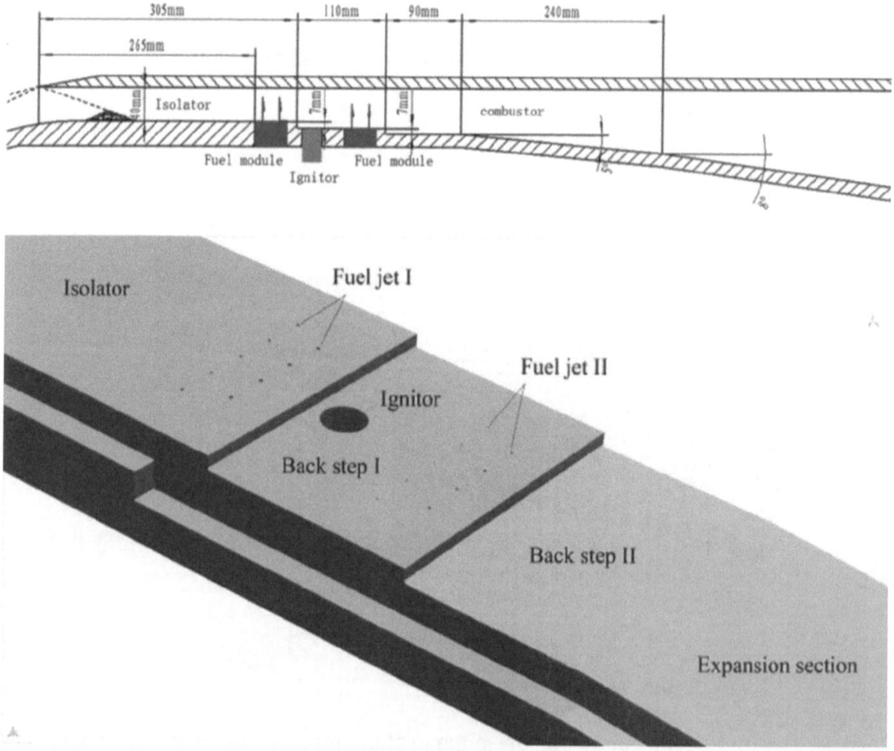

Figure 8.36 Two-backward-step combustor configuration of the model.

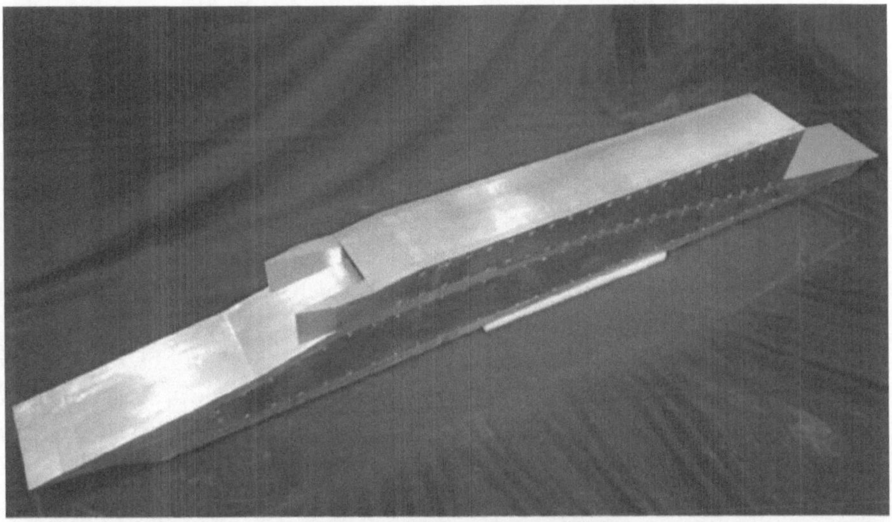

(a) Photo of the scramjet model

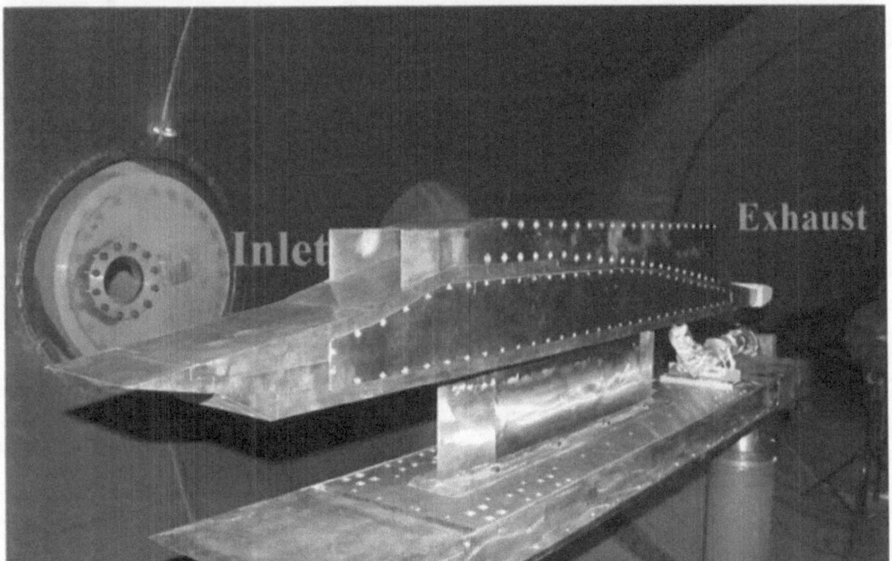

(b) The installation of the model in the wind tunnel

Figure 8.37 The scramjet test model and its installation in the JF-12 Shock tunnel.

distributions on the bottom wall are measured by NS-2 pressure transducers and the exhaust of the scramjet model is recorded by a Fastcam high-speed camera with 3600 fps in 1024×1024 pixels.

8.5.3 Test Facility and Test Conditions

The test facility is the JF-12 shock tunnel of the Institute of Mechanics, Chinese Academy of Sciences. The JF-12 shock tunnel is composed of a driver section, driven section, nozzle, test section, and the dump tank vacuum section. The diameter of the test section is 3.5 m. There are two nozzles with diameters of 2.5 m and 1.5 m. The air test medium of the shock tunnel is compressed and heated by a strong shock wave produced by the detonation wave in the driver section. The backward-detonation driver mode is adopted and the detonation driver is initially filled with a detonable gas mixture of H_2, O_2, and N_2. Typical p_5 pressure histories are shown in Figure 8.38, which shows that the shock tunnel provides stable high-pressure and high-temperature air in the driven section. The typical profiles of pitot pressure and Mach number at the exit of the nozzle are shown in Figure 8.39, which shows good uniformity of the test flow in the shock tunnel test section.

Several JF-12 shock tunnel test conditions in which the scramjet tests are conducted are listed in Table 8.8. The fuel equivalence ratio is kept at about 1.0 by adjusting the fuel pressure. The scramjet was ignited successfully and the combustion was stable in all their test conditions.

Table 8.8 The test conditions of the scramjet test in the JF-12 shock tunnel

	Reservoir gas parameters		Test gas parameters in the test section				
No.	p_0 (MPa)	T_0 (K)	p_t (kPa)	M	q_∞ (Pa)	T_∞ (K)	p_∞ (Pa)
0804	2.10	2,240	54.92	5.754	29,191.8	333.3	1,260.0
0813	3.26	1,871	86.16	5.854	45,864.5	265.1	1,911.9
0820	2.17	2,170	56.90	5.774	30,254.3	319.9	1,296.5
1225	2.54	2,155	61.78	5.880	32,868	307	1,358

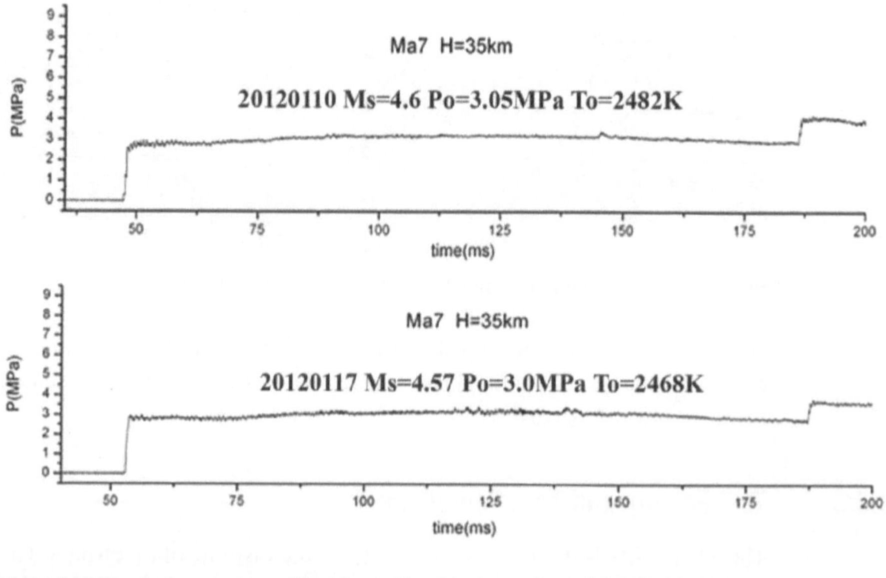

Figure 8.38 Typical p_5 pressure histories in the driven section of the JF-12 shock tunnel.

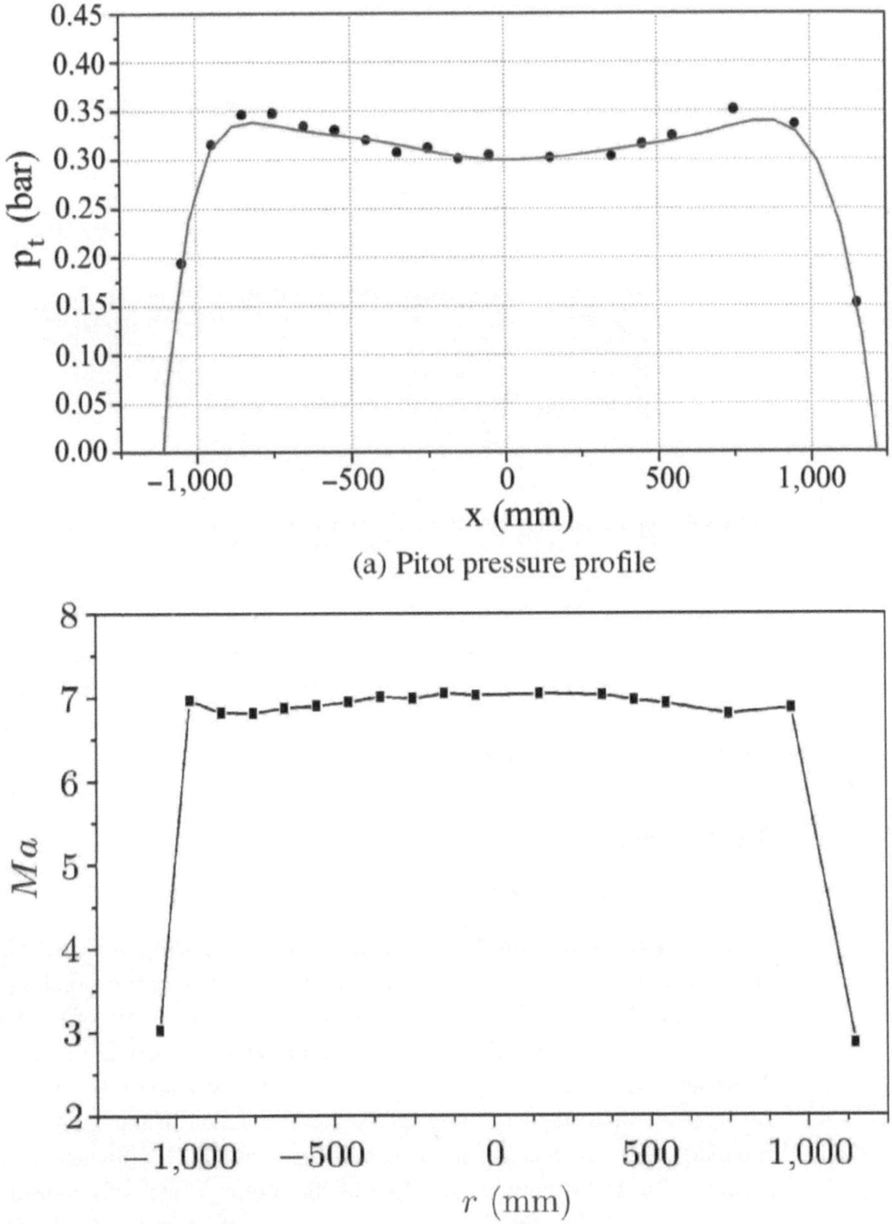

(a) Pitot pressure profile

(b) Mach number profile

Figure 8.39 Typical profiles of pitot pressure and Mach number at the exit of the nozzle of the JF-12 shock tunnel.

Figure 8.40 shows the typical pressure histories and the test photos recorded by a high-speed camera. After the beginning events in the shock tunnel (ignition in the detonation driver), there is a short duration before the startup of the nozzle. Then, the nozzle starts up and the cold state of the scramjet is established. After a certain delay,

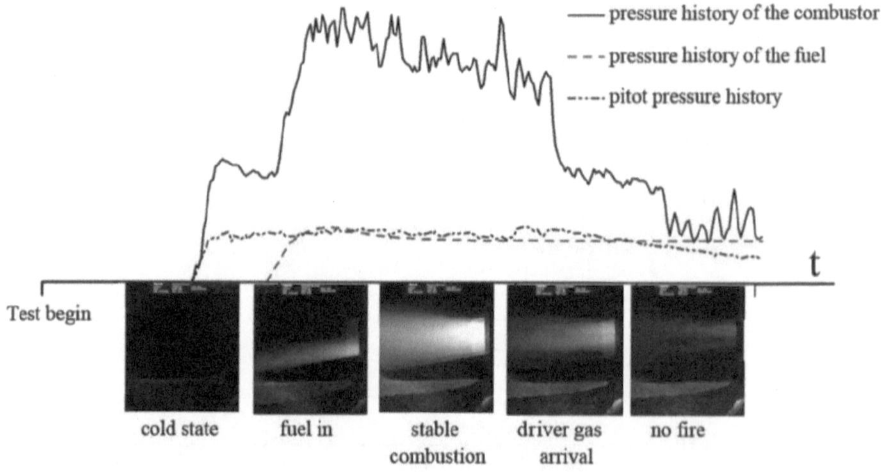

Figure 8.40 Typical pressure histories and test photos recorded by a high-speed camera.

the hydrogen fuel is injected into the scramjet model. Fuel mixing and ignition take place, stable combustion can be observed by the photo of the exhaust of the engine. Once the driver gas arrives, there is no oxygen in the test flow, then combustion stops and the flame is extinguished.

8.5.4 Results and Discussion

(a) Stable combustion of the scramjet

Figure 8.41 shows the non-dimensionalized pressure distribution on the bottom wall of the scramjet in the cold state and combustion state, corresponding to the test condition of No. 0813 in Table 8.8, and the equivalence ratio of the fuel is about 1.06. The p_∞ in the vertical axis represents the static pressure of the test flow ahead of the scramjet model. As shown in the cold state, the wall pressure of the scramjet undergoes a sharp increase induced by the interaction of the shock wave and the boundary layer, and then the pressure decreases gradually with the expansion of the channel area. In the combustion state of the scramjet, the wall pressure increases to a pressure peak as the fuel jet is injected into the supersonic chamber. Due to the combustion of the fuel, the pressure in the supersonic combustor during the combustion state is far higher than that in the cold state. At the downstream of the supersonic combustor, the pressure decreases gradually with the expansion of the channel area. Also, it can be found that the average pressure in the hot state is about 2–3 times that in the cold state, especially in the stable range of the pressure profile stage. Figure 8.42 shows the photos of the exhaust of the scramjet engine in the test condition of No. 0813. The strong combustion phenomenon can be observed in the photo of the exhaust of the scramjet engine.

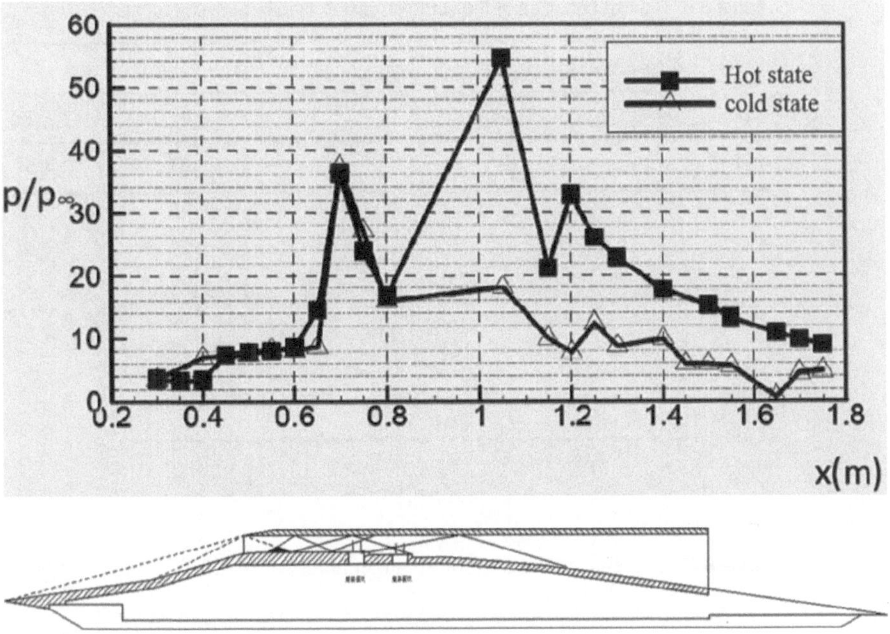

Figure 8.41 Wall pressure distributions in cold and hot states of the scramjet engine.

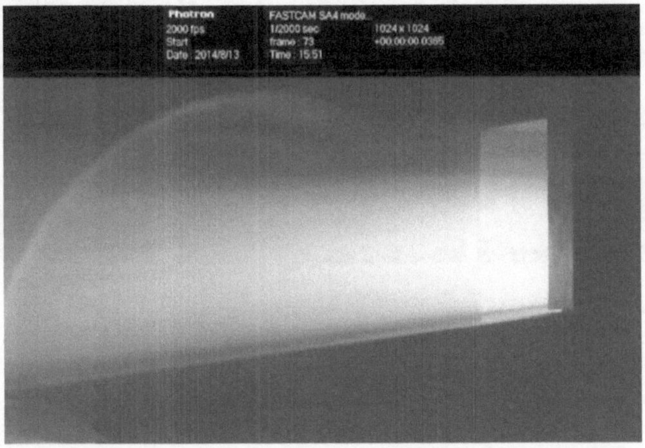

Figure 8.42 Photo of the exhaust of the scramjet engine.

(b) Unstable combustion of the scramjet

In certain test conditions, unstable combustion of the scramjet engine is observed when the fuel equivalence ratio is increased. The test condition is listed in Table 8.9, the fuel equivalence ratio is about 1.10.

Figure 8.43 shows the pressure history recorded by the pressure transducer at the entrance of the supersonic combustor. As can be seen, periodical pressure oscillations

Table 8.9 The test condition of the scramjet test in the JF-12 shock tunnel

	Reservoir gas parameters		Test gas parameters in the test section				
No.	p_0 (MPa)	T_0 (K)	p_t (kPa)	M	q_∞ (Pa)	T_∞ (K)	p_∞ (Pa)
1115	2.43	2,133	57.30	5.932	30,479	299	1,237

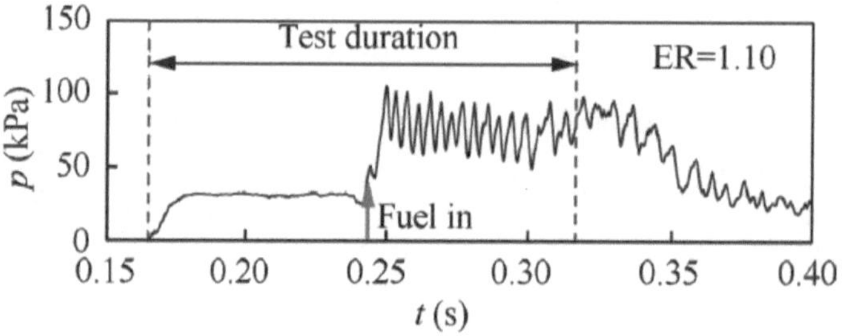

Figure 8.43 Unstable combustion of Scramjet model of No.1115.

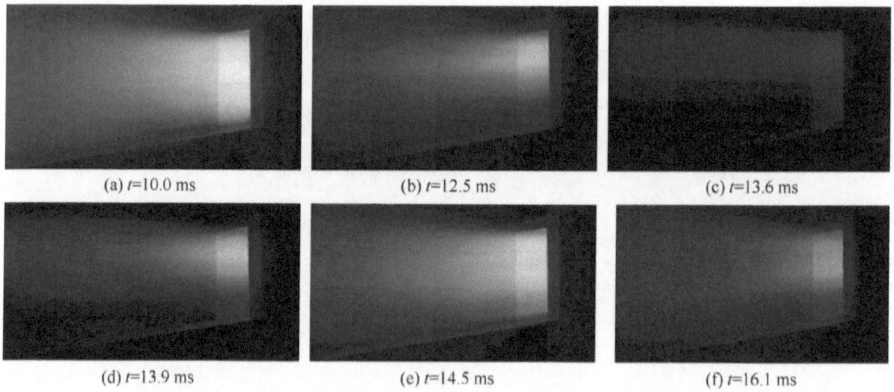

Figure 8.44 Time-sequential photos taken with high-speed cameras during engine surge and six frames showing about one cycle.

exist in the supersonic combustion chamber of the scramjet model. The average frequency is about 200 Hz by counting the peaks of the oscillatory pressure.

The time-sequential photos are presented in Figure 8.44 and these frames were taken with high-speed cameras during the Performance Test Engine (PTE) model operation. It can be observed in the figure that the combustion dies out and is reignited periodically. By checking the time of the frames, it appears that the combustion dies out more slowly than it reignites. The equivalence ratio is the only parameter that is different from the corresponding stable combustion case. This important phenomenon could be defined as engine surge, which is a critical issue for scramjet operation control because no practical engine would be allowed to operate in such an unstable mode.

Generally speaking, there exists a critical operation point for any scramjet engine at certain inflow conditions. Once the heat release is over the critical operation point, upstream-traveling compression waves will get stronger and stronger, and then a shock wave is generated and accelerated due to the temperature gradient distribution ahead. Therefore, thermal choking in the engine takes place once the upstream-traveling shock wave reaches the inlet entrance, the inlet goes into unstart and the combustion flame dies out. Once the flame dies out, the upstream-traveling shock wave will decay rapidly so that the inlet would restart and the scramjet model operates again. This phenomenon is taken as the hypersonic engine surge that is very interesting for the understanding of combustion physics, but a quite tough problem for practical engine development. There must be a balance point at which the inflow Mach number matches with the reaction heat release, and then the shock wave becomes stationary in the combustor.

Nomenclature

B	bias
B_{C_A}	bias limit of C_A
C_A	axial force coefficient
C_D	drag force coefficient
C_N	normal force coefficient
C_M	moment coefficient
c	specific heat capacity
C	Chapman–Rubesin parameter
c_p	specific heat at constant pressure
D	diameter
H	enthalpy
h	static enthalpy
k	the thermal conductivity
Ma	Mach number
Mx	moment coefficients about the x-axis
My	moment coefficients about the y-axis
Mz	moment coefficients about the z-axis
p	pressure
P	precision
Pr	Prandtl number
q	heat flux; dynamic pressure
$\dot{q}_s$	surface heat flux
r	radial coordinate
Re	Reynolds number
$\overline{R}$	averaged value of the aerodynamic coefficient
R_N	nose radius
S	area

$S(T)$	Seebeck coefficient
St	Stanton number
t	time
T	temperature
u	velocity in boundary layer
U_{C_A}	relative uncertainty in C_A
U_{C_N}	relative uncertainty in C_N
U_{C_M}	relative uncertainty in C_M
U_{C_D}	relative uncertainty in C_D
U	velocity in the x-direction
V	velocity in the y-direction
X_{cp}	center of pressure location
X	axial force component in force balance (Figure 8.2)
Y	normal force component in force balance (Figure 8.2)
Z	side force component in force balance (Figure 8.2)
∇V	gradient of voltage
∇T	gradient in temperature
α	angle of attack
σ	standard deviation
ϕ	sensor diameter
τ	ignition delay time; shear stress
τ_{res}	response time of the coaxial thermocouple
ρ	density
μ	viscosity coefficient
δ	boundary layer thickness

Subscripts:

∞	freestream condition
0	total or stagnation condition
5	reflected shock region
e	boundary-layer edge condition
d	diameter
L	length
r	recover state
t	pitot condition
w	wall

References

AIAA. (2003). Recommended Practice: Calibration and Use of Internal Strain-Gage Balances with Application to Wind Tunnel Testing. *AIAA Paper 2003-R-091.*

AIAA-Guide. (2003). Assessing Experimental Uncertainty. *AIAA S-071A-1999 (AIAA Paper 2003-G-045).*

AIAA-Standard. (1999). Assessment of Experimental Uncertainty with Application to Wind Tunnel Testing. *AIAA Paper No. 99-S-071A*.

Alexander, F. (2011). Transition and Stability of High-Speed Boundary Layers. *Annual Review of Fluid Mechanics*, **43**, 79–95.

Anderson, J. D. (2006). *Hypersonic and High-Temperature Gas Dynamics*, 2nd ed. Reston, VA: AIAA.

Arrington, P. J., Joiner, R. J., and Henderson, A. J. (1954). Longitudinal Characteristics of Several Configurations at Hypersonic Mach Numbers in Conical and Contoured Nozzles. *NASA TN D-2489*.

Bernstein, L. (1975). Force Measurement in Short-duration Hypersonic Facilities. *AGARDograph No. 214*.

Bertin, J. J. and Cummings, R. M. (2003). Fifty Years of Hypersonics: Where We've Been, Where We're Going. *Progress in Aerospace Sciences*, **39**, 511–536.

Bertin, J. J. and Cummings, R. M. (2006). Critical Hypersonic Aerothermodynamic Phenomena. *Annual Review of Fluid Mechanics*, **38**, 129–157.

Buttsworth, D. R. (2001). Assessment of Effective Thermal Product of Surface Junction Thermocouples on Millisecond and Microsecond Time Scales. *Experimental Thermal and Fluid Science*, **25**, 409–420.

Casper, K. M., Beresh, S. J., and Schneider, S. P. (2014). Pressure Fluctuations Beneath Instability Wave Packets and Turbulent Spots in a Hypersonic Boundary Layer. *Journal of Fluid Mechanics*, **756**, 1058–1091.

Chen, H. (2000). Investigation on the Link-up between the Aerodynamic Force Data Measured in Supersonic and Hypersonic Wind Tunnel. *ACTA Aerodynamica Sinica*, **18**(3), 345–349.

Cheng, H. K. (1959). Hypersonic Shock-layer Theory of a Yawed Cone and Other Three-dimensional Pointed Bodies. *WADC TN 59–335*.

Chynoweth, B. C., Edelman, J., Gray, K., Mckiernan, G., and Schneider, S. P. (2017). Measurements in the Boeing/AFOSR Mach-6 Quiet Tunnel on Hypersonic Boundary-Layer Transition. *47th AIAA Fluid Dynamics Conference*.

Chynoweth, B. C., Schneider, S. P., Hader, C., Fasel, H., Batista, A., Kuehl, J., et al. (2019). History and Progress of Boundary-Layer Transition on a Mach-6 Flared Cone. *Journal of Spacecraft Rockets*, **56**(2), 333–346.

Comte-Bellot, G. (1976). Hot-Wire Anemometry. *Annual Review of Fluid Mechanics*, **8**, 209–231.

De Filippis, F., Savino, R., and Martucci, A. (2005). Numerical-Experimental Correlation of Stagnation Point Heat Flux in High Enthalpy Hypersonic Wind Tunnel. In *AIAA/CIRA 13th International Space Planes and Hypersonics Systems and Technologies Conference*. Capua, *AIAA-2005-3277*.

Fay, J. A. and Riddel, F. R. (1958). Theory of Stagnation Point Heat Transfer in Dissociated Air. *Journal of Aeronautic Science*, **25**(2), 73–85.

Fluid Dynamics Panel Working Group 15. (1994). Quality Assessment for Wind Tunnel Testing. *AGARD-AR-304*.

Gao, Y. L. (2008). Study on Hypervelocity Flow Generation Techniques and Essential Hypersonic Phenomena. Ph.D Thesis, Institute of Mechanics, Beijing, China.

Gray, J. D. (1964). Summary Report on Aerodynamic Characteristics of Standard Models HB-1 and HB-2. AEDC-TDR-64-137.

Hirschel, E. H. (2005). *Basics of Aerothermodynamics*. Berlin: Springer.

Holden, M., Wadhams, T., MacLean, M., and Mundy, E. (2007). Experimental Studies in LENS I and X to Evaluate Real Gas Effects on Hypervelocity Vehicle Performance. *45th AIAA Aerospace Sciences Meeting and Exhibit: 204.*

Hornung, H. G. (2010). Ground Testing for Hypersonic Flow, Capabilities and Limitations. *RTO Technical Report. RTO-EN-AVT-186-1.*

Hu, Z. M., Zhou, K., Peng, J., and Jiang, Z. L. (2017). Hypervelocity Test with a Detonation Driven Expansion Tube. *The 31st International Symposium on Shock Waves*, Nagoya, Japan, Jul. 13–18.

Jiang, Z. and Yu, H. (2014). Experiments and Development of Long-test-duration Hypervelocity Detonation-driven Shock Tunnel (LHDst). *AIAA Paper 2014–1012.*

Jiang, Z. L. (2014) Experiments and Development of Long Test Duration Hypervelocity Detonation-driven Shock Tunnel (LHDst). *AIAA SciTech 2014*, Maryland, USA.

Jiang, Z. L. and Yu, H. R. (2017). Theories and Technologies for Duplicating Hypersonic Flight Conditions for Ground Testing. *National Science Review*, **4**(3), 290–296.

Jiang, Z. L., Hu, Z. M., Wang, Y. P., and Han, G. L. (2020). Advances in Critical Technologies for Hypersonic and High-enthalpy Wind Tunnel. *Chinese Journal of Aeronautics*, https://doi.org /10.1016/ j.cja.2020.04.003.

Jiang, Z. L., Li, J. P., Zhao, W., and Liu, Y. (2012). Investigation into Techniques for Extending the Test-duration of Detonation-driven Shock Tunnels. *Chinese Journal of Theoretical and Applied Mechanics*, **44**(5), 824–831. (in Chinese)

Jiang, Z. L., Wu, B., Gao, Y. L., Zhao, W., and Hu, Z. M. (2015). Development of the Detonation-driven Expansion Tube for Orbital Speed Experiments. *Science China Technology Sciences*, **58**(4), 695–700.

Joarder, R. and Jagadeesh, G. (2003). A New Free Floating Accelerometer Balance System for Force Measurements in Shock Tunnels. *Shock Waves*, **13**, 409–412.

Kammeyer, M. E. (1999). Uncertainty Analysis for Force Testing in Production Wind Tunnels. In *1st International Symposium on Strain Gauge Balances.*

Kuehl, J. J. (2018). Thermoacoustic Interpretation of Second-Mode Instability. *AIAA Journal*, **56**, 1–8.

Ladson, C. and Blackstock, T. (1962). Air-helium simulation of the aerodynamic force Coefficients of Cones at Hypersonic Speeds. *NASA TN D-1473.*

Laurence, S. J. and Karl, S. (2010). An Improved Visualization-Based Force-measurement Technique for Short-duration Hypersonic Facilities. *Experiments in Fluids*, **48**, 949–965.

Li, J. P., Chen, H., Zhang, S. Z., Zhang, X., and Yu, H. (2017). On the Response of Coaxial Thermocouples for Transient Aerodynamic Heating Measurements. *Experimental Thermal and Fluid Science*, **86**, 141–148.

Li, Q., Jiang, T., and Chen, S. Y. (2019). Measurement Technology and Measuring of Boundary Layer Transition in Shock Tunnel. *Acta Aeronautica et Astronautica Sinica*, **40**(7), 122740

Li, S. X., Shi, J., and Guo, X. (2016). Experimental Study of Boundary Layer Features on a Flat Plate in High Speed Flow. *Physics of Gases*, **1**(6), 1–4.

Li, X., Yu, C., and Tong, F. (2018). Direct Numerical Simulation of Hypersonic Boundary-layer Transition of Blunt Cones. Tenth International Conference on Computational Fluid Dynamics Barcelona, Spain.

Liu, M. K., Han, G. L., and Jiang, Z. L. (2019). Experimental Investigation on Plate Boundary Layer Transition in JF-12 Hypersonic Shock Tunnel. *Proceedings of the 32nd International Symposium on Shock Waves (ISSW32 2019).*

Lu, F. K. and Marren, D. E. (2002). Advanced Hypersonic Test Facilities. Progress in Astronautics Vol. 198.

MacWherter, M., Noack, R. W., and Oberkampf, W. L. (1986). Evaluation of Boundary-layer and Parabolized Navier-Stokes Solutions for Reentry Vehicles. *Journal of Spacecraft and Rockets*, **23**(1), 70–78.

Marineau, E.C., Grossir, G., Wagner, A., Leinemann, M., Radespiel, R., Tanno, H., et al. (2019). Analysis of Second-Mode Amplitudes on Sharp Cones in Hypersonic Wind Tunnels. *Journal of Spacecraft Rockets*, **3**(2), 1–12.

Marineau, E., MacLean, M., Mundy, E., and Holden, M. (2012). Force Measurements in Hypervelocity Flows with an Acceleration Compensated Strain Gage Balance. *Journal of Spacecraft and Rockets*, **49**(3), 474–482.

Martinez, B., Bastide, M., and Wey, P. (2014). Free Flight Measurement Technique in Shock Tunnel. In *Proceedings of the 30th Aerodynamic Measurement Technology and Ground Testing Conference (Atlanta, USA)*.

Maus, J. R., Griffith, B. J., Szema, K. Y., and Best, J. T. (1984). Hypersonic Mach Number and Real Gas Eon Space Shuttle Orbiter Aerodynamic. *Journal of Spacecraft and Rockets*, **21**, 136–141.

Mee, D. J., Daniel, W. J. T., and Simmons, J. M. (1996). Three-Component Force Balance for Flows of Millisecond Duration. *AIAA Journal*, **34**(3), 590–595.

Mikhailov, V. V., Neiland, V. Ya., and Sychev, V. V. (1971). The Theory of Viscous Hypersonic Flow. *Annual Review of Fluid Mechanics*, **3**, 371–396.

Morkovin, M., Reshotko, E., and Herbert, T. (1994). Transition in Open Flow System-A Reassessment. *Bulletin of American Physical Society*, **39**, 1882.

Naumann, K. and Ende, H. (1990). A Novel Technique for Aerodynamic Force Measurements in Shock Tubes. *AIP Conference Proceedings*, **208**, 653–658.

Naumann, K., Ende, H., Mathieu, G., and George, A. (1993) Millisecond Aero-dynamic Force Measurement with Side-jet Model in the ISL Shock Tunnel. *AIAA Journal*, **31**, 1068–1074.

Park, C. (2013). Hypersonic Aerothermodynamics: Past, Present and Future. *International Journal of Aeronautical and Space Sciences*, **14**(1), 1–10.

Parziale, N. J., Shepherd, J. E., and Hornung, H. G. (2014). Free-stream Density Perturbations in a Reflected-shock Tunnel. *Experiments in Fluids*, **55**, 1665.

Penland, J. A. (1961). Aerodynamic Force Characteristics of a Series of Lifting Cone and Cone-cylinder Configurations at m = 6.83 and Angles of Attack up to 130 degree. *NASA TN D-840*.

Rasmussen, M. (1994). *Hypersonic Flow*. New York: John Wiley & Sons.

Robinson, M. J., Schramm, J. M., and Hannemann, K. (2011). Design and Implementation of an Internal Stress Wave Force Balance in a Shock Tunnel. *CEAS Space Journal*, **1**, 45–57.

Roediger, T., Knauss, H., Estorf, M., Schneider, S. P., and Smorodsky, B. V. (2009). Hypersonic Instability Waves Measured Using Fast-Response Heat-Flux Gauges. *Journal of Spacecraft Rockets*, **46**, 266–273.

Rose, P. H. (1958). Development of the Calorimeter Heat Transfer Gauge for Use in Shock Tubes. *Review of Scientific Instruments*, **29**, 557.

Sahoo, N., Mahapatra, D. R., Jagadeesh, G., Gopalakrishnan, S., and Reddy, K. P. J. (2003). An Accelerometer Balance System for Measurement of Aerodynamic Force Coefficients Over Blunt Bodies in a Hypersonic Shock Tunnel. *Measurement Science and Technology*, **14**, 260–272.

Sanderson, S. R., Simmons, J. M. and Tuttle, S. L. (1991). A Drag Measurement Technique for Free-piston Shock Tunnels. *AIAA Paper 91–0540*.

Sanderson, S. R. and Sturtevant, B. (2002). Transient Heat Flux Measurement Using a Surface Junction Thermocouple. *Review of Scientific Instruments*, **73**, 2781.

Saravanan, S., Jagadeesh, G., and Reddy, K. P. J. (2009) Aerodynamic Force Measurement Using 3-component Accelerometer Force Balance System in a Hypersonic Shock Tunnel. *Shock Waves*, **18**, 425–435.

Schlichting, H. and Gersten, K. (2017). *Boundary-Layer Theory*. Springer Berlin Heidelberg.

Schultz, D. L. and Jones, T. V. (1973). Heat-transfer Measurements in Short-duration Hypersonic Facilities. *AGARD AG-163*.

Seiler, F., Mathieu, G., George, A., Srulijes, J., and Havermann, M.(2005) Development of a Free Flight Force Measuring Technique (FFM) at the ISL Shock Tube Laboratory. In *25th International Symposium on Shock Waves*, ISSW 25, Bangalore, India. *ISL Report PU 620/2005*.

Spadaccini, L. J. and Colket III, M. B. (1994). Ignition Delay Characteristics of Methane Fuels. *Progress in Energy and Combustion Science*, **20**, 431–460.

Stalker, R. J. (1967). A Study of the Free-piston Shock Tunnel. *AIAA Journal*, **5**, 2160–2165.

Tani, I. (1969). Boundary-Layer Transition. *Annual Review of Fluid Mechanics*, **1**, 169–196.

Tanno, H., Komuro, T., Sato, K., Fujita, K., and Laurence, S.J. (2014). Free-flight measurement technique in the free-piston shock tunnel HIEST. *Review of Scientific Instruments*, **85**, 045112.

Tanno, H., Komuro, T., Sato, K., Itoh, K., and Yamada, T. (2013). Free-flight Tests of Reentry Capsule Models in Free-piston Shock Tunnel. *AIAA Paper 2013–2979*.

Tashiro, Y., Biwa, T., and Yazaki, T. (2005). Calibration of a Thermocouple for Measurement of Oscillating Temperature. *Review of Scientific Instruments*, **76**, 124901.

Wang, Q., Li, J. P., Zhao, W., and Jiang, Z. L. (2016a). Comparative Study on Aerodynamic Heating Under Perfect and Nonequilibrium Hypersonic Flows. *Science China*, **59**, 624701.

Wang, Q., Olivier, H., Einhoff, J., Li, J., and Zhao, W. (2019). Influence of Test Model Material on the Accuracy of Transient Heat Transfer Measurements in Impulse Facilities. *Experimental Thermal and Fluid Science*, **104**, 59–66.

Wang, Y. P., Liu, Y. F., Luo, C. T., and Jiang, Z. L. (2016b). Force Measurement Using Strain-gauge Balance in a Shock Tunnel with Long Test Duration. *Review of scientific instruments*, **87**(5), 55108.

Wey, P., Srulijes, J., Bastide, M., Martinez, B., and Gnemmi, P. (2012). Determination of Aerodynamic Coefficients from Shock Tunnel Free Flight Trajectories. In *Proceedings of the 28th Aerodynamic Measurement Technology*, Ground Testing and Flight Testing Conference 2102, Vol. 2. Reston, VA: AIAA, pp. 949–957.

Willeke, K. and Bershader, D. (1973). An Improved Thin-film Gauge for Shock-tube Thermal Studies. *Review of Scientific Instruments*, **44**, 22.

Woods, W. C, Arrington, J. P., and Hamilton, H. H. (1983). A Review of Preflight Estimates of Real-gas Effects on Space Shuttle Aerodynamic Characteristics. In *Shuttle Performance: Lessons Learned*. NASA Conf. Publ. 2283.

Wu, B. (2012). *Study on the Interaction of Strong Shock Wave and the Hypervelocity Experimental Method*. Ph.D. Thesis. Institute of Mechanics, Beijing, China.

Wu, S., Shu, Y. H., Li, J. P., and Yu, H. R. (2014). An Integral Heat Flux Sensor with High Spatial and Temporal Resolutions. *Chinese Science Bulletin*, **59**, 2484–2489 (in Chinese).

Xiong, Y., Yu, T., Lin, L., Zhao, J. and Wu, J. (2020). Nonlinear Instability Characterization of Hypersonic Laminar Boundary Layer. *AIAA Journal*, **58**(12), 5254–5263.

Yi, S., He, L., Zhao, Y., Tian, L., and Cheng, Z. (2009). A Flow Control Study of a Supersonic Mixing Layer Via NPLS. *Science in China Series G: Physics, Mechanics and Astronomy*, **52**, 2001–2006.

Yu, H., Esser, B., Lenartz, M., and Grönig, H. (1992). Gaseous Detonation Driver for a Shock Tunnel. *Shock Waves*, **2**, 245–254.

Zhang, H. (2004). *Hypersonic Aerodynamic Test*. Beijing, China: National Defense Industry Press.

Zhao, Y., Yi, S., Tian, L., and Cheng, Z. (2009). Supersonic Flow Imaging Via Nanoparticles. *Science in China Series E: Technological Sciences*, **52**, 3640.

Zhong, X. (2009). Numerical Simulation of Hypersonic Boundary Layer Receptivity and Stability on Blunt Circular Cones. *47th AIAA Aerospace Sciences Meeting Including the New Horizons Forum and Aerospace Exposition*.

Zhou, K., Peng, J., Ou, D. B., and Jiang, Z. L. (2020). Experimental Study on Aerodynamic Heat and Wall Catalytic Effects of Hypervelocity Flow. *Scientia Sinica Technologia*, **50**, 1095–1101 (in Chinese).

Zhou, K., Wang, Q., Hu, Z. M., and Jiang, Z. (2016a). Performance Study of a Detonation-driven Expansion Tube. *Acta Aeronautica ET Astronautica Sinica*, **37**, 810–816 (in Chinese).

Zhou, K., Yuan, C. K., Hu, Z. M., and Jiang, Z. (2016b). Test Flow Analysis and Upgrade of the JF-16 Expansion Tube. *Acta Aeronautica ET Astronautica Sinica*, **37**, 3296–3303 (in Chinese).

Zhu, Y., Chen, X., Wu, J., Chen, S., Lee, C., and Gad-El-Hak, M. (2018). Aerodynamic Heating in Transitional Hypersonic Boundary Layers: Role of Second-mode Instability. *Physics of Fluids*, **30**, 011701.

Index